*Adjusting to Policy Failure
in African Economies*

Food Systems and Agrarian Change

Edited by Frederick H. Buttel, Billie R. DeWalt,
and Per Pinstrup-Andersen

A complete list of titles in the series appears at the end of this book.

ADJUSTING TO POLICY FAILURE IN AFRICAN ECONOMIES

EDITED BY

David E. Sahn

Cornell University Press

ITHACA AND LONDON

First published 1994 by Cornell University Press.

Library of Congress Cataloging-in-Publication Data

Adjusting to policy failure in African economies
     p.   cm. — (Food systems and agrarian change)
  Includes bibliographical references and index.
  ISBN 0-8014-2906-4. — ISBN 0-8014-8136-8 (paper)
  1. Structural adjustment (Economic policy)—African, Sub-Saharan—
  Case studies. 2. Africa, Sub-Saharan—Economic conditions—1960—
  Case studies. I. Sahn, David E. II. Series.
  HC800 .A55257 1994
  338.967—dc20   93-29987

Printed in the United States of America

Contents

v

Tables and Figures

FIGURES

Acknowledgments

The work contained in this volume was prepared with funding under a Cooperative Agreement between the Africa Bureau of the U.S. Agency for International Development (AID) and the Cornell Food and Nutrition Policy Program (CFNPP). The support of AID is greatly appreciated, especially that of Jerome Wolgin and Leonard Rosenberg.

Many of the chapters draw on work found in CFNPP Monographs and Working Papers. Several research support specialists were instrumental in gathering and organizing data and deserve credit for their contribution. These include René Bernier (for Madagascar and Niger), Kajal Budhwar (for Niger), Mattias Lundberg (for Gambia), Lemma Merid (for Tanzania), and Elizabeth Stephenson (for Guinea). Furthermore, the cooperation and assistance of the governments and individuals therein of the countries discussed in this book are appreciated.

D. E. S.

Washington, D.C.

Contributors

HAROLD ALDERMAN, formerly a senior research associate with the Cornell Food and Nutrition Policy Program, is an economist with the World Bank in Washington, D.C.

JEHAN ARULPRAGASAM is a research support specialist with the Cornell Food and Nutrition Policy Program.

RENÉ BERNIER is a research support specialist with the Cornell Food and Nutrition Policy Program.

DAVID BLANDFORD, formerly affiliated with the Cornell Food and Nutrition Policy Program and a professor in the Department of Agricultural Economics at Cornell University, is with the Organization for Economic Cooperation and Development (OECD) in Paris.

PAUL DOROSH is a senior research associate with the Cornell Food and Nutrition Policy Program.

DEBORAH FRIEDMAN is a consultant to the Cornell Food and Nutrition Policy Program.

PETER GLICK is a research associate with the Cornell Food and Nutrition Policy Program.

CATHY JABARA, formerly a senior research associate with the Cornell Food and Nutrition Policy Program, is an economist with the U.S. International Trade Commission.

STEVEN KYLE is an assistant professor in the Department of Agricultural Economics at Cornell University and is affiliated with the Cornell Food and Nutrition Policy Program.

SARAH LYNCH, formerly a graduate assistant in the Department of Agricultural Economics, Cornell University, is currently a consultant.

NATASHA MUKHERJEE is a consultant to the Cornell Food and Nutrition Policy Program.

DAVID E. SAHN is the director of the Cornell Food and Nutrition Policy Program and an associate professor of development economics at Cornell University.

ALEXANDER H. SARRIS is a senior research associate with the Cornell Food and Nutrition Policy Program and a professor of economics at the University of Athens.

ERIK THORBECKE is H. Edward Babcock Professor of Economics and Food Economics at Cornell University and is affiliated with the Cornell Food and Nutrition Policy Program.

WA BILENGA TSHISHIMBI is a research support specialist with the Cornell Food and Nutrition Policy Program.

ROGIER VAN DEN BRINK, formerly a research associate with the Cornell Food and Nutrition Policy Program, is an economist with the World Bank in Washington, D.C.

*Adjusting to Policy Failure
in African Economies*

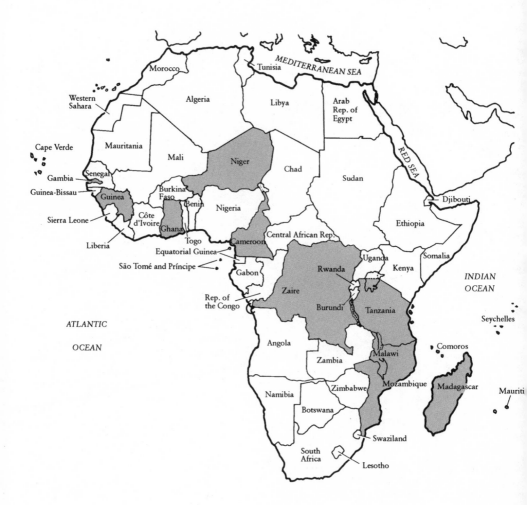

I

Economic Crisis and Policy Reform in Africa: An Introduction

David E. Sahn

Africa in the 1990s remains in crisis. There is some evidence of recovery from a decade or more of economic turmoil and stagnation. Nonetheless, most countries in sub-Saharan Africa are still suffering the harsh consequences of ill-advised policies and an inhospitable external environment that caused havoc with their economies, especially during the late 1970s and the first half of the 1980s. Although few countries have avoided the ravages of economic decline, there is considerable variability in past and present economic performance and in future prospects of countries in sub-Saharan Africa. It is also the case that the great majority of countries in the region have been forced by hardships to examine the fundamentals of policy formation that have significantly contributed to their economic problems. Reform to rectify errors of the past and weaknesses in institutional structures has become the key operational concept of economic (and related social) policy during the latter half of the 1980s and the early 1990s.

It is our purpose in this book to examine the early efforts at economic reform that took place in sub-Saharan Africa in the years after countries found themselves facing unsustainable deficits in the internal and external accounts. It was these balance-of-payments and budget deficits, rather than the tragic human dimensions of Africa's economic crisis, that provided the compelling impetus for reform.

Efforts at reform were most often taken under the sponsorship of, and with financing provided by, the International Monetary Fund (IMF), World Bank, and bilateral donors. Such leveraging of reform programs with foreign financing has become the hallmark of policymaking throughout sub-

Saharan Africa. These programs, often referred to as stabilization and structural adjustment programs, have become controversial. The controversy reflects the divergence of views as to whether the policy prescriptions basic to most economic reform and recovery programs are appropriate and effective as well as equitable.

To provide insight into the divergent opinions regarding the genesis of Africa's economic crisis, and the growing controversy concerning the appropriateness of the subsequent response, in this book we explore the external and domestic factors that precipitated Africa's crisis as well as the nature and consequence of the process of economic reform. Chapters 2 through 11 describe the experiences of ten diverse countries: Cameroon, Gambia, Ghana, Guinea, Madagascar, Malawi, Mozambique, Niger, Tanzania, and Zaire. These countries vary markedly in the general structure and characteristics of their economies, the stock of physical and human capital resources, and their experience and success in undertaking reforms. Some basic statistics on the ten countries are highlighted in Table 1.1. The differences are considerable, in terms of the range of population size, from Gambia, one of the smallest countries in Africa, to Zaire, one of the largest; in terms of GNP per capita, ranging from one of the wealthiest countries in the region, Cameroon, to the poorest country in the region and world, Mozambique; in terms of population density, ranging from land-scarce Malawi to its land-abundant neighbor, Tanzania; in terms of the structure of merchandise exports, heavily concentrated in minerals in Zaire and Guinea but primarily agricultural Gambia and Malawi; in terms of agriculture's share of GDP, which ranges from over 60 percent in Mozambique and Tanzania to less than half that in Guinea and Zaire; and

Table 1.1. Basic indicators, 1989

	Population (millions)	GNP per capita (US$)	Population density (per sq. mi.)	Structure of merchandise exports (%)		Agriculture % of GDP	Adult literac (%)
				Fuels and minerals	Other primary commodities		
Cameroon	11.6	1,000	24.4	48	49	27	48.0
Gambia	0.8	240	77.2	—	—	—	20.3
Ghana	14.4	390	60.3	29	63	49	52.8
Guinea	5.6	430	22.8	83	6	30	16.8
Madagascar	11.3	230	19.3	6	85	31	76.9
Malawi	8.2	180	69.4	0	94	35	41.7
Mozambique	15.3	80	19.1	9	43	64	27.6
Niger	7.4	290	5.8	—	—	36	21.5
Tanzania	23.8	130	25.2	4	84	66	52.0
Zaire	34.5	260	14.7	85	6	30	65.9

Sources: World Bank 1991d; UNDP 1991.

in terms of social indicators, such as adult literacy, which ranges from a low of 17 percent in Guinea to a high of 77 percent in Madagascar.

In recounting the factors that contributed to economic decline and the accomplishments and setbacks of these countries in undertaking reforms, we focus on the institutional and endogenous policy-making process that determines economic performance. Particular attention is given to agriculture because of its sizable contribution to value added and exports and its importance in employment generation.

In these chapters we explore the role of the state in African economies and how its shortcomings contributed to economic decline. Prominent among these shortcomings are the failure of the state to foster economic growth, especially in agriculture, as a consequence of its direct involvement in production and allocation of goods and services, and neglect of essential roles such as infrastructure development, agricultural research, and other "crowding-in" investments whereby government spending complements, and thus encourages, investments by private sector agents. The recounting of this shared experience of failure among the ten countries is complemented by an analysis of the obstacles to implementing economic reforms and of the reasons the response is not always in keeping with expectation. The difficulty of transforming the agricultural economy as well as other productive sectors through quick fixes such as state disengagement and getting prices right is emphasized.

Dimensions of Africa's Crisis

To set the context for the discussion of the determinants of economic decline and the magnitude of the economic crisis in the sub-Saharan African countries to be discussed, it is useful to examine a few key economic aggregates. Figure 1.1 presents data on GDP and GDP per capita in Africa from 1976 to 1990. Growth rates were generally in decline from 1976 to 1983 and then followed an oscillating pattern in 1985–1990 that did not indicate any sustained improvement in performance. Over the entire period, the growth rate for middle-income countries in sub-Saharan Africa was approximately 50 percent higher than the growth rate recorded for low-income countries. When the exceedingly rapid rate of population increase is taken into account, the average growth rate of GDP per capita was in fact negative over the entire period for all sub-Saharan Africa.

Among the countries included as case studies in this book, significant differences in the pattern and pace of aggregate growth are evident (Table 1.2). For most of the countries (the obvious exception being oil-exporting Cameroon), the early 1980s showed the worst growth performance. The

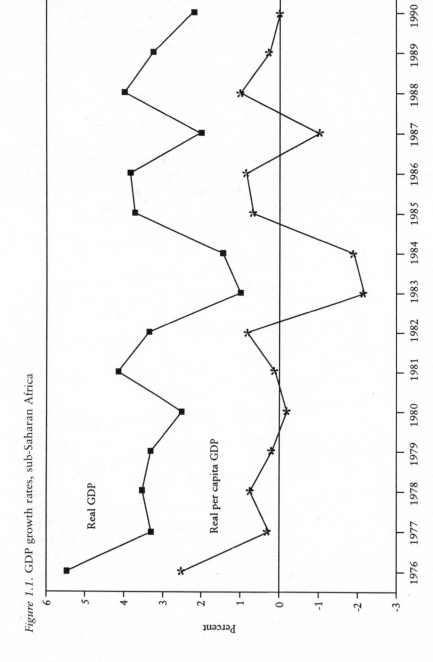

Figure 1.1. GDP growth rates, sub-Saharan Africa

Sources: World Bank [undated data diskettes] Africa Tables; World Bank 1991e.
Note: The figure was derived from an unweighted average of annual growth rates of thirty-nine countries.

Table 1.2. GDP and GDP per capita average annual growth rate

	GDP growth rate (% change)						GDP per capita growth rates (% change)					
	1975–79	1980–83	1984–85	1986–87	1988–89	1990	1975–79	1980–83	1984–85	1986–87	1988–89	1990
Cameroon	8.0	9.7	6.7	0.8	-5.6	-1.2	4.9	6.6	3.6	-2.5	-8.8	-4.4
Gambia	4.5	2.0	0.6	5.5	5.2	2.1	2.5	-0.3	-2.6	2.1	1.9	-1.3
Ghana	-2.1	-2.4	3.7	4.8	5.6	2.7	-3.9	-5.8	0.5	2.3	1.5	0.1
Guinea	—	1.2	1.2	4.9	5.1	4.1	—	-1.1	-1.2	2.3	2.2	1.3
Madagascar	1.5	-2.4	1.5	1.5	3.8	3.5	-0.9	-4.9	-1.5	-2.8	0.6	0.3
Malawi	5.5	0.2	4.4	1.6	3.8	4.7	2.5	-2.9	1.0	-1.8	0.4	1.3
Mozambique[a]	2.9	-7.4	-4.6	2.6	3.8	—	0.4	-10.0	-7.2	-0.1	1.1	—
Niger	5.3	0.7	-6.9	2.0	0.7	—	2.1	-2.7	-10.3	-1.3	-2.3	—
Tanzania	2.0	-0.1	3.0	3.4	4.3	3.9	-1.3	-2.9	0.2	0.6	1.5	1.1
Zaire	-2.9	0.9	2.6	2.7	0.5	—	-5.6	-2.0	-0.6	-0.3	-0.5	—

Sources: World Bank 1991; Government Finance Statistics (IMF, various years); and various sources cited in the following chapters.

[a] These are growth rates for global social product, an eastern bloc–inspired measure of aggregate production. This measure is not the same as GDP but tends to track it closely. See Chapter 8 for a further discussion.

Figure 1.2. Government budget surpluses, sub-Saharan Africa

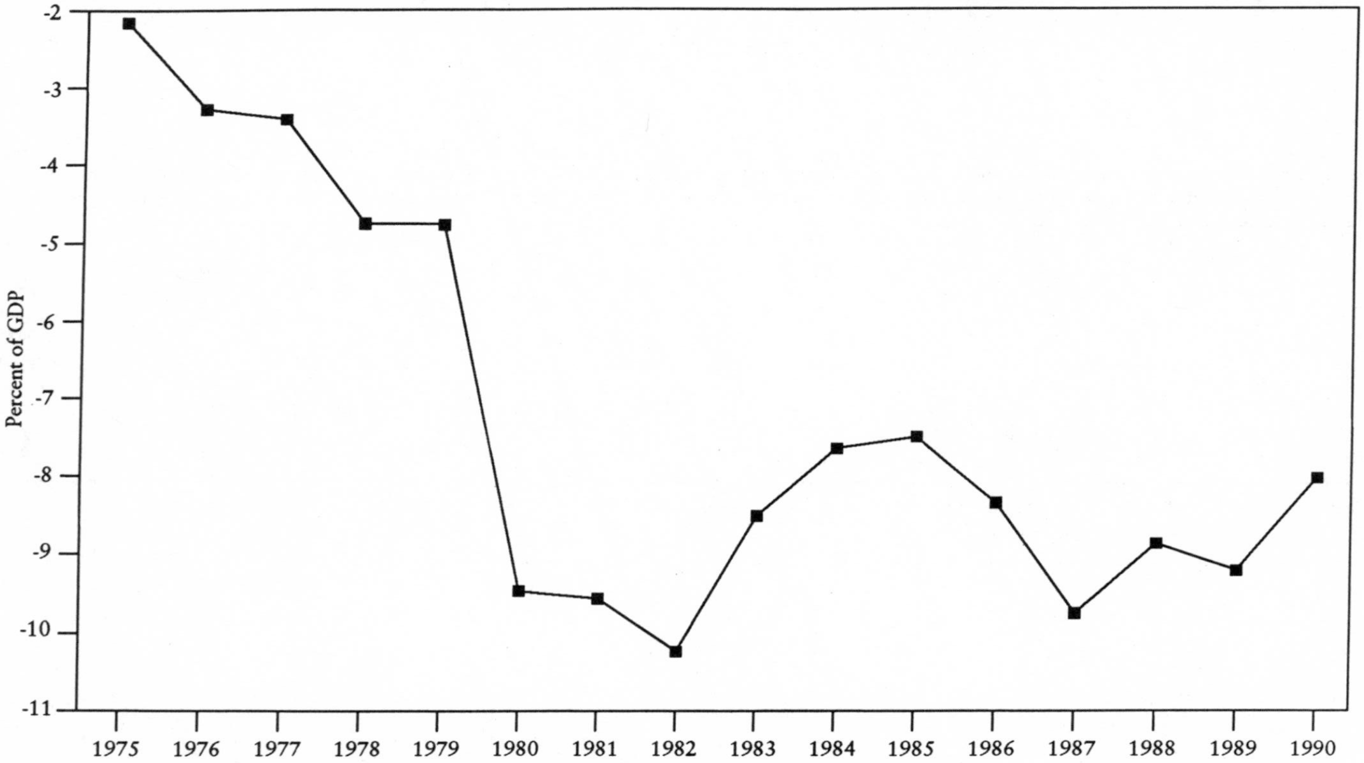

Sources: World Bank [undated data diskettes] Africa Tables; World Bank 1991e.

beginning of World Bank–sponsored adjustment operations coincided or followed shortly after these periods of economic decline. By the late 1980s, when adjustment programs were in place, growth rates were showing some sign of improvement over those observed earlier in the decade, an indication that economic recovery had commenced, albeit slowly.

Aggregate growth, however, was not all that faltered during Africa's years of economic crisis. During this period, internal and external account balances in Africa worsened, ultimately leading to the necessity for undertaking adjustment programs designed to restore financial stability. To illustrate, Figure 1.2 shows how government budget deficits as a share of GDP increased dramatically between 1977 and 1982 and subsequently showed some, albeit modest, improvement. Although the experiences of the individual countries included in this book differ, it was generally the case that the deficit as a share of GDP had reached unsustainable levels by the beginning of the 1980s (Table 1.3).

The worsening budget deficits and balance-of-payments situation in Africa reflected the heavy reliance on foreign borrowing, which temporarily boosted incomes and domestic investment but did little to increase productive capital in the economy. This borrowing was in large part a result of cheap credit made possible from the recycling of petrodollars in the world financial system, coupled with the prevailing view during the late 1970s that encouraged countries to borrow as part of a big push for development. In practice, however, the foreign borrowing did little to improve the performance of the productive sectors of the economies. This low growth in productivity was in turn manifested in the declining volume of exports during the late 1970s and early 1980s. The worsening account imbalances also reflected the concurrent increase in government spending, often to finance unproductive capital investments. One consequence of this spending was the mounting external debt in the region, a trend that began in the early 1970s but continued in earnest during the 1980s. A look at the countries included in this book indicates that total debt as a percentage of GNP grew significantly during the course of the 1980s (Table 1.4). This is especially so in Madagascar, Tanzania, and Zaire. Debt service as a percentage of exports of goods and services, however, did not increase as rapidly as debt as a share of GNP.

Explaining the failing of Africa's economies in general terms is difficult, especially given the diversity. The details of the resource base, economic structure, political economy, and ways these combined to determine economic performance are discussed in the chapters that follow, but there is little question that in general the deterioration of economic performance is explained by a combination of exogenous factors and misguided policies.

Negative external shocks are reflected in the declining terms of trade that

Table 1.3. Budget surplus/deficit and terms of trade

	Budget surplus/deficit (% of GDP)						Terms of trade (1980=100)				
	1975–79	1980–83	1984–85	1986–87	1988–89	1990	1975–79	1980–83	1984–85	1986–87	1988–89
Cameroon	0.8[a]	-1.5	1.1	-1.5	-3.4	—	99.3	104.8	104.8	69.0	68.2
Gambia	-8.0	-16.1	-13.0	-22.0	-7.0	—	110.0	86.8	75.5	63.0	66.4
Ghana	-9.1	-5.2	-1.5	0.8	1.1	0.6	99.0	73.0	74.0	37.0	46.0
Guinea	—	-4.5	-14.2	-4.3	-6.2	-5.7	120.0	98.0	107.0	94.0[b]	—
Madagascar	-5.6	-9.7	-3.6	-3.4	-5.6	-0.8	125.0	94.0	102.0	122.0	115.0
Malawi	-6.0	-8.0	-6.9	-6.0	1.8	—	143.0	115.0	110.0	87.0	80.0
Mozambique	—	-13.6	-17.2	-19.9	29.2[c]	—	—	—	94.0[d,e]	100.0[f,e]	91.0[g,e]
Niger	-4.8	-12.0	-8.3	-9.5	-10.1	—	133.0	112.0	115.0	92.0	80.0[c]
Tanzania	-12.4	-11.5	-8.1	-11.4	-12.1	—	135.7	87.6	97.9	88.5	92.1[c]
Zaire	-5.8	-6.1	-2.6	-7.4	-21.0	—	104.5	86.5	83.5	81.6	75.9

Sources: World Bank 1991e; Government Finance Statistics (IMF, various years); and various sources cited in the following chapters.
[a]1977–79 only.
[b]1986 only.
[c]1988 only.
[d]1985 only.
[e]These are 1987=100. No reliable earlier figures are available.
[f]1987 only.
[g]1989 only.

Table 1.4 External debt, ODA, and cancellation of ODA debt

	Total external debt (% of GNP)		Cancellation of ODA debt (million US$)ᵃ		ODA as % of			
					GDP		GDI	
	1980	1989	1980–87	1988–90	1980	1987	1980	1987
Cameroon	36.8	44.2	—	—	4	2	18.7	9.2
Gambia	—	—	4.2	1.2	25	38	92.3	197.9
Ghana	29.7	54.9	72.0	418.3	4	7	76.6	68.4
Guinea	—	85.3	26.4	242.0	6	10	36.1	61.2
Madagascar	31.5	154.1	17.6	604.5	7	17	27.8	117.4
Malawi	72.1	91.4	106.7	54.8	11	21	46.7	174.0
Mozambique	—	426.8	8.4	270.9	7	49	38.4	238.4
Niger	34.5	79.4	101.9	269.9	7	16	18.4	177.8
Tanzania	50.2	186.1	488.9	145.2	13	25	56.6	149.0
Zaire	33.5	96.6	1.1	737.8	4	8	28.0	83.0

Sources: World Bank 1990g; World Bank 1989a; World Bank 1990f.
ᵃDoes not include France's cancellation of official nonconcessional debt.

began during the 1970s in all but the oil-exporting countries (Figure 1.3). Twenty-four of thirty-two countries in Africa saw their terms of trade decline by more than 3 percent annually between 1970 and 1980, with low-income countries facing the greatest deterioration. Among the ten countries discussed here, there is once again considerable variation, although with the exception of oil-exporting Cameroon all witnessed a decline from the period 1975–1979 to the period 1980–1983. For most countries, the terms of trade continued to fall for the remainder of the decade (Table 1.3).

The state is crucial in setting the policy that conditions the economy's ability to respond to, and therefore cope with, changing exogenous circumstances. In the face of external fluctuations and downturns in the terms of trade such as sub-Saharan Africa experienced, the state is all the more significant. This point is addressed in each of the case studies in some detail. The general picture that emerges is of government policymakers directly contributing to the difficulties. Economic institutions fostered by government policy, often directly under the control of the state, were not sufficiently strong or flexible to respond to changing exogenous factors.

Instead of state intervention promoting growth and fulfilling the development functions to which policy was ostensibly directed, policy failure was felt on various fronts. These generally involved the mismanagement of state-controlled resources, the failure of the state to make enabling or "crowding-in" capital investments that encouraged complementary investments by private sector agents, as well as market failures and repression of private sector and indigenous institutions. Although the country experi-

Figure 1.3. Terms of trade indices, sub-Saharan Africa

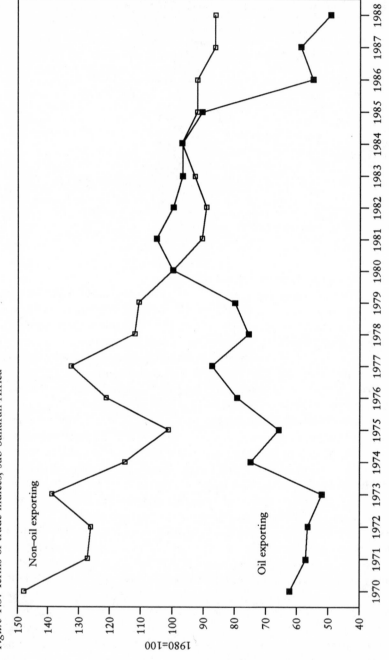

Sources: World Bank [undated data diskettes] Africa Tables; World Bank 1991e.

ences detailed in this book provide ample evidence of the role of the state in contributing to economic decline, the divergence in response to crisis—of government, indigenous and informal civil institutions, and the international community—is also clear.

Responding to Economic Crisis

A major theme of the case studies is that the economic crisis, manifested in account imbalances and stagnation, compelled governments to work together with bilateral donors and international financial institutions, albeit at times in a less than harmonious fashion, to address the failures of the developmental state. The objective was to move toward equilibrium in the current account of the balance of payments. This strategy requires either that steady long-term financing from abroad be provided, that fiscal and monetary restraint be employed to reduce the sum of consumption, government spending, and investment (which are together referred to as absorption), or that market reforms and the incentive structure be changed to foster more efficient use of factors and to raise output so that aggregate incomes and aggregate expenditures become more aligned.

In practice, the cooperation between the international financial institutions and host governments in the process of economic reform assumed two basic forms: stabilization and structural adjustment. Both are designed to restore external and closely related internal financial account balances, although the means differ. Stabilization programs, which are most often associated with the IMF, generally combine the provision of short-term financing with efforts at demand reduction to restore equilibrium. Structural adjustment, in contrast, is in theory designed to increase the efficiency of resource allocation and investment. The removal of distortions in product and factor markets is thus the centerpiece of the structural adjustment programs of the World Bank, which also provide medium- and long-term quick-dispersing financing in exchange for government promises to adopt new incentive structures to encourage increased production and marketing. This financing, coupled with the growth that is supposed to be associated with the realignment of relative prices, is intended to lessen the hardships of restoring balance between aggregate demand and supply. Put simply, the intention is that the new structure of incentives induce a large enough supply-side response to render any contraction in demand unnecessary.

In practice, the terminology of structural adjustment and stabilization suffers from imprecision and emotive connotation. This is partially a consequence of the identification of stabilization with the IMF and of adjustment with the World Bank, institutions that have long been criticized for promulgating demand-reducing policies that are particularly harmful to the poor. The distributional aspect of reform is the subject of a complementary volume under preparation; the equally salient issue addressed in this volume is the characteristics of reform: has it been primarily expenditure-reducing, bringing supply and demand into balance through contractionary income-lowering policies, or expenditure-switching, shifting relative prices and thereby restoring incentives to producers in order to generate a positive supply response? Put in other terms, has reform been associated with tangible measures that have brought about increases in output or, instead, only declines in absorption?

The extent to which economic reform programs emphasize stabilization versus structural adjustment, and the extent to which either pathway to restoring account balances is successful, differs dramatically from one country to the next. In fact, the nature of adjustment is largely dependent on prior conditions—especially, in this context, the degree to which the 1970s resulted in unsustainable investment and foreign borrowing. But what is similar is that, in most of the countries discussed in this book, stabilization and structural adjustment go hand in hand. It is nonetheless arguable that demand reduction is a more central feature of the process of economic reform in some countries, whereas growth-oriented structural adjustment is the predominant feature of the process of reform in others.

The practical distinction between stabilization and structural adjustment is often tenuous, for several reasons. First, in any given country, policy orientation changes over time, so that an initial emphasis on fiscal and monetary restraint may be quickly followed by a shift toward increasing efficiency of resource allocation and investment. Second, many of the major types of policy instruments employed in stabilization and structural adjustment are the same. Depreciation of the currency, monetary contraction and interest rate reform, budgetary restraint, and tax adjustment can all arguably fall in either domain, although others such as trade and market liberalization and related institutional reforms of public enterprises are more closely identified with structural adjustment than with stabilization.

A key point in examining and evaluating the process of reform is that stabilization and structural adjustment programs can be mutually reinforcing. Structural adjustment is more likely to succeed in a relatively stable economic environment, and many structural adjustment polices (e.g., improving efficiency of public sector enterprises) promote the objectives of a stabilization program. Yet certain policies that fall under the domain of structural adjustment may tend to worsen internal and external account balances in the short term. For example, removing trade restrictions on all imports may be advisable on grounds of efficiency. If tariff revenues are an important part of the budget, however, deficits can rise in the short term. There may also be justification for adopting an expansionary fiscal policy to counter a possible reduction of economic activity in the transitional period due to trade liberalization, contributing to an overall worsening of existing imbalances. Thus, whereas stabilization and structural adjustment are in practice highly interdependent and generally mutually reinforcing, in some cases decisionmakers confront trade-offs among policies aimed at restoring equilibrium in the account balances and those designed to foster greater long-term economic growth.

The process of adjustment has led to a variety of policy reforms (and subsequent policy reversals), which in turn have had differing degrees of success in turning countries' economic performance around. But one thing constant across all countries reviewed in this volume is the increasingly important role of donors in trying to restore growth to sub-Saharan Africa by reducing the large and burdensome financing gap. This assistance has taken several forms, including debt cancellations, donor financing, and the related conditionality that accompanies such financing.

In regard to the issue of the growing debt burden, the end of the 1980s has witnessed considerable progress in the cancelling of Official Development Assistance (ODA) debt (Table 1.4) in addition to major efforts at rescheduling. Nonetheless, there remains much to be done. Concerning the issue of financing, the World Bank, IMF, and bilateral donors have generally assumed a major role in transferring significant resources as an accompaniment to policy-based reform. In fact, these financial flows have actually been as important a part of the story of economic reform as the conditionality the loans were intended to support.

Overall, the 1980s saw dramatic growth in ODA to sub-Saharan Africa. For example, net ODA in the form of International Development Association (IDA) loans from the World Bank increased by an average annual rate of 15 percent during the period 1980–85, with the rate of increase accelerating to 49 percent in 1986 before moderating to 10 percent in 1987. But the most dramatic aspect of the burgeoning importance of ODA is seen when it is examined as a share of recipient GDP. In Africa as a whole, this portion doubled from 13.4 percent in 1980 to 26.7 percent in 1987, although in the countries surveyed here there is considerable variability. At one extreme, ODA represented nearly half the size of Mozambique's GDP in 1987, up from just 7 percent in 1980, whereas in Cameroon the importance of ODA actually fell during the course of the decade (Table 1.4). This fall is, however, anomalous: all other countries showed significant increases.

ODA's share of gross domestic investment (GDI) increased even more rapidly, owing to the virtual disappearance of the domestically financed investment budget in much of sub-Saharan Africa. Excluding from consideration Cameroon, we see at the low end of the spectrum that Guinea's ODA as a percentage of GDI did not quite double to reach 61 percent at the end of the decade. This figure is in contrast to the case of Niger, where the percentage increased nearly ten times (Table 1.4), reflecting an increase in aid coupled with a fall in gross investment in the post-adjustment years. Thus, there is little question that, if nothing else, the years of adjustment in Africa have brought about an increasing role for ODA disbursements from bilateral and multilateral agencies, with the foreign savings implied representing a sizable share of GDP and often most of the investment budget.

In fact, the growth of ODA is attributable largely to financing to promote economic reform programs. This case is illustrated in Table 1.5, which shows the levels of financial commitment from the World Bank in support of reform programs. There was a dramatic rise in World Bank policy-based lending, although no such clear trend is observed in IMF financing. The increase in World Bank policy-based lending for the countries included here parallels the trend observed for sub-Saharan Africa as a whole, where the US$135.0 million in 1980 is but a small fraction of the $1,544.7 in 1989.

Early recipients of World Bank financing were Malawi and Tanzania. Whereas Malawi remained on good terms with the Bank and was perceived to be serious about the process of reform, the opposite was initially the case for Tanzania, although this changed in the late 1980s when financial flows in the form of policy-based lending were renewed. Ghana is not only the star economic performer among the countries included here but also received considerably more financial commitment in the form of adjustment loans from the World Bank. At the bottom end of financial commitment to support reform is Gambia, with its population of only 800,000 persons. Nonetheless, population size was not the key determinant of the level of financial support. Mozambique, for example, with a population comparable to that of Ghana, received less than half Ghana's amount of policy-based lending during the decade. Instead of simply population, a combination of criteria were used, including the perceived need for donor support, the opportunity for external financing to make a useful contribution to the development process, the willingness on the part of donors and recipients to negotiate a set of conditions, the assessment by the Bank as to whether the adherence to the earlier conditions warrants further loan agreements, and a variety of other subjective judgments.

Although the issue of financial flow is clearly of importance, of greater concern in this volume is the conditionality that accompanied the money, the degree to which it was adhered to, and the consequences of the reforms undertaken. In this regard, the contributors to this book concentrate on

Table 1.5. World Bank financial commitment in support of reform programs (million US$)

	1980	1981	1982	1983	1984	1985	1986	1987	1988	1989	Cumulative
Cameroon	—	—	—	—	—	—	—	—	—	150.0	150.0
Gambia	—	—	—	—	—	—	—	16.5	—	23.0	39.5
Ghana	—	—	—	40.0	76.0	87.0	53.5	115.0	100.0	126.6	598.1
Guinea	—	—	—	—	—	—	42.0	—	65.0	—	107.0
Madagascar	—	—	—	—	—	60.0	93.0	83.0	125.0	1.4	362.4
Malawi	—	45.0	—	5.0	55.0	—	70.0	10.0	70.0	5.2	260.4
Mozambique	—	—	—	—	—	45.0	—	—	88.6	90.0	223.6
Niger	—	—	—	—	—	—	60.0	80.0	—	—	140.0
Tanzania	—	50.0	—	—	—	—	—	96.2	30.0	147.5	323.7
Zaire	—	—	—	—	—	—	80.0	149.3	—	—	229.3
TOTALS	0.0	95.0	0.0	45.0	131.0	192.0	398.5	550.0	478.6	543.7	2,434.0
TOTAL SUB-SAHARA	135.0	170.0	220.0	336.5	818.2	299.9	1,023.3	1,221.4	1,005.6	1,544.7	6,774.6

Source: World Bank.

Table 1.6. Government expenditure and lending minus repayment (percentage of GDP)

	1977	1978	1979	1980	1981	1982	1983	1984	1985	1986	1987	1988	19
Cameroon	16.0	16.0	15.0	14.2	18.8	19.3	22.9	22.7	22.3	21.6	24.3	22.3	20
Gambia	24.2	36.1	27.7	31.1	32.0	32.0	29.9	31.4	30.5	27.7	43.3	40.8	28
Ghana	20.0	15.7	15.6	11.1	11.0	11.2	8.2	10.1	14.0	14.3	14.4	14.2	14
Guinea	—	—	—	32.5	27.4	49.6	33.5	27.0	42.4	23.0	20.2	21.8	20
Madagascar	18.2	21.6	26.4	25.1	24.3	19.4	17.1	18.0	17.0	16.0	18.5	17.1	22
Malawi	23.5	30.4	34.5	39.4	35.7	29.4	28.7	27.5	32.1	34.1	31.2	24.1	22
Niger	13.5	15.7	16.6	19.1	23.6	18.4	19.0	19.2	20.2	20.8	20.7	19.0	—
Tanzania	25.6	28.4	35.2	28.7	27.7	31.6	26.2	23.5	21.5	20.3	21.4	21.0	—
Zaire	16.9	16.4	16.1	12.1	13.5	12.9	11.8	15.9	15.3	16.7	20.0	17.5	—

Source: Data from World Bank 1991d.

how the role of the state in the economy has been transformed by economic adjustment. Broadly speaking, two major areas are discussed. One is the area of fiscal policy, including public expenditures and generation of revenues. Given the objective of reducing the fiscal deficit, adjustment programs are often concentrated on the complementary objectives of reducing the size of government and raising and rationalizing taxation. As can be seen in Table 1.6, however, fiscal contraction was observed in only a few countries (Tanzania, Guinea, and Cameroon) but did not occur after the beginning of adjustment programs in the other cases examined. In some cases, such as Ghana, government actually increased in size relative to GDP.

A second area of concern, one of greater relevance to the case studies that follow, is how the role of the state in terms of intervening in markets, attempting to allocate goods and resources, setting macro and sectoral prices, including those for foreign exchange and agricultural products, has evolved under adjustment. Rather than focus on the size of government, then, we concentrate on the behavior of government, on its influence on markets and civil economic institutions, both formal and informal. Particular attention is paid to the process whereby, after independence, the state attempted to redefine the rules by which the economy operated, seeking to impose new institutional structures to replace informal and indigenous ones. Occasionally those institutions born during the colonial period were targeted for elimination, although in many instances colonial structures and institutions, albeit marginally transformed, became the basis for new institutions—for example, parastatals controlling export crops or state-run factories and trading companies. This discussion of the evolution of new institutional structures is followed by the analysis of one of the key elements of adjustment: the liberalization of markets and the disengagement of the state. Special consideration is given to public enterprises, which have become so closely identified with the fostering of inefficiencies and rent seeking. In addition, the state's role in the agricultural sector, which is predominant in the economy both in terms of its contribution to GDP and as a source of employment and incomes, is accorded thorough scrutiny. The role of the state in other sectors, especially manufacturing, is discussed in keeping with its importance in the economies discussed.

The experiences of the countries examined differ dramatically with regard to the need for, and nature and degree of, institutional and market reforms. The extent and means employed by the state to directly control and repress markets before reform is in fact the key element in determining the appropriate response to the economic crisis; that is, the appropriateness and adequacy of the adjustment process is conditioned by prior conditions such as the degree and method of taxation of farmers (directly as well as indirectly through the exchange rate); the extent to which marketed output was controlled by parastatals; the pervasiveness and enforcement of systems of rationing, including for foreign exchange, agricultural inputs and food, credit, and other scarce resources; the degree to which parallel markets arose in response to market controls; and the level of transaction costs involved in these parallel markets. The consequences of, and proper responses to, these conditions, which are a result of intervention by the state coupled with external factors, are discussed in each of the chapters.

We also explore the effects of attempts to address the economic crisis on the various economies and different sectors within them. We examine changes in trade policy and macro prices, such as the cost of foreign exchange, as well as in consumer and producer prices in order to determine the process by which, and direction in which, they have evolved as adjustment proceeds. We review as well the effect of reform on price incentives within the context of the role of the state and its effect on the functioning of markets.

Indeed, the importance of prices is amply illustrated and discussed throughout this volume. An analysis of prices, how they are formed, and the limitations of price policy is reinforced in the case studies by a discussion of the role of institutions, both formal and informal. This discussion includes, first, an exploration of the emergence after independence of officially sanctioned institutions such as parastatals which assumed a vast array of roles, leading to corruption, inefficiencies, and rent seeking. Of greater importance, however, is the emergence of informal and indigenous institutions to compensate or cope with ineffectual state control. Included in the chapters, therefore, is considerable discussion of parallel markets and other coping mechanisms to respond to the state's attempts to impose new institutional structures that actively intervene in the economy.

Plan of the Book

After this contextual introduction, we begin in Chapters 2 and 3 with the West African nations of Ghana and Guinea, the two countries where reform has perhaps proceeded most briskly. Harold Alderman notes in Chapter 2 that Ghana witnessed a dramatic decline in its economic position as one of the wealthiest countries in sub-Saharan Africa at independence. Attempts at an import-substitution strategy of development and neglect of agriculture, including a heavy taxation of cocoa producers, eventually led to a virtual collapse of the state. Since the beginning of the reform program in 1983, Ghana has been one of the star performers in Africa. In fact, it is most often cited as the model of what adjustment can accomplish, and it illustrates the potential benefits of vigorous pursuit of economic reform. In addition, the Ghana case provides some compelling evidence that adjustment is not necessarily synonymous with the reduction of the role of the state; quite the contrary, government was prominent in promoting recovery and providing goods and services that were unavailable before the reforms. In addition, despite the increase in agricultural output, especially from the cocoa sector, the response of agriculture to economic restructuring has been slower than that of other sectors—thus illustrating structural problems that go beyond issues of price policy, including an overdependence on tree crops and a lack of institutional requirements to promote technical change.

The excesses of the state in Guinea, the subject of Chapter 3, authored by Jehan Arulpragasam and David E. Sahn, was perhaps unparalleled throughout sub-Saharan Africa in terms of its pervasiveness in all aspects of economic life and its calamitous implications. By the time the process of economic reform commenced in 1985, the economy had deteriorated in all respects, ranging from the elimination of legal agricultural exports, to a domestic banking system that was insolvent, to a system of corruption and rent seeking that was the basis for economic survival in a complex of state control enterprises no longer capable of delivering the most basic services. Despite this deterioration, the enclave mining sector sustained the state from total collapse until the mid-1980s. Since then, however, Guinea has made major efforts toward reform. Dramatic changes in most aspects of economic policy, ranging from macroeconomic management to institutional reform, make for an interesting story on the scope and limitations of adjustment in restoring economic growth.

Standing in stark contrast to the concerted effort at reform in Ghana and Guinea is Zaire in central Africa. Zaire's large size and good resource base did little to protect it from the abuses of state control and extraction of rents that debilitated the economy. In fact, the discussion of Zaire in Chapter 4 by wa Bilenga Tshishimbi demonstrates the deleterious consequences of adjustment that is chaotic, not sustained, and subject to policy reversals, presumably in response to greed and the immediate political imperatives of the day. Nonetheless, the evidence suggests that in the first two years after the initial 1983 reforms, when the reforms were allowed to proceed, the economy responded positively. Still, subsequent fiscal and monetary goals were not met, and by the end of the 1980s the reform process was completely off-track. The Zaire story, although anomalous among the other countries included in this book in terms of the unsustainability of reforms, is not unique in Africa, as witnessed by similar stories in Somalia and Zambia, for example. The case of Zaire raises important issues of political economy and the sustainability of reform. We are left asking how adjustment can be crafted to be more politically acceptable, and under what circumstances the political environment is so hostile and unstable as to preclude a successful economic recovery program.

Among the other countries included in this book are two CFA countries also located in West Africa, Cameroon and Niger. Cameroon, the subject of Chapter 5, authored by David Blandford and colleagues, is an oil exporter and properly classified as a middle-income country. Cameroon's economic performance was considerably better than the other countries' in this volume because of its advantageous resource endowment as well as its pursuit of relatively sound economic policies. This was so both before and during the oil boom. In fact, Cameroon was able to avoid many of the most harmful manifestations of Dutch disease (see Chapter 5, note 3), in contrast to neighboring Nigeria. This success may be attributed in part to Cameroon's membership in the franc zone and all the benefits and constraints that involves. Nonetheless, when the dollar-denominated prices of Cameroon's exports, especially oil, fell dramatically during the second half of the 1980s, Cameroon had little recourse but to look to the IMF and World Bank. Thus a process of structural adjustment began, albeit later than the other countries discussed in this book and in response to different circumstances. Adjustment's short history in Cameroon makes it difficult to judge whether the measures taken are sufficient to restore growth. At best, however, the pace of reform has been slow and characterized by greater austerity than in the other countries.

In contrast to Cameroon, with its ample resource endowment, is Niger. Its landlocked location in the Sahel, irregular rainfall, noticeable degradation of the fragile soils, and extreme dependence on uranium exports are among the most important factors behind Niger's precarious economic status, reflected in the financial crisis that precipitated the need for adjustment. As discussed by Paul Dorosh in Chapter 6, factors exacerbating Niger's attempts to cope with these structural weaknesses include its inability to adjust the exchange rate and the substantial dependence of the country's economic situation on policy in neighboring Nigeria. But perhaps of greater importance, Niger illustrates the limits of policymaking and state intervention in an economy where the vast majority of the population has little contact with formal sector institutions.

After Niger, the country included here with the most limited resource base is likely the other landlocked country, Malawi. Economic adjustment began early in Malawi, which has been viewed by donors and international financial institutions as a country committed to reform. Despite its characterization as a country that has undertaken an ambitious and comprehensive reform program, David E. Sahn and Jehan Arulpragasam casts some doubt in chapter 7 on the degree to which far-reaching structural change has occurred and suggest that the process of reform may have been much less comprehensive than usually portrayed. This is especially the case in the key sector, agriculture, where through the enforced duality between the export sector and the smallholder sector the state has actively contributed to technological stagnation and low productivity in both sectors.

The scarcity of cultivable land and primary product resources in Malawi stands in contrast to the other southern African country in this book, neighboring Mozambique. But in Chapter 8 Steven Kyle indicates that the favorable resource base, including good ports and plentiful land and minerals, has done little to foster development in Mozambique. The constant state of war that has ravaged the economy since independence, coupled with the faith in the command economy in the early years, so thoroughly distorted the structure of incentives that in the three years 1982–85 overall production fell by 30 percent. It was in this context that the government of Mozambique began its economic reform program in 1987. Policy changes have been tangible, and the response satisfactory. Nonetheless, security problems continue to plague the economy, and progress in rehabilitating infrastructure has been limited. Furthermore, the growing dependence on aid, especially for food, is not likely to be alleviated in the near future, even under the most optimistic scenarios.

Mozambique's neighbor to the north, Tanzania, is perhaps best identified with its quest to achieve equity through African socialism. As discussed in Chapter 9 by Alexander H. Sarris and Rogier Van den Brink, it too witnessed economic failure in the 1970s and the first half of the 1980s. The crisis eventually became so severe as to precipitate a serious effort at reform during the late 1980s. Looking back, the authors note that poor economic management, coupled with unfavorable external conditions, brought about stagnation in the rural sector, decay of infrastructure and social services, and a collapse of the industrial base. These problems initially led to the call for reform in the early 1980s, but the feeble attempts of an uncommitted bureaucracy were unmitigated failures. And though the recovery program that was introduced in the second half of the 1980s was more extensive, it could nonetheless best be described as gradualist. Despite the resistance to a rapid pace of reform, there is evidence that the Tanzanian economy is responding, mainly driven by the agricultural sector. Threatening continued recovery, however, is the difficulty of bringing about true institutional and administrative reforms in an economy heavily encumbered by state controls through parastatals and public enterprises.

The final two countries included in this book are Gambia in West Africa and the island nation of Madagascar. Surrounded by the dominating economy of Senegal, the degrees of freedom open to policymaking in Gambia are fewer than in the other countries included in this volume. Still, the need for embarking on an economic recovery program was no less compelling as Gambia witnessed a deterioration in its economy in the late 1970s and early 1980s because of a combination of worsening terms of trade, fiscal deficits, and distortions in domestic pricing and the exchange rate. The most important reforms undertaken as part of the economic recovery program concern the groundnut sector, in particular, bringing domestic prices of groundnut in line with international prices. As discussed by Cathy Jabara in Chapter 10, a series of institutional reforms regarding marketing of groundnuts and agricultural inputs have commenced, and the limited progress in these areas has been important in defining the need for further reform efforts to achieve the types of sustainable growth of the economy desired for the 1990s.

Madagascar distinguishes itself in that its rice-based economy is in many ways more analogous to an Asian country than to its African neighbors. Nonetheless, the genesis of its economic crisis and the process of economic reform, as outlined in Chapter 11 by Paul Dorosh and René Bernier, elucidates some themes common to many other countries included in this volume. In particular, it was Madagascar's heavy investment push beginning in the late 1970s that initially led to the country's financial crisis. In response, the government first concentrated its reform program on stabilization measures, an effort that successfully reduced large government budget and balance of payment deficits. Other major policy reforms to address structural constraints to growth, ranging from price policy to institutional reforms of marketing structures, were postponed for several years. Arguably this sequencing of reforms retarded the important, albeit belated supply response that the economy has finally shown since 1986.

In the closing chapter of this book, David E. Sahn distills some of the salient lessons about the burgeoning economic crisis that gripped Africa during the late 1970s and early 1980s and the early attempts to rectify the situation. These lessons, we hope, will guide policymakers in the years ahead as Africa enters its second decade of adjustment facing challenges that seem only marginally less daunting than those at the commencement of the process of economic reform.

2

Ghana: Adjustment's Star Pupil?

Harold Alderman

A recently published travel guide on West Africa claims that Ghanaians are a favorite of travelers to the region. Though this hypothesis is possibly testable, it will most certainly have its challengers. A similar claim, however, can be made that Ghana is a favorite of economists. The first sub-Saharan colony to gain its independence, in 1957, Ghana has often tested the validity of theories of economic development. For example, as Killick (1978) illustrates, Ghana consulted some of the most noted economists of the period and received tacit endorsement for its ambitious development plan in the early 1960s. Killick also indicates how unrealistic the plan was in the face of political constraints and poor implementation. In other contexts as well, Ghana's economic planning experiences, mainly misguided efforts, have been used to draw general conclusions about institutions and the development process; an example is Bates's (1981) essay on the tendency of interest groups to capture programs and policies.

Ghana's recent economic performance is also presented as a case study of the impact of economic policies often, but not always, in a favorable light. As is now well known, Ghana was unable to sustain positive economic growth in the 1970s or early 1980s; GNP declined at a rate of 1 percent per annum between 1975 and 1983. In contrast, the growth in GNP and GNP per capita was positive every year between 1983 and 1989, with the

This essay is based on my monograph *Downturn and Economic Recovery in Ghana: Impacts on the Poor,* Cornell Food and Nutrition Policy Program, Monograph no. 10 (Ithaca, N.Y.: CFNPP, 1991). I thank Paul Higgins and Gerald Shively for assistance in drafting that monograph and the Cornell Food and Nutrition Policy Program for permission to reproduce material here.

Table 2.1. Value added to GDP, by sector, Ghana (percentage change)

	1974	1975	1976	1977	1978	1979	1980	1981	1982	1983	1984	1985	1986	1987	1988	1989	1990
GNP	—	-12.4	-3.7	2.7	8.6	-2.7	0.5	-3.5	-6.9	-4.6	8.6	5.1	5.2	4.8	5.6	5.1	3.0
Agriculture	—	-19.9	-1.6	-5.4	19.1	3.8	2.2	-2.6	-5.5	-7.0	9.7	0.7	3.3	0.0	3.6	4.2	-2.4
Agri./livestock	-21.3	—	—	—	—	—	0.1	-0.8	-4.2	-8.0	15.5	-1.9	0.2	-0.3	6.0	5.1	-4.7
Cocoa	-26.1	—	—	—	—	—	9.5	-4.4	-17.2	-14.2	-8.4	13.2	18.2	3.3	-6.4	3.2	3.4
Forestry	-1.7	—	—	—	—	—	2.0	-12.2	7.8	7.4	1.4	0.2	1.1	1.5	3.4	1.2	3.7
Fishing	7.4	—	—	—	—	—	9.1	1.1	-4.2	3.4	0.7	12.0	14.0	-10.1	2.2	0.6	1.7
Industry	—	-1.0	-2.6	4.1	-7.5	-14.2	0.3	-16.0	-17.0	-12.0	9.1	17.6	7.6	11.3	7.4	4.2	4.3
Mining	—	-5.8	-4.2	-9.9	-6.9	-13.1	-3.1	-7.3	-7.6	-14.4	13.5	6.5	-3.0	7.9	17.8	10.0	10.5
Construction	—	-21.3	1.6	12.9	-17.8	-10.9	4.3	-15.6	-13.5	-14.5	2.3	2.8	-2.7	15.0	9.2	4.2	5.2
Manufacturing	—	9.1	-4.4	3.3	-3.5	-16.8	-1.4	-19.3	-20.5	-11.2	12.0	24.3	11.0	10.0	5.1	3.0	2.6
Electricity/water	—	-5.5	13.2	-0.5	-11.7	17.3	12.9	11.9	-8.1	-7.0	-6.1	20.7	18.0	18.7	12.9	7.7	8.8
Services	—	-5.0	-4.0	14.0	7.8	-1.8	-2.3	3.3	-3.6	2.3	6.6	7.5	6.5	9.4	7.8	5.8	8.6
Transp./commun.	—	1.3	-18.5	13.6	-5.4	5.8	-13.2	6.8	1.1	7.3	12.8	8.5	5.6	10.9	10.2	7.9	12.0
Trade	—	-17.2	-7.4	2.0	-3.1	3.5	-8.6	-1.9	-10.4	5.3	10.1	13.7	9.0	17.4	7.4	7.5	12.5
Finance	—	-15.2	6.6	12.2	9.3	-0.7	3.9	4.5	3.3	3.3	9.3	2.6	7.7	5.5	6.7	3.9	7.3
Government	—	22.5	1.3	31.8	22.9	-8.6	3.5	6.2	-3.4	5.7	1.3	6.0	4.2	5.1	8.1	4.8	4.9

Sources: World Bank 1989d, 1989e, 1991a; Government of Ghana, Quarterly Digest of Statistics (Accra: Ghana Statistical Service, various years).

former measure averaging 5.4 percent per year between 1983 and 1988 (Table 2.1). It is natural, then, to ask which policies contributed to this turnaround.

Ghana has attracted attention, however, not solely because it managed to achieve an economic recovery, but also because it attempted to do so without creating a class of newly poor. In addition to policies designed to correct structural imbalances, the government initiated the Programme of Actions to Mitigate the Social Costs of Adjustment (PAMSCAD) with bilateral and multilateral donor support. It is, therefore, also natural to inquire what social costs were incurred as well as how effective compensatory programs have been.

Adjustment and poverty alleviation may, of course, not be distinct concepts. Not only does a growing economy make the task of reducing poverty easier, the absence of a large disaffected group may contribute to the sustainability of the programs that constitute the economic restructuring. This latter issue is of concern because Ghana has had several abortive attempts at economic reform in its short history (Leith 1974; Rimmer 1992).

Ghana's First Quarter Century: Black Star Setting

One cannot ignore Ghana's history when evaluating the options open to the economy at any given time.[1] For Ghana, a key to its subsequent economic path was the introduction of cocoa in the late nineteenth century. From that time until the middle of the 1970s, Ghana dominated the world cocoa market. To a large degree, cocoa dominated Ghana as well. Although a small part of the entire economy (cocoa comprised roughly 15 percent of agriculture by value and less than 7 percent of GNP in 1990), cocoa dominates export earnings and government revenues. In 1980, for example, after a decade of decline of cocoa production, it accounted for 64 percent of export earnings and 18 percent of government revenues. The instability inherent in relying on one export for revenues is indicated by the fact that in 1979 the share of revenues from cocoa duties was twice that of 1980 and in 1981 cocoa provided a net drain on the treasury.[2]

Partially in response to this price volatility, the colonial government established the Cocoa Marketing Board (CMB). There was an additional, largely unstated, presumption that the state was more efficient than the

[1] Path dependency in economic history has been explored by several researchers (e.g., David 1985).
[2] As is discussed in Garcia and Llamas 1988, export booms, if not impeccably managed, can be as dangerous as busts.

producer at price forecasting and at saving. In actuality, after independence, surpluses accumulated during periods of high prices were viewed as
part of the general budget and not as funds held in reserves for income
stabilization. Thus, as early as 1959, when cocoa revenues were pooled in
the general development budget, the CMB had no real role in the stabilization of incomes of cocoa producers (Rimmer 1992).

By the mid-1960s, the CMB added long-run supply management to its
stated objectives and in so doing became a manifestation of the growing
role of the state in virtually all aspects of the economy. In addition to
attempting to manage exports, in the 1960s the Nkrumah government
embarked on a development program that attempted to increase the nation's capital stock rapidly with major public investments guided by a
national plan. This "big push," however, eventually resulted in investment
without incremental output. Whereas some large projects such as the Volta
dam went more or less as planned, investment generally proceeded in
virtual disregard to planning. Killick (1978) presents examples, widely
quoted in subsequent articles and books, of factories being built with no
consideration for the source of raw materials or with capacity far in excess
of any market absorption potential.

It is noteworthy that the national savings rate, a key feature of economic
growth in most economic models of the period, was quite high during the
first half of the 1960s. Investment absorbed over 20 percent of GDP between 1963 and 1965 (Rimmer 1992). The financing of this misguided
investment was in part through foreign debt. Direct foreign investment and
aid combined provided only 10 percent of fixed investment, mostly for the
Volta dam and the attendant aluminum smelting project. An appreciable
share, on the other hand, came through retained earnings and reduced
consumption. Also noteworthy is the fact that, despite an ideology that
favored the public sector, the relative growth in state-owned enterprises
was sought at the margin—that is, through investment rather than expro-
priations. The net effect, however, was similar: with foreign investment
actively discouraged for most of the period before 1983 and with foreign
exchange and the inputs it could obtain heavily regulated, the legacy of this
strategy has been a stunted private industrial sector.

Nevertheless, the most pervasive feature of economic mismanagement
was not the ill-advised investment strategy but the inability to recognize
that real exchange rates are not set by fiat. The investment push not only
depleted the appreciable foreign reserves with which Ghana was favored at
independence, it contributed to macroeconomic mismanagement that
eroded the real exchange rate. Devaluation, a logical though temporary
solution to the resulting trade imbalances, was rarely viewed as a viable
option; a devaluation in December 1971 was followed by a coup the

following month (Table 2.2). Whether there was a causal connection or not, subsequent governments failed to devalue the currency until the parallel exchange rate was eight times the official. By that time (1978), the distortions in export incentives had undermined the cocoa industry.[3] The government may not have even noticed the direct linkage of the sectoral decline and exchange distortions; as Krueger et al. (1988) point out, on paper the cocoa sector received a net nominal subsidy, although the exchange rate distortion resulted in a large implicit tax. Falling production, masked in the mid-1970s by high world prices, as well as increased incentives for growers to smuggle their output to neighboring countries (May 1985), led to reduced government revenues.

The fiscal consequences of foreign exchange misalignment were also affected by the licensing of imports. Since the smuggling of output usually also implies the smuggling in of commodities, import duties were reduced. The overall inability to mobilize resources contributed to the government's deficit (averaging 9 percent of GDP between 1976 and 1979; see Table 2.3). This led to subsequent inflation and the erosion of the inadequate devaluation.[4] In turn, the eroding exchange rate limited the prices paid for export crops and encouraged additional smuggling.

Here we touch on a general issue relevant to any analysis of the Ghanaian economy or of general welfare. The economic crisis in the latter half of the 1970s and the first few years of the 1980s was of sufficient magnitude and duration to have reoriented most forms of economic relations. Many observers write of the *kalabule,* or underground, economy (Chazan 1983). Although this term often carries a negative connotation similar to the English word *profiteering,* it is also indicative of an economy reverting to barter as well as smuggling. Such barter transactions are more reflective of the scarcity value of goods and labor than of cash transactions at official (controlled) wages and prices.

There is an economic cost to the type of rent seeking that characterized Ghana in recent years; resources devoted to subverting controls are a deadweight loss to society in general (Azam and Besley 1989). Moreover, as difficult as it is to gain a picture of the overall level of production and trade during the period of economic disassociation, it is even more difficult to explore the distribution of economic rents. To some degree the policies of deregulation that are central to economic restructuring have as great an impact on this distribution of rents as on the overall level of the economy. These, however difficult to measure, may account for the economic and political consequences of policy reforms.

---

[3] Over twenty-five years Ghana's share of world production had declined from over 40 percent to barely above 10 percent (Bateman et al. 1990).

[4] The effectiveness index of Ghanaian devaluations is discussed in Alderman 1991a.

Table 2.2. Nominal and real exchange rates, Ghana

	Official exchange rate (C/$, 1984=100)			Parallel exchange rates (C/$, 1984=100)			
	Official nominal rate	Index of official rate	Index of real official	Parallel nominal rate	Index of parallel rate	Index of real parallel	Paralle nomin. (%)
1964	0.714	1.96	246.70	1.18	1.22	153.86	165
1965	0.714	1.96	202.06	1.41	1.46	150.59	197
1966	0.714	1.96	183.15	1.38	1.43	133.59	193
1967.1	0.714	1.96	201.60	1.77	1.83	188.60	248
1967.2	0.714	1.96	194.87	1.77	1.83	182.31	248
1967.3	1.020	2.80	288.24	1.77	1.83	188.76	174
1967.4	1.020	2.80	284.61	1.77	1.83	186.38	174
1968	1.020	2.80	253.84	1.86	1.92	174.69	182
1970	1.020	2.80	244.91	1.69	1.75	153.14	166
1971	1.020	2.80	247.09	1.52	1.57	138.96	149
1972.1	1.820	4.99	460.89	1.68	1.74	160.55	92
1972.2	1.280	3.51	300.58	1.68	1.74	148.88	131
1972.3	1.280	3.51	310.97	1.68	1.74	154.03	131
1972.4	1.280	3.51	312.44	1.68	1.74	154.76	131
1973.1	1.210	3.32	301.52	1.50	1.55	141.06	124
1973.2	1.150	3.15	278.11	1.50	1.55	136.90	130
1973.3	1.150	3.15	271.16	1.50	1.55	133.47	130
1973.4	1.150	3.15	291.53	1.50	1.55	143.50	130
1974	1.150	3.15	279.03	1.73	1.79	158.41	150
1975	1.150	3.15	240.34	1.99	2.06	156.95	173
1976	1.150	3.15	154.12	2.91	3.01	141.91	253
1977	1.150	3.15	83.61	9.20	9.52	250.18	800
1978.1	1.150	3.15	65.78	6.98	7.22	150.67	607
1978.2	1.175	3.22	56.91	8.84	9.14	161.59	752
1978.3	1.979	5.42	93.07	10.00	10.34	177.48	505
1978.4	2.750	7.54	106.33	10.00	10.34	145.91	364
1979	2.750	7.54	96.14	15.56	16.10	209.89	566
1980	2.750	7.54	75.76	15.87	16.41	165.20	577
1981	2.750	7.54	33.78	26.25	27.15	120.23	955
1982	2.750	7.54	25.67	61.67	63.79	209.61	2,242
1983.1	2.750	7.54	16.93	76.67	79.31	178.17	2,788
1983.2	24.690	67.68	96.32	60.67	62.76	89.32	246
1983.3	24.690	67.68	90.17	78.83	81.03	107.96	317
1983.4	30.000	82.24	97.08	90.67	93.80	110.73	302
1984.1	30.300	83.06	85.44	93.00	96.21	98.96	307
1984.2	35.000	95.94	92.70	97.00	100.34	96.96	277
1984.3	38.500	105.54	107.31	97.00	100.34	102.03	252
1984.4	42.120	115.46	115.36	99.67	103.11	103.02	237
1985.1	50.000	137.06	117.90	128.33	132.75	114.20	257
1985.2	53.000	145.28	127.65	129.00	133.45	117.25	243
1985.3	57.000	156.25	145.60	131.67	136.21	126.93	231
1985.4	60.000	164.47	159.90	136.00	140.69	136.78	227
1986.1	90.000	246.71	225.92	—	—	—	—
1986.2	90.000	246.71	218.53	—	—	—	—
1986.3	90.000	246.71	222.38	200.00	206.89	186.50	222
1986.4	149.417	409.58	335.54	192.50	199.14	163.14	129
1987.1	152.636	418.40	321.92	201.82	208.78	160.63	132

(continued

ble 2.2. (Continued)

	Official exchange rate (C/$, 1984=100)			Parallel exchange rates (C/$, 1984=100)			
	Official nominal rate	Index of official rate	Index of real official	Parallel nominal rate	Index of parallel rate	Index of real parallel	Parallel/ nominal (%)
)87.2	158.364	434.10	308.08	226.36	234.17	166.18	143
)87.3	165.077	452.50	307.61	251.54	260.21	176.89	152
)87.4	174.333	477.88	342.37	260.00	268.96	192.70	149
)88.1	180.667	495.24	326.63	—	—	—	—
●88.2	185.833	509.40	302.14	—	—	—	—
)88.3	217.231	595.47	327.65	—	—	—	—
)88.4	230.083	630.70	365.27	—	—	—	—
)89.1	245.350	672.55	351.99	—	—	—	—
)89.2	266.350	730.11	330.91	—	—	—	—
)89.3	275.590	755.44	334.97	—	—	—	—
)89.4	292.770	802.40	350.64	—	—	—	—
)90.1	307.420	842.69	—	—	—	—	—
)90.2	321.380	880.96	—	—	—	—	—
)90.3	334.040	915.66	—	—	—	—	—
)90.4	342.490	938.82	—	—	—	—	—
)991.1	351.42	963.30	—	—	—	—	—

Sources: Nominal official Ghanaian exchange rate through 1986.3, as well as trade statistics, the Ghanaian PI, and price indices and official nominal exchange rates of trading partners, are from IMF, various years (b). arallel exchange rate through 1986.3 from International Currency Analysis, various years.

Notes: All annual rates are arithmetic averages of reported quarterly figures. Beginning with 1986.4, quar- rly official nominal exchange rate is calculated as the arithmetic average of the relevant weekly rates estab- shed in the official foreign exchange auction. Similarly, the nominal parallel rate after 1986.3 is the average of e weekly rates established in the (legal) private foreign exchange bureau. Real exchange rates are calculated as e product of the nominal rate and the ratio of a trade-weighted world wholesale price index to the Ghanaian PI. Weights are constructed using Ghana's six largest trading partners, exclusive of Nigeria, during the period 980–84 (see Edwards 1989, chap. 4, for a discussion of an alternative definition).

Manifestations of Economic Disequilibrium: Black Star's Nadir

The reverting of the exchange economy to *kalabule* which occurred over a period as long as a decade was basically a result of economic mismanage- ment. Nevertheless, successive exogenous shocks between 1981 and 1983 can be viewed as contributing factors to economic collapse. In 1981 and 1982, world real cocoa prices dropped to a third of their 1977 peak (Table 2.4). For two years the government failed to collect any cocoa duties (Table 2.5). Subsequently, as earnings from this key commodity began to rise, a severe drought followed by bush fires led to a 12 percent real decline in the value of agricultural output and a much larger decline in production of food. Moreover, the low rainfall reduced the capacity to generate electricity and foreign exchange earnings from aluminum smelting. Also, in 1983,

Table 2.3. Deficit finance, Ghana (million cedis)

	1974	1975	1976	1977	1978	1979	1980	1981	1982	1983	1984	1985	1986	1987	1988	1989
GNP	4,629	5,241	6,478	11,123	20,938	28,123	42,670	72,404	86,226	182,398	266,918	337,280	498,797	725,476	1,024,670	1,389,026
Deficit (surplus)	196	401	736	1,057	1,897	1,800	1,808	4,707	4,848	4,933	4,843	7,579	(299)	(4,059)	(8,916)	(10,350)
Foreign grants	—	1	—	1	—	—	45	48	52	57	914	1,620	3,868	6,037	11,553	21,343
Revenue gap	196	401	736	1,058	1,897	1,800	1,853	4,755	4,990	4,990	5,757	9,199	3,569	1,978	2,637	10,993
Deficit/GNP	4.2	7.7	11.4	9.5	9.1	6.4	4.2	6.5	5.6	2.7	1.8	2.2	-0.1	-0.6	-0.8	-0.7
Revenue gap/ GNP	4.2	7.7	11.4	9.5	9.1	6.4	4.3	6.6	5.7	2.7	2.2	2.7	0.7	0.3	0.3	0.8
Total financing	196	401	736	1,057	1,897	1,800	1,808	4,707	4,848	4,933	4,843	7,579	(299)	(4,059)	8,916	10,350
Domestic	198	400	734	1,044	1,720	1,800	1,518	4,340	4,421	3,825	3,028	4,043	5,315	(2,879)	(6,166)	(15,279)
Foreign	(2)	1	2	13	67	0	290	367	389	970	1,815	3,522	(5,614)	(1,180)		
Other	0	0	0	0	110	0	0	0	38	138	0	14	0	0	0	0
Internal financing (% of total)	101.0	99.8	99.7	98.8	90.7	100.0	84.0	92.2	91.2	77.5	62.5	53.3	—	—	25.3	47.6

Sources: IMF, various years (b); World Bank 1989d.
Note: Revenue gap is the difference between revenue and expenditure, the deficit when grants are not considered.

Table 2.4. World cocoa prices (1985 U.S. cents/kg)[a]

	Current		Constant	
	Cents/kg	1985=100	Cents/kg	1985=100
1950	63.1	28.0	238.1	105.6
1960	58.9	26.1	191.9	85.1
1961	48.5	21.5	158.5	70.3
1962	45.9	20.4	149.5	66.3
1963	55.3	24.5	180.7	80.1
1964	50.5	22.4	164.5	72.9
1965	36.6	16.2	116.9	51.8
1966	51.8	23.0	160.4	71.1
1967	59.7	26.5	184.3	81.7
1968	72.1	32.0	217.2	96.3
1969	90.4	40.1	262.0	116.2
1970	67.5	29.9	189.1	83.8
1971	53.8	23.9	145.8	64.6
1972	64.4	28.6	166.8	74.0
1973	113.1	50.2	259.4	115.0
1974	156.1	69.2	301.4	133.6
1975	124.6	55.3	220.1	97.6
1976	204.6	90.7	345.0	153.0
1977	379.0	168.1	602.5	267.2
1978	340.4	151.0	502.1	222.6
1979	329.3	146.0	431.6	191.4
1980	260.4	115.5	299.0	132.6
1981	207.9	92.2	218.8	97.0
1982	173.6	77.0	179.2	79.4
1983	212.0	94.0	215.9	95.7
1984	232.1	106.2	238.4	105.7
1985	225.5	100.0	225.5	100.0
1986	207.0	91.8	213.1	94.5
1987	199.4	88.4	200.1	88.8
1988	158.4	70.3	153.1	67.9
1989	124.1	55.0	114.1	50.6

Source: World Bank 1989n.

[a]Annual average, which in turn is the average of the first three positions daily on the terminal markets in New York and London, as reported by the International Cocoa and Chocolate Organization. Real price and real index constructed using the U.S. wholesale price index.

Nigeria expelled up to a million expatriate workers and their families back to Ghana.

The food price increases that resulted from the crop shortfall and increased population correlate with an increase of malnutrition in 1983 (see Figure 2.1). Although this is plausible and consistent with consumer theory, it must be recalled that incomes declined in the same period that prices increased. The direct causal pathway may be ambiguous, but the increase in malnutrition rates are a clear indication of the economic downturn.

As the economy broke down in the period before 1983, neither the

Table 2.5. Government revenue, Ghana (million current cedis)

	1975	1976	1977	1978	1979	1980	1981	1982	1983	1984	1985	1986	1987	1988	1989
GNP[a]	5,241	6,478	11,123	20,938	28,124	42,672	72,294	86,225	182,798	266,918	337,280	498,797	725,476	1,024	670
Total revenue	809	870	1,141	1,392	2,600	2,951	3,234	4,804	10,185	21,728	38,691	69,758	105,009	142,492	193,170
Tax revenue	727	765	1,041	1,251	2,411	2,747	2,970	4,182	8,459	17,931	31,918	61,923	93,917	126,033	174,522
Individual income	69	84	94	171	210	308	388	694	999	1,654	3,058	5,242	8,191	11,096	12,323
Cocoa duties	180	269	278	835	920	535	—	—	2,800	4,509	8,861	13,901	26,996	24,464	31,415
Import duties	139	131	177	241	355	361	472	570	1,792	3,159	6,591	14,236	17,784	25,166	44,843
Sales and excise	136	242	351	409	422	832	1,263	1,884	1,727	5,620	8,592	19,778	26,545	36,318	52,069
Nontax revenue	82	105	100	141	189	204	264	622	1,726	3,797	6,773	7,835	11,092	16,459	18,648
Foreign grants	—	—	1	—	—	45	48	52	57	914	1,620	3,868	6,037	11,553	21,343
Govt. expenditures	997[c]	1,308[c]	2,137	3,165	4,296	4,668	7,719	9,530	14,755	26,694	45,763	70,660	106,987[d]	144,875	204,163
Tax/GNP	14	12	9	6	9	6	4	5	5	7	9	12	13	12	13
Tax/tot. rev.	90	88	91	90	93	93	92	87	83	83	82	89	89	88	90
Income tax/tot. rev.	9	10	8	12	8	10	12	14	10	8	8	8	8	8	6
Cocoa duties/tot. rev.[b]	22	31	24	60	35	18	0	0	27	21	23	20	26	17	16
Import duties/tot. rev.	17	15	16	17	14	12	15	12	18	15	17	20	17	18	23
Sales & exc./tot. rev.	17	28	31	29	16	28	39	39	17	26	22	28	25	26	27
Nontax/tot. rev.	10	12	9	10	7	7	8	13	17	17	18	11	11	12	10
Grants/tot. rev.	0	0	0	0	0	2	1	1	1	4	4	6	6	8	11
Govt. expend./GNP	19	20	19	15	15	11	11	11	8	10	14	14	15[d]	14	15
Indices (1979=100)															
Tax/GNP	162	138	109	70	100	75	48	57	54	78	110	145	151	133	144
Tax/tot. rev	97	95	98	97	100	100	99	94	90	89	89	96	96	95	97
Income tax/tot. rev.	106	120	102	152	100	129	149	179	121	94	98	93	97	96	79
Cocoa duties/tot. rev.[b]	63	87	69	170	100	52	0	0	78	59	65	56	73	49	47
Import duties/tot. rev.	126	110	114	127	100	90	107	87	129	106	125	149	124	126	164
Sales & exc./tot. rev.	104	171	190	181	100	174	241	242	104	159	137	175	156	159	168
Nontax/tot. rev.	139	166	121	139	100	95	112	178	233	240	241	155	145	166	138
Govt. expend./GNP	125	132	126	99	100	72	70	72	53	65	89	93	99[d]	93	98

Sources: IMF, various years (b); World Bank 1989e, 1991a; Government of Ghana, Quarterly Digest of Statistics (Accra: Ghana Statistical Service, various years). All non-notated data are derived from IMF and GFS.
[a] IMF.
[b] Cocoa duties from World Bank 1989e, except 1980 from Government of Ghana 1989.
[c] Government expenditures in 1975 and 1976 are from World Bank.
[d] Provisional.

Figure 2.1. Relative price of food and prevalence of underweight children from clinics, deseasonalized, Ghana

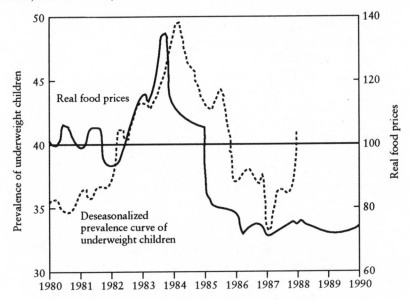

Source: Data from United Nations, Subcommittee on Nutrition 1989.

private sector nor the government maintained capital stocks. Private consumption, roughly 75 percent of GDP in the early 1970s, rose to 94 percent a decade later (Rimmer 1992). Although calculations of relative shares of private and public consumption are somewhat sensitive to cost deflators (private consumption is mainly goods whereas government consumption is primarily services), government consumption can be shown to have exhausted most of the remaining resources available. The absence of stock maintenance, never mind net additional investment, was felt in virtually all sectors. For example, gold production—the country's second largest source of foreign exchange—slumped as machinery broke down without replacement; production in 1983 was 20 percent below production three years earlier and over 40 percent below that of 1977.[5] Similarly, government services contracted. For example, per capita spending on education dropped from $20 in 1972 to $1 in 1983. As real salaries decline, qualified teachers migrated abroad. Those who stayed augmented their salaries with

[5] This may, however, also reflect increased smuggling as the exchange rate became cumulatively distorted.

secondary employment, including tutoring their students on subjects normally in the curriculum (Morna 1989). A decline in quality, then, rather than in quantity, likely characterized the school system at the time.[6]

Wages failed to adjust in keeping with annual inflation rates of over 100 percent (Table 2.6). By the early 1980s real wages for government workers were less than 20 percent of those in 1975. These wages were surely insufficient to live on. As a result, absenteeism, increases in the number of ghost workers, and a variety of other comparatively more unsavory practices became commonplace (Chazan 1983).

One of the symptoms, although not the cause, of the breakdown, then, was too little government, not too much. This fact implies that the manifestation of economic disarticulation, if not counteracted, becomes the cause for the next problems. It also implies that the economic restructuring Ghana required differs from that which might be advocated for an overheated economy.

As mentioned, at the time of restructuring there were few services left to cut; in 1983 total government expenditures as a share of GNP were half what they were in 1979 and 40 percent of the share in 1975.[7] Moreover, this share masks a portion of the reduction, since real GNP had declined 10 percent during this eight-year period. Similarly, Ghana did not have a significant trade imbalance at the onset of the recovery. With no appreciable foreign reserves and with poor credit prospects, Ghana had little choice but to curtail imports in parallel with its export slump.

Scobie (1989) is essentially correct in arguing that there is no clear distinction between stabilization and structural adjustment in the sense that the same set of policy instruments and adjustments are typically applied when an economy is restructuring or when it is stabilizing. Nevertheless, the impact on the general public or subgroups of the population may depend on the trajectory of the economy before the policy reform. Figure 2.2 provides a graphic illustration of the overall trend in the Ghanaian economy before and after the current phase of economic reforms initiated in 1983 and contrasts this picture with selected Latin American economies. Unlike the typical (and simplified) adjustment program in Latin America in the early 1980s, as well as in Ghana in 1971, the shock that precipitated the crisis in 1983 did not hit an overheated economy. Rather than living beyond its means, Ghana had already contracted most services and investment before entering the adjustment program. Thus, the stabilization policies of Latin America and the criticisms thereof, even where valid, cannot be automatically applied to Ghana or several other sub-Saharan countries.

---

[6] Official statistics show only a slight dip in the number of schools or enrollment, although the universities were not in session in 1983/84.

[7] These reduced expenditures reflect three successive declines in the federal deficit. The deficit in 1983 was only 2.7 percent of GNP.

Table 2.6. Price and wage indices, Ghana (1985=100)

	1975	1980	1981	1982	1983	1984	1985	1986	1987	1988	1989
CPI	0.8	11.0	23.8	29.1	64.9	90.7	100.0	124.6	174.2	228.8	286.5
Minimum wage[a], 1985 cedis/day	2	5	12	12	25	35	70	90	90	120	170
Nominal wage index	3	8	17	17	36	50	100	129	129	171	243
Real wage index	352	69	72	59	55	55	100	103	74	76	84.8
Unskilled labor costs											
Nominal cost index	3	11	18	18	25	47	100	138	161	209	259
Real cost index[b]	329	100	77	63	39	51	100	111	93	91	90
Skilled labor costs[a]											
Nominal cost index	3	13	19	19	26	47	100	153	184	237	301
Real cost index	396	121	80	66	40	52	100	123	106	104	105
Average monthly earnings per employee[c]											
Public sector[a]	—	4,009	2,431	2,084	1,650	2,341	3,461	5,886	4,210	5,568	—
Agriculture/forestry	—	2,927	2,058	1,864	1,283	2,426	2,675	4,618	3,303	4,537	—
Mining/quarrying	—	5,564	3,620	2,963	3,411	3,700	10,534	9,864	7,055	4,953	—
Manufacturing	—	4,791	2,650	2,434	1,935	2,646	4,108	6,731	4,815	8,763	—
Construction	—	2,982	2,167	1,779	1,373	1,725	2,474	4,074	2,914	3,862	—
Wholesale/retail trade	—	3,791	2,020	1,981	1,323	1,550	2,681	4,447	3,181	5,417	—
Transport/commun.	—	4,818	2,423	2,452	1,562	2,540	3,503	8,378	5,992	7,781	—
Private sector	—	5,282	2,822	2,980	2,004	3,618	4,702	6,060	4,334	8,156	—
Agriculture/forestry	—	3,782	2,868	2,520	1,807	3,279	3,264	4,337	3,102	5,847	—
Mining/quarrying	—	6,145	3,296	5,548	3,249	4,494	5,998	5,202	3,721	6,309	—
Manufacturing	—	5,155	2,994	2,888	2,072	4,346	5,535	6,297	4,504	9,509	—
Construction	—	4,745	1,961	1,830	1,294	1,781	3,032	3,702	8,648	3,716	—
Wholesale/retail trade	—	7,300	2,557	4,110	2,356	3,263	4,781	5,494	3,929	7,785	—
Transport/commun.	—	5,482	4,010	3,430	2,621	2,946	6,509	9,510	6,802	9,691	—

Sources: World Bank 1989e; IMF, various years (b); Government of Ghana, Quarterly Digest of Statistics (Accra: Ghana Statistical Service, various years).

[a] The minimum wages given for 1983 and 1984 were not imposed until April. The minimum wage given for 1989 refers only to the first quarter.
[b] Real wages and earnings deflated using the average annual CPI.
[c] Average monthly earnings, in constant 1985 cedis, are for December of each year.

Figure 2.2. GNP per capita, selected adjusting countries, constant local currency, Ghana

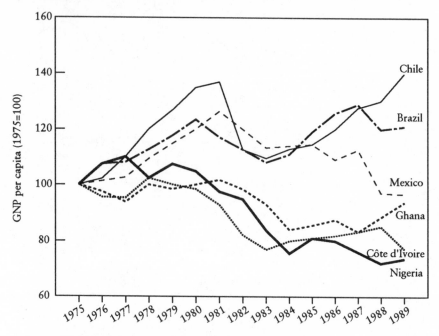

Source: World Bank updates, various years.

The trajectory is useful for more than analyzing the initial economic and distributional impacts of new policies; it also explains why they were risked. Analyzing the political foundation for the recovery program, Kraus (1991) argues that it was widely recognized in Ghana that the downturn left the country with no viable alternatives to the reforms taken. Similarly, Jeffries (1991) uses the extent of the economic disengagement to explain the weaning of Lt. Jerry Rawlings from his leanings toward "a statist brand of economic nationalism, combined with romantic anticapitalism" during the year and a half between the December 31, 1981, coup that returned him to power and the beginning of the economic recovery program.

Initial Policy Reforms: Black Star Ascending

The intertwining of causes and manifestations results in a Gordian knot. Ideologues present reform as a cutting through, but untangling is often a more stepwise procedure. An initial step in Ghana was a massive devaluation in April 1983.[8] More significant than this or the eight other devaluations in the three years between October 1983 and September 1986, however, was the change in the means by which foreign exchange was priced. The foreign exchange auction, inaugurated in September 1986, not only introduced flexibility into the exchange rate determination but also depoliticized the process—or, to the degree that auction rates are determined by central bank policies, at least diffused the political process. This shift was enhanced with the legalization of private trade in currency through licensed foreign exchange bureaus in April 1988.

During the first phase of the auction, certain transactions such as exports receipts for cocoa and the import of petroleum and pharmaceuticals as well as debt service continued at the official fixed rate of exchange. From February 19, 1987, however, both rates were unified at the exchange rate determined by the weekly auction. An additional, parallel but unofficial rate, however, existed throughout this period. This rate remained above the auction-determined rate, as did the bureau rate initially. This gap closed somewhat in the subsequent two years. It remained at 15–25 percent above the auction rate in the fourth quarter of 1989, although it narrowed through the first half of 1990, by which time the rates had essentially converged.[9]

The slow initial rate of narrowing of the gap between the auction rate and parallel rate is not unique to Ghana. The legal private sale of foreign exchange in Somalia remained above the auction rate throughout the existence of that erstwhile auction (Sahn and Alderman 1987). Younger (1991a) points out that the quantity of exchange in the auction increased from US$3.9 million per week in 1987, to $4.9 million in 1988, to $6.2 million in 1989. This increase of funds likely explains the long-run relationship between the auction and the parallel rate.

To a large degree, Ghana has also removed most quotas and import restrictions. Successful auction bids are granted licenses generally within six to eight weeks of the date of the auction. Moreover, restrictions on the use of auction funds have been phased out; service and transfer payments were eligible from March 13, 1987, and consumer goods previously available under special import licenses were phased into the auction between March 1987 and February 1988. The system of special import licenses—a means of importing using retained funds from exporting—was subse-

---

[8] The first devaluation, then, came before the August accord with the IMF.

[9] For example, the auction rate for the dollar was 286 cedis on October 20 and 301 cedis on December 1, 1989. The bureau exchange rate hovered around 350 cedis in the period. By July 1990 the auction rate had risen another 10 percent, whereas the bureau rate remained unchanged.

quently phased out by the beginning of 1989. With a few minor exceptions, the few items that remained prohibited were so for nontrade reasons.

Although I have found no documentation regarding restriction on utilization of auction funds for the import of grains, all major import companies I visited in February 1989 claimed that bids for food items would not
be accepted.[10] Since food imports—including those by government
agencies—were less than 1 percent of all auctioned financed imports up
until third-quarter 1988, this view is credible.

These policies removed the stickiness of exchange rates (Table 2.2), a key
feature of misalignment, but they did not by themselves fully address the
fiscal imbalances that tended to erode the effectiveness of devaluations
(Edwards 1989). Revenues from cocoa duties, however, initially increased
as a result of the devaluation. Similarly, import duties increased with the
volume of trade through official channels.

Other measures were undertaken to increase the efficiency of tax collection. As a result of these measures, the government was able to rapidly
reduce the deficit and even generate a small surplus by 1986 (Table 2.3).
This improvement did not, however, require a reduction of services. Rather,
Ghana achieved its deficit reduction by the traditional—but not necessarily
popular—approach of increasing revenue. Foreign grants (as opposed to
loans) were apparently of minor significance in this recovery. Indeed,
whereas foreign capital was prominent in financing the deficit between
1983 and 1985, capital flow was outward in 1986 and 1987. There was a
net inflow of foreign funds again in 1988, but, given the overall surplus,
they were to retire domestic debt, not strictly to finance a deficit. This
characterization, however, depends critically on what items are listed as on
and off budget.

Looking at the issue from another angle, the public investment–savings
gap was over 6 percent of GNP at the end of the decade.[11] Long-term
sectoral adjustment and infrastructure restoration as well as a variety of
other projects generally occur off budget. Ghana, then, has been able to
maintain government consumption during adjustment, in part because
many expenditures had been cut before adjustment, in part because increased flows of funds from abroad removed the need to make an exclusive
choice between consumption and investment.

Supporting a vulnerable economy during a temporary decline in the
terms of trade is a commendable—indeed, textbook—role for foreign as-

---

[10] In November 1989, at my request, the import manager of a major trading company
approached the Bank of Ghana for clarification on auction restrictions. He was informed that
imports of rice and sugar in particular could not be funded from the auction. No mechanism
for such exclusion is, however, apparent.

[11] Younger (1991b) argues that failure to account for these expenditures in macroeconomic
planning contributes to inflation.

sistance. Nevertheless, the disappearance of the deficit in the face of increasing expectations of adjustment dividends by various groups, including government and state-enterprise employees, is maintained by a high level of per capita aid and credit.[12] An image of Ghana as a prodigal son, enjoying the fatted calf after renouncing its wayward habits, is not, however, accurate. Although foreign assistance increased markedly during the 1980s, such flows were, like so many other indicators for Ghana in this period, dramatic because they were compared to a low base. Between 1983 and 1989 real official development assistance to Ghana increased fivefold. Nevertheless, fourteen of the twenty-four low-income sub-Saharan countries received more assistance in per capita terms than Ghana (World Bank 1991a). Similarly, six of the ten middle-income countries in this region received more in per capita terms.[13]

Let us return to the theme of intertwining reforms. The devaluation enabled the government to raise the real price paid to cocoa farmers by 75 percent between 1982/83 and 1988/89 despite a decline in world prices.[14] Moreover, other reforms such as divestiture of peripheral operations owned by the marketing board Cocobod and the retrenchment of redundant (and often fictitious) workers reduced Cocobod's marketing costs from 29 percent of the f.o.b. price to 17 percent the following year. Divestiture of plantations, however, has moved slowly, with only seven of a targeted fifty-two sales being completed by mid-1990. Cocobod, moreover, remains the only legal buyer of cocoa. The reluctance to liberalize cocoa trade may reflect an awareness of an abortive attempt in Nigeria as well as the memory of a failed initiative during a previous period of reform after Nkrumah was removed from office. That latter attempt was hindered by insufficient rural credit, a problem that has yet to be solved.

Unlike some African countries, Ghana did not directly subsidize food crops, nor were these crops handled by marketing boards. To be sure, some commodity prices were regulated, and enforcement of antihoarding decrees affected private trade. As in many countries, however, *kalabule* prevailed over regulation and most markets cleared by supply and demand, limited as both may have been. Consequently, a quick-fix solution to low productivity by getting prices right was not possible. Indeed, the price

[12] Total aid commitments in 1988 were US$763 million, 40 percent of which were multilateral (World Bank 1989e). This amount is an increase of $151 million from the previous year and $411 million greater than 1986. The grant component has remained roughly a third. Given the acceleration of commitments, annual disbursement levels have been below that of new commitments.

[13] In per capita terms, the aid Ghana received in 1987 was 20 percent below the sub-Saharan average, exclusive of Nigeria. In 1989 it was 7 percent above the average.

[14] Franco (1981) points out that, at the official exchange rate in the late 1970s, it would not have been possible to reach an economically efficient level of production without a sizable subsidy to producers.

policy that most directly affected farmers of grains and root crops was the gradual elimination of input subsidies. Fertilizer prices, for example, were raised in steps; the subsidy was not totally removed until 1990. Similarly, those farmers who used tractor services paid more for their fuel as the prices for all petroleum products were gradually raised. A significant step toward liberalization of fertilizer trade, however, came in late 1988 with the opening of retail trade in selected areas to private dealers.

With the exception of export crops, agriculture suffered less from over-government than undergovernment during the downturn. With the collapse of government revenues, sectoral support evaporated. The financially constrained administration was simply unable to either service or control the majority of farmers. Although a rapid recovery is predicted by models that claim the necessity of getting governments off the backs of farmers, this approach has limited application in Ghana; it may be a prerequisite but not a prescription.[15] Indeed, the need for increased government involvement is implicitly recognized in recent budgets, which devote appreciable resources to rehabilitation of extension services, feeder roads, and other rural infrastructure.[16]

Overgovernment has, however, been an issue with respect to both parastatal enterprises and government overstaffing. The government has an expressed policy of divesting many of the state-owned ventures, but the process has been slow. Many enterprises are unprofitable; they are characterized by obsolete and underutilized capacity and long-term pension and severance contracts unlikely to attract investors. Profitable ventures, on the other hand, are difficult to price. This is especially true with respect to foreign ventures. The sale of, say, the diamond mining corporation to a foreign investor leaves the government open to the criticism that it has auctioned the country's heritage.

Mining has, however, been the most attractive sector for foreign investment, and, as indicated in Table 2.1, it has responded to the investments. For example, the government and Lonhro, assisted by the International Finance Corporation (IFC), put together a $160 million investment package in 1984 aimed at revitalizing gold mining. Similar packages have followed, with five of the six IFC investments through 1989 in mining or oil exploration (97 percent in value terms). Mining has, in addition, been an early beneficiary of deregulation and market liberalizing. Although the

[15] The limitation of a pricism strategy for agricultural development has been recognized by several researchers (e.g., Lipton 1989; Streeten 1987; Shapiro and Berg 1988).

[16] Global 2000, a program of in-kind credit coupled with extension, has been a visible new initiative during the period of economic restructuring. It need not be discussed in detail here, however, less because its long-run viability is uncertain than because it attempts nothing that could not have been tried in the policy environment of Ghana a decade previously; Global 2000 relies neither on exchange rate reform nor on liberalization of internal markets.

state remains the sole legal buyer for gold and diamonds, individuals and small corporations can obtain licenses and are guaranteed a price that reflects world prices.

Two types of reform have more or less counteracted each other with respect to the government's ability to reduce its wage bill. First, the government has redeployed roughly 40,000 employees, in addition to those laid off from Cocobod.[17] These workers received a benefits package including a cash grant—four months' gross salary plus two additional months' salary for each year of uninterrupted service—and training in another profession. Other benefits such as career counseling, subsidized tools, and food for work and input loans for those workers who chose to work in agriculture were also available, albeit a few years after the first round of retrenchment.

Retrenchment did not necessarily reduce the number of workers on the government payroll. In the Ghana Educational Service, for example, non-teaching staff were replaced by new teachers. Moreover, those government employees who remained received sizable salary increases in the first years of the economic recovery program. Even with significant retrenchment of workers, the government wage bill rose from under 25 percent of all government expenditures in 1982 and 1983 to over 35 percent in 1986 and 1987. Moreover, the government adopted a policy to increase the pay differentials between the lowest and highest grades. This differential was 2 to 1 in 1983 and increased to 9 to 1 in 1990. This seemingly dramatic reform was actually less than the target of 13 to 1.

Increases in the wages offered by the government likely put pressure on the private sector as well. Moreover, the minimum wage, which by 1983 had eroded to only 35 percent in real terms from 1977 and 15 percent relative to 1975, was revised several times between 1983 and 1989. To some degree the minimum wage is more important as a gauge watched by laborers than as a determinant of wage floors. The majority of wage earners in Ghana receive wages above the minimum. For example, in 1987/88 only 8 percent of the 965 who reported a wage for their primary employment in the Ghana Living Standards Study received less than the minimum wage at that time.[18]

The package of benefits for redeployed workers was part of the PAMSCAD program. PAMSCAD, however, was not designed with redeployment as its sole focus. Rather, it was intended to provide a range of services, from employment generation to improvement of schools. The diverse group of projects proposed in the initial design (1987) ranged from the

---

[17] The Ministry of Social Welfare and Public Mobilization has computerized records, but omissions and duplications make it impossible to obtain exact figures.

[18] Note, however, that the wage market in Ghana is rather thin; only a small percentage of the population are engaged in either agricultural or nonagricultural wage employment.

purchase of baskets for shea nut gatherers to compensation of redeployed government works. Implementation began only after two years of negotiations with donors who proved reluctant to fund a few of the components of the $84 million scheme (including redeployment compensation).[19] PAM-SCAD, however, managed to achieve much of its desired visibility even before implementation began.

This visibility goal is a partial explanation for the wide range of projects, widely dispersed. A related objective of PAMSCAD is to support community initiatives in each of the 110 districts in the country. This reliance on community initiative and self-help for public works as well as for education is in keeping with other efforts to decentralize social administration and to enhance local councils.

Economic and Social Impacts

The regularity of growth in the period 1984–89, as much as its level, is unprecedented in Ghana's recent history. Even bad weather and declining terms of trade in 1990 did not reverse the recovery. Initial doubts about the sustainability of the recovery (e.g., Green 1987) seem unwarranted. Indeed, in May 1991 the government of Ghana and donors gave serious (and perhaps optimistic) public consideration to aiming for double-digit growth rates.

There is, however, a related concern about the potential for fatigue and for growing political pressures for tangible personal gains to keep pace with the overall economic indicators, possibly to the detriment of long-run growth. One illustration of the medium-term task can be derived by noting that GNP per capita declined at an average annual rate of 1.6 percent between 1965 and 1987 (World Bank 1989m). The growth in the subsequent two years merely brought the average decline from 1965–89 to 1.5 percent (World Bank 1991a). As an alternative illustration, assuming a population growth rate of 3 percent (a figure that is somewhat lower than the average for 1980–1987), it would require an average GNP growth rate of 5.8 percent to restore GNP per capita to its 1965 level by the end of the century. There is nothing magic about the year 2000, nor is this illustration robust to alternative starting points. It does, however, provide an indication of the scope of the real task of recovery and puts short-term measures of success into perspective.

Similarly, it is useful to avoid exaggerating the short-term impacts of policy reforms, not to denigrate them but to explain them better. Clearly,

---

[19] Moreover, not all the funds raised were additional.

the recovery of agriculture (54 percent of GNP in 1984[20]) was largely weather-induced. In like manner, production of electricity and aluminum increased due to external factors as the Volta Lake refilled, although the value obtained was affected by exchange rate reforms. A portion of the increase in exports in 1984 also reflects the reopening of the railroad to Sekondi and Takoradi, the fruition of a World Bank–financed project initiated well before any major policy initiatives.

Furthermore, Tabatabai (1988) argues convincingly that the forced repatriation of workers from Nigeria, which contributed to the country's problems in 1983, also contributed to the recovery of agriculture in the following year. The remigration has both statistical and structural consequences. To the degree that the population was overestimated through underestimation of migration in the early part of the decade, Tabatabai argues, production per capita and food availability per capita were underestimated. Moreover, because the workers who returned from Nigeria largely went to rural areas, they provided a reservoir of labor that contributed to a real increase of agricultural output.

An analogous argument can be made concerning the recovery of export crops. Some apparent increases in production may reflect greater applications of pesticides and labor. A portion of the increase, however, may merely be a shift of trade that had previously gone through Togo and Côte d'Ivoire. Such a shift would enhance government revenues but would not have as much impact on real GNP or even on the real capacity to import as it would on official statistics.

To what degree, then, did exchange rate and other trade policies achieve a rationalization of signals in order to influence production and consumption? Rimmer (1992), for example, contends that the devaluations in 1967 and 1978 had a larger impact on the distribution of rents for holders of import license, or those empowered to grant them, than on domestic prices. A similar argument can be made regarding the devaluations in the early 1980s. If the real rate of exchange is endogenous—as is likely when a country has no foreign reserves to draw down—then a nominal devaluation mainly makes explicit what various rationing and licenses were doing, though the latter is at a cost in terms of allocative efficiency and deadweight loss.

The devaluation has had a clearer impact on exporters than on consumers of imports. Krueger et al. (1988) as well as Stryker (1988) observe that the main source of disincentives to agricultural export production in most developing countries is through exchange rate distortions as opposed to direct intervention. Once these disincentives were removed, production

---

[20] Growing slower than the economy in general, its share dropped to 44 percent in 1990.

responded over time; cocoa volumes as well as traditional exports of gold and timber increased. Moreover, nontraditional exports expanded, albeit from a low base.

The impact on producer incomes was immediate; net payments to cocoa farmers increased from just under 5 billion cedis in 1983/84 to over 13 billion cedis in 1987/88 (in 1985 constant prices).[21] Ecology dictates that such gains will be distributed unevenly across regions, with over half of all payments going to farmers in the western and Ashanti regions.[22] The regions that benefited from the rising cocoa prices were, generally speaking, also the regions that benefited from the higher prices for timber and gold. In the latter case, however, backward linkages are likely relatively low.

Another group that received immediate observable benefits were wage earners. The sharp revision of the minimum wage in 1985 restored its real value to 64 percent of that in 1977 (from 35 percent in the previous year), and another revision in 1986 prevented inflation from eliminating that gain. Subsequently, however, the minimum wage stagnated as the government tried to link further increases to productivity. Consequently, the minimum wage fell in real value between 1986 and 1987 and managed only to keep pace with inflation from 1987 to 1990. An indication of the purchasing power of this wage can be found by comparing it to the price of maize (Alderman 1991b). The minimum wage can purchase around 4 kilograms of maize. Although maize is not the cheapest energy source, this is, nevertheless, a low grain-wage ratio by international standards. On the other hand, the claim by union leaders that a family of four requires 2,000 cedis a day for food alone (Ephson 1990) is clearly bargaining rhetoric.

Although the minimum wage is only indicative of the wages received by the average worker, it is also the case that inflation has kept many wage earners from improving their real earnings after their gains in the first years of the recovery period. This inflation, however, cannot be attributed directly to exchange rate policies. Using time series (autoregressive moving average) procedure, Younger (1991b) finds that a 100 percent devaluation results in only a 10–12 percent increase in food prices, mainly in the first two months after devaluation. A similar conclusion regarding the low transmittal of devaluations to domestic prices was reached by Chhibber and Shafik (1990), who used a different technique of macroeconomic modeling. Both studies are consistent with the conventional wisdom that consumer prices reflect scarcity costs no matter what the import price calculated at official exchange rates.

[21] Calculated from World Bank 1989e.
[22] Extrapolating from work by Boateng et al. (1989), Alderman (1991a) calculates that the increment to incomes of the poor in these regions is on the order of 10 percent of total earnings.

In actuality, the real price of food staples appears to have declined since the recovery program began, even if the main drought year is excluded (Alderman and Shively 1991).[23] This is indicated by the coefficients on the two time trend variables in the regressions reported in Table 2.7. The rate of price decline was slowest for rice, which is imported at the margin and comprises only 2 percent of consumers' budgets.[24] The analysis of prices also confirms that food prices in Ghana remain quite variable, with seasonal increases of up to 100 percent of the harvest period price not uncommon. In some years, such as 1988/89, there has been no interseasonal price rise; grain held in storage in such years must be sold at a loss. Although the government has embarked on a program to increase its storage capacity from 40,000 to over 70,000 tons (with plans for up to 120,000 tons), at no time has the state trading company, Ghana Food Distribution Company, sold more than 20,000 tons of maize, or 3–4 percent of total production. Despite increases in storage capacity, the government is not in a position to defend a price floor for maize, having insufficient liquidity for such a venture. The government recognized this in 1990, introducing a policy of purchasing and selling at market-determined prices.

The downward, if variable, trend in food prices may be interpreted as a shift in rural-urban terms of trade, although such an interpretation disregards the income-increasing effects of any productivity gains. Moreover, many poor, even rural poor, are consumers of the foods that have declined in price and therefore also benefit from this change in relative prices.

Inflation and policies to prevent it, however, remain a brake on the economy. Much of the increase of foreign exchange from official flows as well as from private sources[25] is being sterilized (being used for imports), but some of these funds contribute to increase the money supply. Even with tight credit ceilings and the fact that the government is a net creditor to the banking system, money supply (M2) has been growing at 40 percent a year (Younger 1991a,b). It is not clear that either monetary policy or deficit reductions can be designed to reduce this inflationary pressure without a major contraction in services and investments. Moreover, attempts to moderate this pressure by restricting credit impede the liberalization of a banking industry already burdened by a large portfolio of nonperforming loans. One result is a credit squeeze for private investment.

[23] Although the government occasionally has harassed traders for hoarding and imposed ineffectual price ceilings at various times, there were few explicit subsidies on food items. Hence there is little reason to expect a direct impact of fiscal reforms on food prices.

[24] Most of the price decline occurred immediately after the initial adjustment reforms. It is clear that real prices are less than in the pre-drought period, although the post-recovery period has been too short to model rates of deceleration or acceleration with confidence.

[25] National accounting indicates an increase in remittances. It is not, however, possible to determine how much represents a net increase induced by the devaluation and easing of restrictions and how much represents accounting of previously smuggled funds.

Table 2.7. Regressions indicating price trends of food prices, Ghana

Independent variable	Wholesale price							Retail price	
	Maize[a]	Sorghum	Millet	Cassava	Gari	Yam	Rice	Maize[a]	Gari
Constant	3.163 (108.173)	3.910 (84.590)	3.724 (72.619)	2.065 (18.155)	3.548 (56.361)	4.132 (85.035)	4.382 (114.524)	3.237 (42.872)	3.042 (38.797)
Time trend pre-1983	-5.918×10^{-4} (−3.001)	-5.505×10^{-4} (−1.834)	-1.582×10^{-3} (−4.220)	7.283×10^{-4} (0.955)	-1.034×10^{-3} (−2.498)	-7.542×10^{-4} (−2.549)	-3.260×10^{-3} (−13.356)	-9.717×10^{-4} (−2.468)	1.922×10^{-4} (0.413)
Time trend post-1984	-1.349×10^{-3} (−12.664)[b]	-1.710×10^{-3} (−9.665)[b]	-2.099×10^{-3} (−9.918)[b]	-2.647×10^{-5} (−0.053)[b]	-1.659×10^{-3} (−7.644)[b]	-8.656×10^{-4} (−4.820)	-1.445×10^{-3} (−10.585)[c]	-1.908×10^{-4} (−9.261)[b]	-1.193×10^{-3} (−4.325)
Urban	4.875×10^{-3} (0.398)	0.019 (1.292)	5.811×10^{-3} (0.321)	0.089 (2.900)	−0.052 (−1.792)	0.181 (9.988)	6.589×10^{-3} (0.448)	−0.023 (−0.476)	0.166 (5.057)
Upper region	−0.023 (−1.671)	−0.165 (−9.671)	—	0.569 (7.055)	−0.224 (−1.965)	0.071 (3.396)	−0.101 (−6.077)	0.045 (1.452)	0.369 (10.233)
Northern region	−0.185 (−8.374)	−0.279 (−12.473)	−0.043 (−2.033)	0.769 (9.937)	—	−0.303 (−10.990)	−0.057 (−2.251)	−0.151 (−3.474)	—
January	0.235 (8.107)	−0.212 (−6.050)	−0.059 (−1.442)	0.061 (0.857)	0.145 (2.191)	0.133 (3.185)	−0.126 (−3.654)	0.225 (3.919)	0.117 (1.799)
February	0.249 (8.619)	−0.150 (−4.279)	-8.794×10^{-3} (−0.214)	0.045 (0.638)	0.097 (1.494)	0.121 (2.861)	−0.086 (−2.462)	0.247 (4.310)	0.084 (1.297)
March	0.345 (11.828)	−0.149 (−4.223)	0.020 (0.490)	-2.122×10^{-3} (−0.030)	0.122 (1.847)	0.134 (3.182)	−0.081 (−2.275)	0.334 (5.782)	0.064 (0.987)
April	0.428 (14.584)	−0.116 (−3.294)	0.076 (1.823)	0.057 (0.800)	0.132 (1.952)	0.216 (5.120)	−0.062 (−1.743)	0.435 (7.531)	0.083 (1.267)
May	0.486 (16.669)	−0.058 (−1.652)	0.126 (3.003)	0.144 (2.038)	0.130 (1.940)	0.317 (7.308)	−0.051 (−1.415)	0.495 (8.773)	0.154 (2.363)

June	0.493 (16.900)	−0.036 (−1.027)	0.146 (3.446)	0.079 (1.124)	0.140 (2.127)	0.337 (7.594)	−0.053 (−1.482)	0.497 (8.709)	0.142 (2.178)
July	0.354 (12.034)	−0.025 (−0.714)	0.121 (2.796)	0.034 (0.489)	0.075 (1.142)	0.287 (6.392)	−0.027 (−0.740)	0.409 (7.138)	0.071 (1.092)
August	0.126 (4.288)	0.010 (0.287)	0.086 (2.000)	0.074 (1.061)	0.015 (0.226)	0.132 (3.038)	5.887×10^{-3} (0.164)	0.159 (2.745)	0.078 (1.182)
October	0.025 (0.841)	−0.045 (−1.241)	−0.047 (−1.075)	−0.026 (−0.365)	−0.014 (−0.205)	−0.044 (−1.039)	−0.055 (−1.552)	8.806×10^{-3} (0.152)	0.018 (0.279)
November	0.083 (2.777)	−0.123 (−3.440)	−0.091 (−2.077)	−0.052 (−0.738)	−0.047 (−0.706)	0.055 (1.303)	−0.105 (−2.899)	0.110 (1.884)	−0.061 (−0.930)
December	0.098 (3.264)	−0.283 (−7.842)	−0.190 (−4.452)	−0.040 (−0.558)	−0.016 (−0.230)	0.064 (1.508)	−0.165 (−4.573)	0.096 (1.630)	−0.089 (−1.307)
Imported (pre-1983)	—	—	—	—	—	—	−0.118 (−2.806)	—	—
Imported (post-1983)	—	—	—	—	—	—	0.280c (5.829)	—	—
R²	0.295	0.251	0.267	0.131	0.189	0.242	0.134	0.364	0.323
N	3,202	2,085	1,232d	1,341	664	1,782	2,244	850	770

Source: GOG, Ministry of Agriculture, n.d.

aThe maize series includes data through July 1990 for selected markets; all other commodity series include data through April 1990 only.

bThe coefficient is significantly less than the corresponding coefficient for the earlier period (p, .01 two-tailed test).

cThe coefficient is significantly greater than the corresponding coefficient for the earlier period (p, .01 two-tailed test).

dUpper and northern regions only. Similar trends and significance are observed for the prices from the full sample, although the volume of sales are clearly concentrated in the savannah regions.

Note: T-statistics are in parentheses.

Figure 2.3. Expenditures on health and education as a percentage of GDP, Ghana

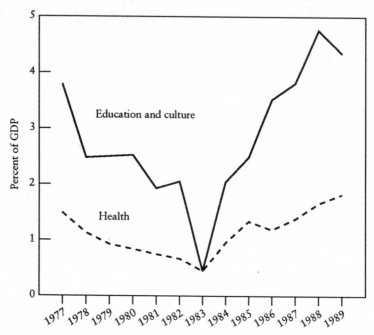

Sources: IMF, various years (b); World Bank 1989m.

As mentioned, the recovery has meant increases in all aspects of the government's budget, including services such as health and education. The sharp recovery indicated in Figure 2.3 is not inclusive of PAMSCAD programs. Although this program adds appreciable resources to human services, concerns have been raised that the overall PAMSCAD program was only superficially planned as a poverty reduction intervention. Given the view expressed elsewhere in this essay that, with the possible exception of retrenched workers who PAMSCAD aims to reach, there are few groups who are absolute (as opposed to relative) losers under the economic recovery program, the mitigation in its title may be a misnomer for a broad populist program. As an illustration, one PAMSCAD program is to deworm children. There are many reasons to suspect improved welfare from such a program but few reasons to expect that the need for the program is an outcome of the structural adjustment process.

In another context, PAMSCAD includes a program to deliver supplementary feeding to a target of 24,000 mothers and young children in priority districts in each region. There is not, however, any strong evidence that supplementary feeding has an appreciable impact on malnutrition in Ghana.[26] It is at least arguable that malnutrition in Ghana is more linked to health and sanitation than to food delivery, although a well-designed nutrition program can improve the former, especially if linked to prenatal health care (Alderman 1990). Again, visibility rather than sustainability appears to be a feature of this initiative.

With a few possible exceptions, such as PAMSCAD's support of small-scale gold miners, however, the program does aim to transfer services and infrastructure to the poor. There is obvious value to such programs, but it is not clear that *crash* programs best serve the objective.[27] Nevertheless, the program's obvious political nature does not rule out the possibility that it may simultaneously serve social and economic objectives.

Conclusions

Hirschman's well-known tunnel metaphor is applicable to contemporary Ghana. He asks one to imagine oneself on one of two lanes entering a tunnel. An obstruction closes the tunnel. One is disappointed but, after all, these things happen. After some time traffic begins moving, but only in the other lane. Originally, this is encouraging to the person stalled, but when one's own turn is slow in coming, one becomes furious. There is no doubt that economic growth was totally stalled by the time Ghana's recovery program began. There is also little doubt that things have begun to move again. The challenge is to distribute progress sufficiently to avoid an appreciable portion of the population feeling that the recovery is only for others.

To be sure, the availability of goods has increased relative to the beginning of the decade, and food prices in particular have fallen since 1983. Moreover, the trajectory of the economy is radically different from that prevailing a decade previously. Individuals, however, have short memories, and few easily conjure up a counterfactual to evaluate their current position relative to what might have been had Rawlings remained an internationally isolated populist.

This issue of growth and equity is, of course, one that goes beyond the concerns of stabilization. Stabilization policies per se have little to do with poverty in Ghana, equitable growth strategies virtually everything. Unlike

---

[26] See Kennedy and Alderman 1987 and the references within for an overall assessment of supplementary feeding.

[27] Initial assessments indicate that the government lacked the capacity to implement all subprojects. Under such constraints, remote (and poorer) areas are less likely to receive services.

many examples from Latin America, there are few individuals who are newly poor due to recovery policies. There remain, however, millions of people who await economic growth and the expansion of government services to bring them out of poverty. Are there any generalizations to be derived from the experience of the 1980s that can guide planners either in Ghana or elsewhere for poverty alleviation in the next decade?

First, the downturn confirms what basic textbooks maintain but specialized disciplines often forget: that macro and micro policies are linked, both economically and politically. Monetary and fiscal policy errors in the mid-1970s helped fuel inflation in the following years. This inflation, in turn, led to progressive misalignment of the exchange rate and subsequent distortion of price signals. The political unpalatability of devaluation led to a reliance on quotas and prices controls, which exacerbated the distortions that eventually led to the collapse of the government's ability to maintain services and infrastructure.

So one of the symptoms of the breakdown was too little government. Recovery, consequently, includes a strengthening of the government's capacity to deliver services—as is manifested in the tripling of public savings as a percentage of GNP concurrent with a rise in the government's wage bill. Government strengthening is also indicated in the steep fall and subsequent dramatic increase of health and education expenditures.

Such improved government capacity does not come cheaply. Although one should not underplay the domestic leadership and commitment necessary to revamp disastrous trade and investment policies, foreign capital has been a sine qua non of the recovery. Ghana has been able to eliminate the fiscal·deficit, increase imports in the face of devaluation, pay off most foreign arrears as well as a portion of domestic debt, provide massive increases in real funding to education and health, all while other investments grew. Sound policies might have established the conditions for these investments to be fruitful, but the funds came from decisions made in Europe, Tokyo, and the United States.

It is unreasonable, however, to expect that domestic resource mobilization alone can initiate major structural adjustment. Moreover, though private incentives are important, so too are public investments. Eight years into a recovery program that appears successful by several measures, private savings rates as a share of GNP have not exceeded preadjustment rates. Similarly, the resource flows into Ghana have largely been through development agencies; private foreign investment—mining excepted—remains limited. Part of the slow response stems from the fact that, in addition to price signals, investors generally respond to the level of uncertainty. Part of investors' uncertainty may be directly economic in origin: for example, high inflation and nominal interest rates in the absence of indexing increases risks for investments with long gestation. More important, successive governments in Ghana manifested distrust of entrepreneurs and traders. Investors are reminded of this legacy whenever an inquiry is launched into the source of investment equity or whenever a local tribunal levies a fine for which there is no appeal in a higher court. A climate that counteracts this legacy cannot be established by a single decree, although judicial reform can help.

Until private sector confidence is established and capital markets can respond to that sector's needs, Ghana's recovery will be led by the public sector, even as it attempts to divest this role. A central concern, then, is the government's ability both to generate returns from its investments and to mobilize these returns. A key to the latter is improvement in the collection of taxes—the first, and comparatively easy, steps toward which Ghana has achieved. A key to the former is the recognition that public as well as private investments need to respond to price signals.

The essential element of the establishment of meaningful price signals is the reform of the means by which the exchange rate is determined. The reform has prevented the real exchange rate from unplanned appreciation due to inflation. Moreover, with the availability of foreign exchange at the bureaus, individual importers can respond to international market prices as well as make comparatively rapid purchases of spare parts, inputs, and inventories. Low liquidity and low internal demand dampen the response of the economy to these new signals, but the revamping of trade policy remains the essential policy reform with the potential to reorient the economy.

Export values respond immediately to such changes in exchange rates, so the government has been able to both raise producer prices for cocoa and increase revenue collection while the world price declined with a corresponding production response.[28] A portion of the supply response, however, is attributable to increased availability of inputs and to improved transportation and utilities. Similarly, the nascent productivity increases in food crops do not reflect price policy but the rehabilitation of support services, including agricultural credit supplied by nongovernment organizations. Ghana's agricultural growth is no less promising for its being due to incremental increases in infrastructure and services. This does, however, imply that the change is not revolutionary—that it is established by footwork more than decree and is in need of continued effort to maintain progress.

Since Ghana subsidized few consumer goods and then only to a privi-

[28] Real producer prices, however, declined in 1989 and 1990, although they rose as a percentage of the world price in those years. Moreover, long-term prospects for cocoa earnings are seriously constrained by global overproduction and corresponding slumping prices.

leged few, impacts of budget cuts on consumers were small. An exception is energy prices. Although devaluation has the potential to increase the cost of living across the economic spectrum, there is little evidence that this has happened. Devaluation came only after imports had already dried up. Marginal consumer prices were unlikely influenced by official exchange rates, hence devaluation affected rents, not the general cost of living.

For many people, however, real wages have stagnated after an initial post-recovery rise. This is particularly true for government employees, but also true for wage earners receiving the minimum wage. Such groups, allied with students who have lost their appreciable boarding subsidies and with the relatively few government workers who have been involuntarily retrenched, are central and visible. They contribute both to journalistic stories of recovery austerity and to political pressures to modify the reform program.

If growth persists, such pressures may dissipate. To a degree, programs like PAMSCAD seek to buy such time. These programs, however political in origin, also have the potential to deliver services and increase employment. PAMSCAD is a component of, but not a substitute for, more pervasive measures to improve health, housing, and other basic services. As indicated by budgetary allocations, the government is undertaking such investments. Largely freed from having to choose between physical and human capital by the level of foreign capital inflows, Ghana has invested in both. In general, social indicators respond to inputs only after some lag. They require, moreover, somewhat more time to collect than price or even GNP figures. The absence of uncontroversial evidence on improvement in health and nutrition, then, is hardly proof that the recovery program has failed in this regard. Here, too, a foundation for incremental improvement through continued investments has been established. The restoration of economic and social indications to pre-downturn levels is a noteworthy achievement. It remains, however, only a prologue to economic growth and the alleviation of poverty.

3

Policy Failure and the Limits
of Rapid Reform: Lessons from Guinea

Jehan Arulpragasam and David E. Sahn

For many countries in sub-Saharan Africa, external factors were important in precipitating the onset of economic crisis in the 1980s. This was not
so in Guinea. Guinea has not experienced the droughts of Sudan and
Ethiopia, the wars of Angola and Mozambique, the transport shock incurred by Malawi, the terms-of-trade shock in oil-exporting Nigeria and
Gabon, or the plain lack of resources of Sahelian countries such as Mauri-
tania and Mali. Indeed, Guinea is extremely resource-rich. The country's
agriculture-based economy boasts a diversity of agroclimatic regions favorable to the strong production of a wide variety of crops. In addition to the
natural resources for agriculture, Guinea is also abundant in mineral resources.

Yet growth faltered in Guinea during the two decades before 1985 (Table
3.1). Real per capita GDP declined by 11 percent between 1960 and 1965
and then declined by 1 percent between 1965 and 1970. Average growth in
real per capita GDP between 1970 and 1980 was only approximately 1
percent annually. Guinea's performance with respect to traditional social
indicators, moreover, was just as troubling as the broad picture of economic stagnation borne out by GDP data. Guinea has been ranked third from
the bottom among countries worldwide on the basis of the human development index, a composite measure of national income, adult literacy, and
life expectancy (UNDP 1991).

In Guinea the onset of economic crisis is attributable primarily to the
history of a politically repressive state and to economic policy failure. The
state's attempts to eliminate all vestiges of price-clearing markets and replace them instead with state-controlled institutions charged with price

Table 3.1. GDP, Guinea

	Nominal GDP		Real GDP	
	Aggregate (bn. current GF)[a]	Per capita (current GF)[a]	Aggregate (bn. 1986 GF)	Per capita (1986 GF)
1960	11.90	3,840.00	—	—
1965	12.00	3,660.00	—	—
1970	15.50	5,000.00	—	—
1975	22.50	5,260.00	—	—
1980	33.52	7,703.16	594.54	136,630.92
1981	36.70	8,253.54	596.11	134,059.60
1982	39.18	8,611.72	609.06	133,870.30
1983	44.17	9,477.36	615.91	132,154.00
1984	48.14	10,088.97	599.62	125,665.75
1985	51.53	10,536.88	629.47	128,713.97
1986	671.15	133,979.91	671.15	133,979.91
1987	845.58	164,632.99	692.04	134,739.01
1988	1,081.59	204,847.99	733.83	138,983.91
1989	1,444.58	266,144.52	764.12	140,778.88
1990	1,820.41	326,251.14	795.54	142,575.48

Sources: 1986–89 data from GOG 1990b. 1980–85 GDP from World Bank 1989a and GOG 1986. 1960–75 data from World Bank 1990e. GOG 1991a.

[a]Until 1986 the currency was denominated in syli, which was replaced on par by the Guinean franc in January 1986.

determination and quantity rationing led to egregious economic distortions and pervasive corruption. By the time Guinea began its reform program in the early 1980s, the official monetized economy had virtually shut down, bankrupt in both financial and philosophical terms. Although parallel markets were pervasive, they were fragmented, badly integrated, inefficient, and characterized by risk and scarcity premia. The result was pervasive shortages, high prices for consumers, and low remuneration for producers.

Guinea stands distinct from most other countries, moreover, not only in the relative importance of misdirected policies in leading to crisis, but also in its policy reform program undertaken since. The policy changes in Guinea have been dramatic, taking the country from African socialism to free markets. Exchange rate policy has been reformed; domestic marketing has been liberalized; commercial banking has been privatized; state-owned enterprises have been liquidated; and government budgeting procedures have been made rigorous.

The Guinea story can thus stand as a textbook case study not only of how policy failure can lead to crisis but also of how policy reform can address economic crisis. There have been some clear efficiency gains with adjustment. At the same time, however, one is forced to examine the shortfalls of reform as, indeed, the crisis in Guinea persists.

The Evolution of Economic and Social Stagnation

The paradox of Guinea's good natural resource potential and its poor economic and living-standards performance is a consequence of the nation's traumatic political history and its associated economic policies. Under the leadership of Sekou Touré and his Parti Démocratique de Guinée (PDG), in 1958 Guinea voted no to General de Gaulle's proposed French Community and became one of the first African countries to gain independence. The break with France, however, was traumatic. Within forty-eight hours, all French expatriates were withdrawn. With their administrative and technical expertise also went equipment, medication, and even lightbulbs. Much of the infrastructure that remained was either inoperable or unserviceable. Both budgetary assistance and favored nation status of Guinean exports to France were terminated.

The severance from France completed the PDG's consolidation of power. In the years that followed, the PDG instituted a one-party government that controlled the political, administrative, and economic spheres. Party officials exercised control over everything from employment in state enterprises to access to foreign exchange at official rates. The patronage system institutionalized as a result is a legacy of state dirigisme that continues to affect Guinean life and the workings of the country's economy.

Exchange and Trade Policies

Guinea's trade and exchange systems were primarily responsible for the stagnation and structural distortions that plagued the country since independence. The link between the official value of the Guinean currency and supply-and-demand considerations was severed early. On independence in March 1960, Guinea withdrew from the franc zone and established its own currency. This marked the beginning of economic mismanagement that was to manifest itself in grievous distortions in foreign exchange markets.

Between 1975 and 1986, the syli was pegged to the SDR (special drawing right) at a constant rate. In the face of inflationary monetary policy undertaken during this period of low real growth, Guinea's exchange regime resulted in an increasing overvaluation of the syli and its sharp appreciation relative to the parallel rate. The parallel rate depreciated rapidly with an uncontrolled expansion in the money supply, much of the economy electing to maintain savings in foreign exchange.

Parastatals and the public sector officials who had access to imports at the official rate had their consumption subsidized, to the detriment of domestic production. Imported beer at hotels, for example, was one-third the price of local beer (Guillaumont 1985). Meanwhile, private sector exporters were heavily taxed through the overvalued official exchange rate. Domestic production and marketing consequently declined, with what little trade taking place occurring illicitly. Moreover, in the face of falling foreign exchange earnings, quantitative restrictions on trade, and high tariffs, imports also declined. By 1985 the parallel exchange rate had risen to average GS 311 per U.S. dollar while the official rate remained at GS 20.46 per dollar.

Despite the grossly overvalued exchange rate, Guinea's trade balance remained positive, due largely to the strong performance of the enclave mining sector in exporting bauxite and alumina.[1] But, at the same time, net services and private transfers, reflecting the repatriation of mining salaries and profits, more than counterbalanced the trade surpluses. As a result, Guinea experienced current account deficits throughout the period before 1986.

Agricultural Policies

Whereas exchange policy transmitted distortions to the rural sector, state dirigisme in Guinea was also reflected at every level of agriculture. With independence in Guinea, the responsibility of land assignment and reassignment, although traditionally held by the village chief or elder, became the charge of the party's PRL (village-level) council. Government control went beyond delineating land rights, however, dictating also the nature and organization of agricultural production in Guinea. Large state-owned farms, or FAPAs (district agro-pastoral farms), were designed to be the technical vanguard of Guinean agriculture.

In light of their special role, the FAPAs had privileged access to inputs. Some FAPA lands were appropriated from colonial plantations, and other land was appropriated through the PRL's privileges in assigning land tenure rights. FAPAs were also the target of heavy capital and mechanical investment by the government.[2] They were staffed by civil servants, many of whom were agricultural graduates the government was obliged to employ. In 1982 the total FAPA wage bill was approximately 6 percent of the government's budgeted recurrent expenditure. Total resources expended on the FAPAs were estimated at 27 percent of the total agricultural sector investment in the Fourth Development Plan (World Bank 1984a).

[1] It should be noted, however, that the mining sector operated under rules that were very different from those that applied to the rest of the economy. Much of the export receipts accruing to the private mining companies were retained and utilized outside Guinea (see below).

[2] In addition to initial working capital of GS 250,000 per FAPA, over 500 tractors, 450 plows, and 270 motor pumps had been made available to the FAPAs by 1984 (World Bank 1984a).

The FAPAs were an unmitigated failure. Rather than learning from the FAPAs, smallholders had to "volunteer" to assist FAPA farmers and accounted for up to 60 percent of the labor on the large farms. Nevertheless, FAPA yields were generally half that of smallholder yields on paddy. Moreover, the mechanization of FAPAs also backfired. By 1982 only one-quarter of the FAPA factory fleet was operational. In total, annual losses by the 360 FAPAs was estimated to be on the order of GS 576 million (World Bank 1984a).

Another production unit that formed part of the First Republic's agricultural policy was the FAC (ferme agricole commune), or agricultural cooperative.[3] Once again PRLs set aside communal land at the village level. This land was to be worked by mandatory labor. Revenue from production was to go toward the village budget for community services. Although not much is known about the operation of communal farms, indications are that, in the case of these farms too, output generally did not cover the investment made (Nellum and Associates 1980).

Smallholders, finally, provided and continue to provide the backbone of agricultural production in Guinea. Under the First Republic, however, the state apparatus permeated smallholder agriculture. Production and marketing decisions were affected by a web of repressive rules and regulations. Regional party authorities were given crop and livestock production and marketing targets by the central government, and village councils were supposed to make sure that they were attained. Each active rural household member was required to market a determined amount of agricultural output at official prices (Thenevin 1988). Official marketing channels in Guinea were run solely by public entities, with private trade considered antisocial behavior, being severely restricted ever since independence and completely prohibited until 1981.

In an effort to hold down prices for domestic consumers and to ensure that the state gained a large share of the value of exports, the First Republic depressed official producer prices, the only legal selling prices for farmers. The nominal price of rice millet and cassava, for example, was not altered at all over the seven-year period from 1975 to 1982. Neither was that of export crops palm kernels, plantains, pineapples, or mangoes (Table 3.2). Coffee prices were increased only once. Given a rapidly appreciating parallel exchange rate, official producer prices coupled with forced sales to the state implied the imposition of large explicit and implicit producer taxes on farmers.

A policy of low, administered producer prices had negative welfare effects on farmers; it also had a disastrous impact on production, marketing, exports, and the balance of payments. Available data show production

---

[3] The FACs were established in 1979 as a consolidation of the preexisting production brigade collectives.

Table 3.2. Nominal producer price of select crops, Guinea

	Export crop (GF/kg)								Food crop (GF/kg)							
	Coffee robusta (net)	Coffee arabica (net)	Groundnuts	Palm kernels	Palm oil	Plantains	Pineapple	Mango	Rice (net)	Maize	Dry Cassava	Fonio (net)	Millet	Cowpea	Onion	Potato
1975	—	—	6	6	—	4.5	—	—	15	6	5	—	7	—	—	—
1976	—	—	7	—	—	4.5	—	—	15	6	5	—	7	—	—	—
1977	34	35	8	6	—	4.5	9	8	15	6	5	—	7	—	—	—
1978	34	35	9	6	—	4.5	9	8	15	6	5	—	7	—	—	—
1979	40	41	9	6	—	4.5	9	8	15	6	5	—	7	—	—	—
1980	40	41	9	6	—	4.5	9	8	15	6	5	13	7	—	—	—
1981	40	41	9	6	28	4.5	9	8	15	7	5	13	7	7	12	15
1982	40	41	9	6	28	4.5	9	8	15	10	7	15	7	7	12	15
1983	45	46	12	7	40	14	9	8	20	—	7	15	10	10	18	20
1984	55	55	12	10	40	14	15	15	20	—	7	15	10	10	18	20
1985	70	70	20	13	60	14	15	15	25	—	—	—	10	10	20	20
1986	400	400	—	60	—	—	15	15	81	80	50	80	—	—	—	—
1987	450	450	110	40	—	14	45	15	96	—	—	—	—	—	—	—
1988	500	—	—	—	—	—	—	—	105	—	—	—	—	—	—	—
1989	290	—	—	—	—	—	—	—	105	—	—	—	—	—	—	—
1990	233	—	—	—	—	—	—	—	—	—	—	—	—	—	—	—
1991	251	—	—	—	—	—	—	—	—	—	—	—	—	—	—	—

Sources: World Bank 1989, 138; MICA (Ministère d'Industrie, Commerce, Artisanat) Service Prix, unpublished data; World Bank 1984a; Weaver 1987. Rice prices from Thenovin 1988; Filippi-Wilhelm 1988; Caputo 1991; GOG, undated (a).

Table 3.3. Crop production estimates, Guinea (thousand metric tons)

	Paddy	Maize	Sorghum	Roots, tubers	Cassava	Bananas	Pineapple	Beans	Green coffee	Cocoa beans	Tobacco	Sugarcane	Fonio	Groundnut	Mango
1970	400	—	—	—	—	—	30	—	23	—	—	—	—	—	—
1971	375	—	—	—	—	—	19	—	30	—	—	—	—	—	—
1972	375	—	—	—	—	—	—	—	—	—	—	—	—	—	—
1973	362	66.3	5	585	450	92.9	20	26	13	0	1	0	72.7	77.2	—
1974	391	67.0	5	605	450	93.9	20	27	14	0	1	0	73.5	78	—
1975	367	67.7	5	661	480	94.8	20	27	14	0	1	125	74.2	78.8	—
1976	432	68.4	5	671	490	95.9	20	28	14	0	1	125	75	80	—
1977	414	61.6	4	677	500	96.8	10	30	14	4	1	125	67.5	80.7	—
1978	418	62.3	4	654	475	97.7	144	30	14	4	1	220	68.2	81.5	—
1979	348	47.2	3	737	555	97.7	147	30	15	4	2	220	68.9	82.4	—
1980	281	57.0	5	785	600	—	—	32	15	4	2	220	—	69	—
1981	485	53.0	5	844	650	113	113	32	15	4	2	245	184	94	—
1982	490	50.0	5	654	494	—	—	45	15	4	2	225	—	—	—
1983	495	39.0	3	658	490	—	—	45	15	4	2	225	—	—	—
1984	500	42.0	3	663	500	—	—	47	15	4	2	225	221	180	—
1985	505	40.0	3	663	500	—	—	50	7	4	2	200	221	61	—
1986	504	50.0	4	663	500	—	—	50	7	4	2	200	227	185	—
1987	509	45.0	4	—	—	—	4	—	—	—	—	—	—	—	—
1988	—	—	—	—	—	—	—	—	—	—	—	—	—	—	—
1989	570	108.0	—	—	358	264	36	—	24	—	—	—	112	45	271

Sources: Fourbeau and Meneux 1989; Hanrahan and Block 1988; Hirsch 1986a; AIRD 1989; World Bank 1981c, 1990e; Nellum and Associates 1980; UNDP and World Bank 1989; USAID 1989; FAO, various years (a).

stagnating given pricing and exchange policies (Table 3.3). Production of groundnut, fonio, tobacco, coffee, beans, bananas, cassava, roots and tubers, maize, and rice all remained virtually unchanged through the 1970s, signifying declines in per capita availability. With production costs higher than producer prices, there was an obvious disincentive to produce beyond subsistence levels for the official market. This disincentive was reinforced by the fact that farmers were reluctant to produce a surplus in the fear that it would have to be surrendered as *livraisons obligatoires* (Thenevin 1988).[4] Moreover, the lack of progress on the front of agricultural research also contributed to low levels of technology and agricultural productivity (see SCETAGRI 1986 and Hirsch 1986 on agricultural research). Stagnant yield levels also reflected the virtual absence of any market for agricultural inputs.

The ensuing consequence of such low official producer prices and related institution and market failures was the virtual cessation of production for official markets. In the three years between 1977 and 1980, rice (net) sales to the state fell by 57 percent, fonio (net) by 63 percent, pineapples by 85 percent, and unshelled groundnuts by 85 percent (Associates for International Resources and Development [AIRD] 1989). Consequently, rice marketed to the state was only an estimated 1.8 percent of total production. Similarly, officially marketed fonio was only an estimated 5.5 percent of production and officially marketed groundnuts were approximately 0.8 percent of production. The government was able to ensure only a minimum procurement of commodities through the official window, presumably extracted in large part in the form of the obligatory marketing quota levied on all active household members.

The decline in officially marketed production not only increased the foreign exchange burdens of relying increasingly on food imports but also translated into a dramatic fall in export earning from cash crops (Table 3.4). Between 1970 and 1985, pineapple exports fell by 94 percent, coffee exports by 98 percent, palm kernel exports by 99 percent, and banana exports by 100 percent. The share of agricultural exports in total exports fell by 71 percent to less than 1 percent between 1975 and 1981 (Figure 3.1). Related to unattractive incentive structures, lack of competition, and weak administrative capacity, the bad management of exporting parastatals also had much to do with faltering exports.[5]

---

[4] Despite the fact that required sales were per person, it would appear that those who had a surplus were evidently forced to sell more than those who did not. It is unclear, furthermore, to what degree farmers may have been coerced to "sell" to powerful village-level party officials, at prices below the official price or quantities beyond the legal *normes*.

[5] The parastatal FRUITEX, for example, made only one purchasing agreement with pineapple producers in 1984. And, even then, it failed to follow through, resulting in the wastage of the entire expected shipment (AIRD 1989).

Table 3.4. Volume of agricultural exports, Guinea (thousand metric tons)

	Coffee	Banana	Palm kernel	Mango	Pineapple
1960–65	10.38	43.00	19.73	NA	4.12
1966–70	8.49	31.40	14.64	NA	7.40
1971–75	4.76	9.12	12.06	0.78	8.64
1976–80	1.84	0.05	12.33	0.98	2.06
1980–85	0.58	0.00	6.08	0.18	0.62
1986	4.57	0.00	2.50	0.12	NA
1987	4.60	0.00	4.10	0.13	0.38
1988	5.72	0.00	NA	0.34	0.80
1989	9.95	0.00	7.00	NA	NA
1990	13.25	NA	NA	NA	NA

Sources: 1960–80 data from World Bank 1984a. 1989 data from BCRG 1989. 1990 data from BCRG 1990, "Bulletin trimestriel d'études et de statistiques," no. 15. All other years from AIRD 1989.

Note: NA, not available.

Mining Sector Policies

In contrast to agriculture, the success of Guinea's mining sector has been important in saving an otherwise floundering formal economy. Endowed with approximately one-third of the world's bauxite reserves, Guinea also has extraction potential in gold, diamonds, iron, granite, lead, platinum, silver, zinc, nickel, cobalt, and uranium.

Several policy factors have differentiated the operation of the mining sector from the rest of the Guinean economy since independence. First, most mines have been owned jointly with the private sector. The government has also left the operation of mining in the hands of the private sector. Second, all the large private mine operators in Guinea are foreign. Third, the mining sector has relied primarily on its own infrastructure, rather than on public infrastructure. Private mining companies have been largely responsible for the construction and maintenance of their own services, in areas of the country where the public provision and maintenance of roads, rail, bridges, water, and electricity, for example, would otherwise be scant and subject to the vagaries of a debt-ridden government's public finance. Fourth, the mining companies have been permitted to circumvent some import taxes and restrictions and hence also avoid some of the distortions inherent in the trade regime. Fifth, and most important, the mining companies have been permitted not only to retain much of their foreign exchange earnings but also to retain them in accounts overseas and thus avoid the severe distortions associated with the exchange regime.

In the face of a deteriorating agricultural sector, a steady increase in mining exports has resulted in the mining sector having grown to dominate

Figure 3.1. Composition of exports, Guinea

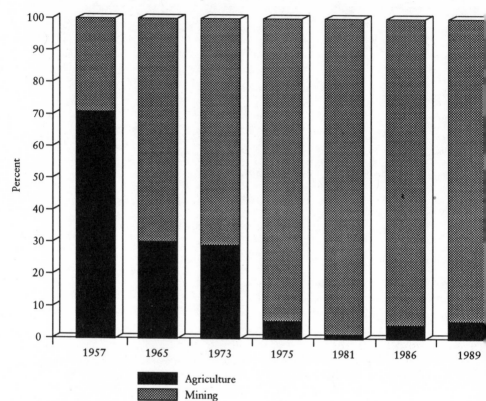

Sources: World Bank 1981c, 1984b, 1990e.

total exports. Whereas mining exports constituted 29 percent of total exports at independence, this share was up to 70 percent by 1965, 94 percent by 1975, and 99.4 percent by 1981 (Figure 3.1). These indications of success, though, must be tempered by the recognition that the profits from these exports do not all accrue to Guinea. The foreign holding companies retain their earnings and profits in accounts overseas.

Guinea has, however, partaken in the success of mining through the accrual of government tax revenue from this sector. In 1986, for example, mining sector tax revenues represented 86 percent of total tax revenue and 69 percent of total government revenues and grants. Guinea's reliance on the mining sector for its foreign exchange is also high. In 1987, 66 percent of all public sector foreign exchange revenue came from mining taxes (World Bank 1990e).

Whereas mining has benefited Guinea indirectly through the channeling of foreign exchange to the government through taxes, it has not proved that beneficial to Guinea in terms of employment creation. Approximately 9,000 people are employed by the formal mining companies, of which several hundred are expatriates (World Bank 1990e). This number of workers would have represented only approximately 0.4 percent of the domestic labor force in 1986.[6] Although the mining sector contributed close to 25 percent of GDP in 1986, there have been few forward or backward linkages with the rest of the economy.

Financial and Banking Policies

Guinea's departure from the West African franc zone in 1960 and its replacement of the CFA franc by the Guinean franc (and subsequently the Guinea syli) meant the launching of an independent monetary policy that was to finance and maintain an inefficient and unsustainable public sector over the course of the next two and a half decades. Moreover, bad financial management, together with corruption, was to be the cause of institutional chaos within the banking system.

The restrictions and pressures to which the banking sector was subject under the First Republic contributed to the system's frailty. First, with one exception, all the banks in Guinea were owned by the government. Each was additionally restricted to a specialized set of operations. Furthermore, banks could generally not even choose their own clients. Rather, as implementing agents of the development plan, they were compelled to lend to public enterprise according to the directives of the plan. Second, all the state-owned banks were initiated with seed capital. Since, however, all profits, reserves, and depreciation allowances had to be transferred to the government, these banks were unable to augment their capital position in step with their extension of credit. Third, lax administrative practices, incompetence, and bad management exacerbated the precarious condition of banks. The resulting accounting chaos, furthermore, was amenable to corruption. These factors reveal themselves in statistics on monetary aggregates (Table 3.5). Domestic credit expanded rapidly, at an average annual rate of 17 percent between 1960 and 1980. Moreover, the public sector was the main destination of this credit, garnering over 90 percent of all domestic credit during the period 1960–80. Money supply expanded at an average annual rate of 41 percent over this period.

Although expansionist policies manifested themselves as added pressure on parallel market prices, they were not evident in the movement of official prices and wages. Consequently, most economic actors took measures to

[6] The labor force figure used is based on the estimate that approximately 50 percent of the population is between the ages of fifteen and sixty (GOG 1989b).

Table 3.5. Financial sector, Guinea (billion GF)

	1960	1965	1970	1975	1980	1986	1987	1988	1989	1990a
Net foreign assets	0.50	-0.80	-1.90	-1.10	-6.70	7.90	25.10	11.10	24.30	58.50
BCRG	—	—	—	—	—	-1.90	10.30	-5.80	8.10	43.80
Commercial banks	—	—	—	—	—	9.80	14.80	16.90	16.20	14.70
Net domestic assets	—	—	—	—	—	42.60	43.40	75.90	85.20	85.80
Domestic credit	1.00	5.30	9.90	12.80	21.80	37.20	42.10	66.00	67.10	85.80
Public sector (net)	1.00	5.10	9.30	11.70	20.60	25.80	17.10	32.40	24.80	22.80
Government (net)	0.40	1.30	4.00	0.90	6.10	37.20	36.40	48.60	38.70	—
State enterprises (net)	0.60	3.80	5.30	10.80	14.50	-11.40	-19.30	-16.20	-13.90	—
Private sector	0.00	0.20	0.60	1.10	1.20	11.40	25.00	33.60	42.30	63.00
Other items (net)a	—	—	—	—	—	5.40	1.30	9.90	18.10	11.70
Money and quasi-money (M1)	1.40	3.80	7.40	11.10	13.00	42.60	59.60	80.20	103.80	123.00
Currency	0.90	1.30	3.20	3.70	3.30	34.60	43.30	55.70	69.40	84.70
Deposits	0.50	2.50	4.20	7.40	9.70	8.00	16.30	24.50	34.40	38.30
Foreign currency deposits	—	—	—	—	—	7.90	8.90	8.00	5.70	5.60
Indices/ratios										
CPI (index 12/86=100)	—	—	—	—	—	100	133.7	168.9	212.81	270.41
CPI (%)	—	—	—	—	—	—	33.70	26.33	26.00	27.07
Interest rates (%)	—	—	—	—	—	11.00	15.00	17.00	22.00	—
GDP (billion GF current)	11.9	12.8	16.8	24.2	33.52	671.27	851.01	1149.98	1534.94	1774.39
Net domestic credit as % of GDP	8.40	41.41	58.93	52.89	65.04	5.54	4.95	5.74	4.37	4.84
M1 as % of GDP	11.76	29.69	44.05	45.87	38.78	6.35	7.00	6.97	6.76	6.93
Deposits as % of M1	35.71	65.79	56.76	66.67	74.62	18.78	27.35	30.55	33.14	31.14
Domestic bank deposits in 1986 GF	—	—	—	—	—	8.00	12.19	14.51	16.16	14.16
% growth in domestic bank deposits	—	—	—	—	—	—	52.39	18.98	11.44	-12.38
% growth in deposits	—	400.00	68.00	76.19	31.08	63.92	215.00	99.39	63.67	27.62
Private share of domestic credit (%)	0.00	3.77	6.06	8.59	5.50	30.65	59.38	50.91	63.04	73.43
Private sector credit in 1986 GF	—	—	—	—	—	11.40	18.70	19.89	19.88	23.30
% growth in private sector credit	—	—	—	—	—	—	64.02	6.39	-0.08	17.21
% growth in net domestic credit (real)	—	—	—	—	—	—	-15.35	24.10	-19.31	0.63

Sources: BCRG, IMP, and World Bank from World Bank 1990e.
aProvisional data from GOG 1991a.

avoid the domestic monetary system. The mining sector, for example, utilized offshore foreign exchange accounts. Others held their savings either in foreign exchange or in real assets. Inflation and the general failure in confidence in the financial system ultimately resulted in large-scale reliance on barter trade throughout the country.

Fiscal and Public Sector Policies

The finances of the First Republic reiterate the story of a state-run economy in crisis. There was no formal-budget document. Proper budgetary processes and expenditure control procedures were nonexistent. Consequently, many expenditures went unrecorded; most government development expenditures, for example, were outside the budget (IMF 1987c). The treasury could not track the government's cash position, for accounts were usually debited without the treasury's knowledge.

The domestic tax system at the termination of the First Republic was "inadequate to meet even the most modest financing requirements" (World Bank 1988c). The fall in the share of nonmining revenue was largely due to an eroding tax base as a result of corruption and the consequent granting of frequent, uncontrolled, and often large-scale discretionary tax exemptions. Just as disconcerting was the fact that the mining sector accounted for approximately 40 percent of all government revenue and grants by 1984, making the government's finances extremely vulnerable to fluctuations in international mineral markets.

The most important elements of Guinea's public finance on the expenditure side were the expenditures necessary to sustain the employment of a large, underpaid, unmotivated civil service, those required to operate an inefficient parastatal sector, and expenditures required to pay for food subsidies. In 1979 about 35 percent of the country's labor force that was not engaged in rural self-employment had a job in the public sector (World Bank 1981d).[7] Between 1981 and 1984, on average, 41 percent of current expenditures went to financing wages and salaries (Arulpragasam and Sahn, forthcoming).

Also important to fiscal accounts was the subsidy inherent in official food prices transmitted through the ration system in place in urban areas under the First Republic. Monthly official entitlements of rice in Conakry implied an estimated subsidy to consumers of 42 percent, or of approximately GS 8 per kilogram (see Arulpragasam and Sahn, forthcoming). This subsidy policy, which discriminated against low-income households, implied a huge drain on the treasury.

---

[7] A more recent public sector census indicated that, excluding military personnel, there were 90,000 public servants in 1986, 72,000 of whom were in administration and 18,000 of whom worked at public enterprises (World Bank 1990e).

Inefficient and corrupt parastatals were an additional drain on fiscal resources. Between 1981 and 1984, on average, 28 percent of all current expenditure was on subsidies and transfers to numerous public enterprises that covered every aspect of economic activity. In fact, there was little formal enterprise other than public sector enterprise: parastatals accounted for approximately 75 percent of all modern sector employment and 92 percent of all domestic credit.

Partly as a consequence of the large current expenditure costs of wages, salaries, and transfers and subsidies to parastatals, capital expenditures levels during the First Republic were low. By 1984 development expenditures accounted for approximately 16 percent of total expenditures, down from 34 percent ten years earlier. Much government investment, moreover, was allocated to unprofitable collectivized agriculture and to the inefficient parastatals, consequently earning low rates of return (World Bank 1990e). Over the course of the First Republic, meanwhile, public infrastructure as well as the physical capital of public enterprises had suffered significant degeneration.

The social sectors also suffered as a consequence of budgetary mismanagement. The Ministry of Health's budget was approximately 1.5 percent of GNP between 1979 and 1984 (World Bank 1986c). Furthermore, the share of the national operating budget allocated to health declined steadily to approximately 5 percent in 1984. Given an inordinate allocation of operating funds for wages and salaries, the health and education sectors suffered shortages in medical and teaching supplies. Moreover, the low levels of capital expenditure led to the dilapidation of hospitals and schools.

Impacts of Reform on the Micro and Sectoral Levels

The death of Sekou Touré in 1984 represented a dramatic turning point for Guinea. In conjunction with political reform, the Second Republic under General Lansana Conte has commenced sweeping economic reform. The new regime has committed itself to an economy in which private ownership and foreign investment are to be accepted and promoted.

Phase one of the PREF (Programme de Redressement Economique et Financier) committed the government of Guinea to (1) undertake a large-scale devaluation and realign its exchange rate, (2) liberalize internal and external trade by eliminating price controls and state marketing, (3) restructure the banking and financial sector by shutting down state banks and promoting commercial banking, (4) reduce the scale and increase the efficiency of the public sector by privatizing, liquidating, and restructuring parastatals as well as by reducing public sector employment, (5) increase and reorient public investment so as to improve productivity, and (6) institute commercial and institutional reforms that would promote private sector savings and investment. In concentrating on effectively pushing forward and adhering to initial reform guidelines, the second phase of the PREF also focuses on (1) strengthening local economic management and policy implementation capabilities, (2) removing sector-specific infrastructural bottlenecks, (3) improving the legal and institutional fabric for the private sector, and (4) developing a social policy which, in protecting vulnerable populations and those most likely to be hurt by reforms, will also facilitate the reform program from a political standpoint.

Trade and Exchange Sector Policy Reform

The severance from a fixed exchange rate system has probably been the most dramatic and important reform undertaken by Guinea's Second Republic, due to both the magnitude of the initial devaluation and the subsequent move to a more flexible exchange rate regime. The nominal exchange rate experienced a seventeen-fold devaluation between the last quarter of 1985 and the first quarter of 1986, bringing the official rate roughly in line with the parallel (Table 3.6).

In real effective terms too, the devaluation has been just as dramatic. The real effective exchange rate (REER) takes into account not only trade shares and exchange rates of Guinea's trading partners but also Guinea's inflation rate relative to the inflation rates of trading partners. Despite the high domestic inflation rates brought about by the nominal devaluation, the REER experience an initial devaluation of approximately 1,205 percent (Arulpragasam and Sahn, forthcoming). The REER (index of Guinean francs per U.S. dollar) in early 1991 wa still over eight times its level in 1985. Moreover, the gap between the parallel and official exchange rate has also narrowed. In 1988 and 1989, the margin was maintained at approximately 8 percent (Table 3.6).

Although exchange rate policy with adjustment has had an overall real impact on the price of foreign exchange in Guinea, the simple REER does not tell the whole story. There has also been considerable liberalization of the trade regime. Most important, the private sector has since been permitted to import all commodities, including food, with the exception of petroleum products (the import of which is also in the process of being liberalized). Import licensing was abolished in 1986. The private sector can now also engage in export activity in all commodities but gold and diamonds. In an important move to liberalize the trade regime, the tariff structure was generally reduced and dramatically simplified in January 1986.

Table 3.6. Exchange rates, Guinea (GS/GF per US$)

	Official rate	Parallel rate	Official/parallel ratio
1970	246.85	404.67	0.61
1971	246.85	1,084.58	0.23
1972	22.74	—	—
1973	20.46	56.00	0.37
1974	20.46	91.25	0.22
1975	20.46	57.08	0.36
1976	20.46	60.83	0.34
1977	20.46	60.92	0.34
1978	20.46	30.58	0.67
1979	20.46	30.58	0.67
1980	20.46	41.70	0.49
1981	20.46	73.08	0.28
1982	20.46	93.33	0.21
1983	20.46	128.08	0.16
1984	20.46	283.42	0.07
1985	20.46	310.67	0.07
1986	365.00	396.25	0.92
1987	429.00	443.33	0.97
1988	475.00	521.00	0.91
1989	593.00	642.79	0.92
1990	661.67	708.75	0.93
1991[a]	724.71	764.14	0.95

Sources: Pick's Currency Yearbook, various years; World Bank 1984b, 1990e; GOG 1989a; BCRG unpublished data for 1990 and 1991.
[a]Data from first seven months of 1991 only.

Despite these commendable policy initiatives, there continues to be an active (albeit now shallower) parallel market for foreign exchange. The fact that some trade does indeed continue on the parallel market reflects both continued capital flight and several disincentives to participating in the official market. First, many traders are discouraged from participating in the banking system: it requires being in Conakry, filling out numerous forms, and paying high banking fees. Participation in the foreign exchange "auction" requires that a trader be registered with the Ministry of Commerce and have a license to participate.[8] The trader must also have an account at a commercial bank. Second, and perhaps most important, by

---

[8] In practice, the foreign exchange rate for the week is announced after the collection of private sector applications for foreign exchange that week. Although referred to as an auction, no price bids are ever made during this process. Demand and supply at the auction do not equate, demand generally being in excess of supply among the private sector. Moreover, the exchange rate is relatively unresponsive to short-term changes in relative demand and supply at the private sector (or auction) allocation. The shortfall in foreign exchange supplied by the private sector is met with public sector foreign exchange revenue from mining taxes and balance-of-payments support.

avoiding official channels traders also avoid filling out import declaration forms and consequently avoid customs and duty payments as well as pre-shipment verification fees. Third, by avoiding official channels, traders can avoid even the short negative list and use the auction to transfer interest and dividend payments overseas, for example. Fourth, circumventing official channels also frequently means circumventing long administrative delays associated with the official process. Fifth, the disincentive to attaining foreign exchange through official channels was significantly enhanced with the policy, effective February 1990, that once again requires advanced deposits of local currency at the Banque Central de la Republique de Guinée (BCRG) by traders who want to import consumption goods. Aimed at limiting the large volume of imports, this measure replaced a 1989 restriction that prohibited the use of credit for the purchase of foreign exchange.

A primary argument for exchange rate devaluation is to increase the competitiveness and sale of export products. In Guinea exchange and trade reform has resulted in a notable increase in export revenues. Between 1986 and 1989, export value increased by 16 percent (Table 3.7). This growth was largely due to the dramatic increase in nonmining exports from its previously low level. As in most other sub-Saharan countries, however, the balance-of-payments problem in Guinea since the commencement of adjustment has been that exports have been unable to keep up with the large increases in imports.

Imports using foreign exchange accessed through the auction have reflected a heavy consumption bias to private sector imports with liberalization (Arulpragasam and Sahn, forthcoming).[9] In 1989, 81 percent of all commodity imports through the auction were accounted for by consumer goods; 51 percent of all foreign exchange sold through the auction was used to import food; and 21 percent was used specifically to import rice. With inclusion of public sector imports through the off-auction channel, the share of commodity imports represented by foodstuff is lower (19 percent) and the shares for capital and intermediate goods increase to 20 and 35 percent, respectively (Government of Guinea [GOG] 1991).[10] Nevertheless, the low share of capital and intermediate goods imports by the private, nonmining sector is a somewhat disheartening indicator of the prospects of early growth in the nonmining sectors.

The rapid increase in imports and consequent trade balance movements

[9] It is unclear to what extent overinvoicing may be used in conjunction with the auction to mask capital flight.

[10] The larger share of capital and intermediate goods in total imports reflects the importance of purchases such as petroleum, equipment, and intermediate goods by the government and mining companies (GOG 1991a).

Table 3.7. Balance of payments, Guinea (million US$)

	1981	1982	1983	1984	1985	1986	1987	1988	1989
Trade balance	67.0	66.3	135.8	103.5	136.5	147.5	120.3	9.9	145.9
Exports, f.o.b.	493.0	443.9	501.4	510.0	512.9	655.2	687.1	650.9	761.5
Bauxite/alumina	—	428.4	490.7	502.0	495.0	467.2	484.8	454.5	530.6
Other	—	15.5	10.7	8.0	18.0	188.0	202.3	196.4	230.9
Imports, c.i.f.	426.0	(377.6)	(365.6)	(406.5)	(376.5)	(507.7)	(566.8)	(641.0)	(615.6)
Public sector	—	(224.1)	(228.9)	(264.0)	(228.5)	(204.3)	(176.9)	(191.1)	(180.3)
PIP imports	—	—	—	—	—	(57.6)	(77.9)	(97.1)	(100.6)
Other	—	—	—	—	—	(146.7)	(99.0)	(94.0)	(79.8)
Mixed mining companies	—	(153.5)	(135.8)	(142.5)	(145.0)	(99.0)	(131.1)	(131.9)	(124.4)
Other private	—	—	—	—	(3.0)	(204.4)	(258.8)	(318.0)	(310.9)
Services/private transf. (net)	—	(183.2)	(161.4)	(165.9)	(201.6)	(249.8)	(298.5)	(314.2)	(375.3)
Nonfactor services, net	—	—	—	—	—	(80.1)	(112.7)	(109.3)	(118.9)
Factor services, net	—	—	—	—	—	(144.1)	(136.1)	(168.6)	(200.9)
Interest on official reserves	—	—	—	—	—	1.0	1.9	2.3	1.9
Interest on MLT debt (incl. USSR)	—	—	—	—	—	(43.8)	(48.6)	(70.3)	(55.8)
Mixed mining companies (excl. OBK)	—	—	—	—	—	(80.6)	(62.6)	(67.4)	(110.9)
Other	—	—	—	—	—	(20.7)	(26.7)	(33.1)	(36.1)
Private transf. (net)	—	—	—	—	—	(25.6)	(49.7)	(36.4)	(55.6)
Mixed mining companies (excl. OBK)	—	—	—	—	—	8.2	(11.0)	(10.5)	(11.6)
Other	—	—	—	—	—	(33.8)	(38.7)	(25.8)	(44.0)
Public sector (net)	—	-64.0	-63.0	-70.0	-109.0	—	—	—	—
Interest on public debt	—	(34.2)	(29.9)	(31.8)	(27.4)	—	—	—	—
Interest on financing gap	—	(13.2)	(17.1)	(20.5)	(20.3)	—	—	—	—
Other (net)	—	(16.6)	(16.0)	(17.4)	(60.9)	—	—	—	—
Mixed mining companies (net)	—	(114.8)	(94.1)	(88.0)	(88.0)	—	—	—	—
Other private (net)	—	(4.4)	(4.3)	(8.2)	(5.0)	—	—	—	—
Public transfers	—	26.5	18.2	51.0	18.0	42.3	82.8	83.5	97.8
Current account deficit (−)	—	(90.4)	(7.5)	(11.4)	(47.2)	(60.1)	(95.4)	(220.8)	(131.6)
Capital movements (net)	—	57.4	31.0	(24.5)	(34.8)	36.9	71.6	73.7	89.5
Public capital, long-term	—	6.6	25.7	(19.5)	(32.0)	49.8	50.4	38.0	70.9
Disbursements, no.[a]	—	86.1	91.9	78.9	70.0	173.0	167.4	220.6	207.3
Amortization	—	(79.5)	(66.3)	(98.4)	(102.0)	(123.2)	(117.0)	(182.6)	(136.4)
Other	—	6.6	(3.2)	—	—	—	—	—	—

Public capital, short-term	—	—	—	—	(15.2)	0.0	0.0	0.0
Amort. due to 1986 resched.	—	—	—	—	—	—	—	—
Mixed mining companies (net)	44.2	8.6	(5.0)	14.2	(2.7)	16.0	29.3	8.6
Other private, incl. direct investment (net)	—	—	—	(17.0)	5.0	5.2	6.5	10.0
Errors and omissions	(29.9)	(127.2)	(59.1)	(7.3)	(28.9)	13.1	24.0	5.7
Overall balance	(62.9)	(103.7)	(95.0)	89.3	(52.0)	(10.7)	(123.1)	(36.4)
Financing items	62.9	103.7	95.0	89.3	52.0	10.7	123.1	36.4
IMF credit	13.2	—	—	—	10.9	14.6	0.0	17.2
Purchases	13.2	—	—	—	17.6	21.9	0.0	22.5
Repurchases	—	—	—	—	(6.7)	(7.3)	0.0	(5.3)
of which: SAF (net)	—	—	—	—	—	—	—	—
Gold	—	—	—	—	0.0	(9.2)	(19.6)	(18.4)
Other reserve movements (net) (increase −)	3.3	62.0	53.0	11.2	(23.3)	(34.0)	42.0	(21.3)
Reduction of arrears (gross)	—	—	—	—	(285.3)	(16.5)	(19.0)	(104.7)
Other liabilities or arrears of which short-term	—	—	—	3.0	(13.5)	0.0	0.0	0.0
IDAs, SAL I, cofinanciers	—	—	—	—	—	—	—	—
Paris Club resched. 1986	—	—	—	—	219.2	6.3	0.0	0.0
Paris Club resched. 1989	—	—	—	—	0.0	0.0	0.0	116.5
Other debt resched. (in process)	—	—	—	—	133.5	25.5	43.7	30.9
Accumulation of arrears	46.4	41.7	42.0	75.1	10.5	24.0	76.0	16.2
Other bilateral (USSR) (net)	—	—	—	42.0	—	—	—	—
Memorandum items								
Current account balance/GDP (%)	—	—	—	—	−3.27	−4.90	−9.77	−5.51
Current account balance (excl. public trans.)/GDP (%)	—	—	—	—	−5.57	−9.15	−13.47	−9.61
Exports/GDP (%)	—	—	—	—	35.63	35.26	28.81	31.90
Exports (excl bauxite/alumina)/GDP (%)	—	—	—	—	10.22	10.38	8.69	9.67
Imports/GDP (%)	—	—	—	—	−27.61	−29.00	−28.37	−25.78
Private imports (excl. mining)/GDP (%)	—	—	—	—	−11.11	−13.28	−14.08	−13.02
Gross official reserves (year end)[b]								
In millions of US$	56.0	22.5	14.7	2.2	21	64	42	82
In months of nonmining, non-PIP imports	—	—	—	—	1	2	1	3
GDP in current prices (US$ M)	—	—	—	—	1,839.1	1,948.5	2,259.2	2,387.5
GDP in current prices (GF B)	—	—	—	—	671	836	1073	1416
Exchange rate (GF/US$)	—	—	—	—	365	429	475	593

Sources: IMF 1987c for 1982–85 data. World Bank 1990e for 1986–89 data.

[a] Disbursement figures include project related loans, IDA and cofinancing, and drawings on USSR special loans.

[b] Reserve figures include gold in data after 1986.

reflect several factors. The increase in imports was caused by a pent-up demand, in addition to a large-scale rechanneling of extant demand that had earlier been satisfied through unofficial channels or financed through own resources. Moreover, balance-of-payments support permitted imports in excess of exports. To the extent that concessional financial inflows contributed to slowing a depreciation of the exchange rate, they may have additionally served to reign in exports while imports increased. Finally, Guinea's terms of trade commenced to experience a sharp fall in 1986, dropping by 12.2 percent that year (World Bank 1986d). Adverse terms-of-trade movement continued to affect Guinea through 1990, with a fall in the price of exported coffee due to the failure of the International Coffee Agreement and the suspension of quotas and with an increase in the price of petroleum due to the Persian Gulf crisis.

There have clearly been some impressive accomplishments in the reform of Guinea's exchange rate policy and management, as well as of trade policy in general. Commercial imports of consumer and other goods have increased dramatically, exports have shown some tentative sign of responding, and prices rather than quantities are beginning to clear markets. Efficiency gains have also been garnered through the removal of gross distortions that contributed to market failures and rent seeking. Urban consumers and traders have likely been the greatest beneficiaries of these reforms, with plentiful supplies of food (especially imported rice) and other consumer goods returning to the markets. Furthermore, new opportunities for commercial activities have also risen.

Still, these observations raise some important concerns about the extent of structural change with reform and about the overall sustainability of the current balance-of-payments scenario. High levels of imports evident in private sector foreign exchange deficits are being counteracted not by export revenues as much as by other sources (Table 3.7). First, mining sector foreign exchange tax revenue continues to be a primary source of financing import needs. Projected declines in mining tax revenues raise an alarm as to the sustainability of the balance-of-payments situation if agricultural exports do not increase. Second, in addition to mining tax revenue, balance-of-payments assistance from donors has also been important on the revenue side of foreign exchange accounts, especially for financing capital and intermediate goods expenditures. Clearly Guinea's balance-of-payments plight, in general, and its continued reliance on current levels of imports, in particular, will continue to depend on high levels of donor support. Third, Guinea's foreign exchange expenditures have also been financed by the drawing down of external reserves and the accumulation of arrears. For example, in 1988 the level of foreign exchange supplied to the auction was made possible only by running down reserves and running up arrears and then again in 1990 by accumulating in arrears on the order of GF 31.13 billion (GOG 1991a). Fourth, financing of the balance-of-payments deficit since the commencement of the formal adjustment program has been made possible by the rescheduling of debt and the extension of IMF credits. For example, total debt relief in 1986 reached close to US$353 million; in 1989, after another round of Paris Club rescheduling, total debt rescheduling was $147 million (Table 3.7). The importance of annual debt relief to meet balance of payments raises additional concerns regarding the sustainability of the country's balance of payments.[11]

Agriculture Sector Policy Reform

In addition to trade and exchange rate reforms that affect agricultural incentives and food prices, direct interventions within the agricultural sector itself have been fundamental in Guinea's efforts to generate economic growth and improve the welfare of producers and consumers. With 74 percent of the national population being rural, agricultural policy is important from an incomes standpoint for producers and from a prices and quantities standpoint for consumers and domestic food security. On the macroeconomic front, agricultural-policies are critical to production, to export growth, and thus to the balance of-payments equation.

Impact on Production, Marketing, and Export. Agricultural reforms have been focused on dismantling the extensive state controls of the First Republic and establishing a policy framework that will foster the development of a market-oriented economy. First, in 1984, production was liberalized with the abolishment of state collective farms. In 1985 producer prices were decontrolled. Second, steps were taken to liberalize marketing. In April 1984 all existing quotas for the forced marketing of agricultural products to the state were terminated. Restrictions that controlled the internal movement of goods, including police barricades, were also eliminated. Commencing in 1986, private traders were allowed to participate at all levels of internal and external trade. State marketing monopolies were eliminated, as were parastatal monopolies on long-distance transportation and storage.

Recent reforms have effectively reduced or eliminated the taxation of producers with respect to the production of both export and food crops. Changes in the level of taxation are reflected in changes in the level of the nominal protection coefficients (NPC) for various crops—that is, the ratio

[11] The need for debt relief underlines the fact that Guinea's debt has been increasing at an average annual rate of growth of 11 percent over the course of the decade (World Bank 1990f). By 1987 total debt (outstanding and disbursed) was approximately US$1.6 billion.

of domestic producer prices to international prices valued at farmgate. Under the First Republic, producers were effectively taxed. Domestic producer prices evaluated at the appropriate parallel exchange rate were lower than border prices less a transport and handling margin. At the parallel rate, NPCs for export crops remained well below 1.0 for the period 1975– 85 (Table 3.8).[12] The producer price for coffee evaluated at the parallel exchange rate was on average approximately one-third of the border price over this period. The average NPCs for groundnuts and palm kernels over the ten years were approximately 0.5 and 0.6, respectively. The nominal producer price increases commencing in 1982 had little impact in stemming the decline in NPCs given the precipitous depreciation of the parallel exchange rate. In 1984, Sekou Touré's last year in power, the NPC of coffee was as low as 0.12, that of groundnuts 0.22, and of palm kernels 0.11. The NPC for rice, the country's main food crop, was 0.33 in 1985.

The liberalization of producer prices has resulted in a dramatic increase in the legal prices to farmers. In the case of export crops, this is most clearly evident in the case of coffee, for which annual producer price data are readily available.[13] In 1986 the revised indicative producer price for coffee, at GF 400 per kilogram, was close to six times higher than the official producer price in the preceding year (Table 3.2). When prices are valued at the parallel exchange rate, an important increase in the NPC for coffee, from 0.12 to 0.53, is noted between 1985 and 1986.[14] The NPC appeared to reach 0.70 in 1988 when an increase in the indicative producer price coincided with the beginning of a fall in world prices for coffee. Since 1989, and then with the collapse of the International Coffee Agreement (and the disruption of the export quota system that went with it) in 1990, actual producer prices have fallen well below the stated indicative price.[15] Nevertheless, at 0.55 in 1991, the NPC for coffee continued to be higher than during the First Republic.

Similarly, policy reform has effectively led to a large increase in legal producer prices for rice. In 1986 and 1987, observed producer prices (GF

---

[12] The fact that NPCs are lower than 1.0 when producer prices are valued at the parallel rate and higher than 1.0 when they are valued at the official rate indicates that the taxation is largely implicit in nature, transmitted by the distortion in the exchange rate rather than by nominal producer price policy per se.

[13] The producer price series for coffee is constructed using indicative official farmgate prices issued by the Ministère d'Industrie, Commerce, Artisanat and producer price information from RC2, a coffee project. Indicative prices were closely adhered to from 1986 through 1989.

[14] The fact that there was a large devaluation of the exchange rate at the same time makes it fallaciously appear that producers started being heavily taxed in 1986, according to the NPCs valued at the official exchange rate (Table 3.8).

[15] Whereas the indicative price for coffee was GF 500 per kilogram in 1990 and 1991, actual producer prices paid to farmers were GF 233 and GF 251 per kilogram for those respective years.

Table 3.8. Nominal protection coefficients, Guinea

	At official exchange rate					At parallel exchange rate				
	Rice	Coffee robusta	Coffee arabica	Groundnuts	Palm kernels	Rice	Coffee robusta	Coffee arabica	Groundnuts	Palm kernels
1975	1.77	—	—	1.15	2.69	0.63	—	—	0.41	0.96
1976	2.12	—	—	1.32	2.18	0.71	—	—	0.44	0.73
1977	2.68	0.77	0.62	1.15	1.48	0.90	0.26	0.21	0.39	0.50
1978	2.14	0.67	0.54	1.12	1.32	1.43	0.45	0.36	0.75	0.89
1979	1.96	1.06	0.97	1.27	0.95	1.31	0.71	0.65	0.85	0.64
1980	1.65	0.88	0.74	1.57	1.54	0.81	0.43	0.36	0.77	0.75
1981	1.47	1.10	1.26	1.21	1.75	0.41	0.31	0.35	0.34	0.49
1982	1.75	1.72	1.31	2.08	2.48	0.36	0.36	0.27	0.43	0.52
1983	3.59	1.49	2.14	2.97	1.63	0.57	0.24	0.34	0.48	0.26
1984	3.86	1.65	1.64	2.99	1.57	0.28	0.12	0.12	0.22	0.11
1985	5.03	1.86	1.85	5.02	—	0.33	0.12	0.12	0.33	—
1986	1.04	0.58	0.36	—	—	0.96	0.53	0.33	—	—
1987	1.09	0.46	0.86	—	1.00	1.05	0.44	0.83	—	0.97
1988	0.80	0.77	0.61	—	—	0.73	0.70	0.55	—	—
1989	0.57	0.42	—	—	—	0.52	0.39	—	—	—
1990	0.62	0.52	—	—	—	0.58	0.49	—	—	—
1991	—	0.58	—	—	—	—	0.55	—	—	—

Sources: MICA unpublished data; USAID 1989. World Bank 1984b, 1987b, 1989b, 1990e; GOG 1989a; Nellum and Associates 1980; Thenovin 1988; UNDP/World Bank 1989; AIRD 1989; IMF, various years (b); FAO, various years (b); Weaver 1987; Caputo 1991.

80–100 per kilogram) were three to four times greater than the official producer prices of two years earlier (GF 25 per kilogram) (Table 3.2). NPCs calculated with the parallel exchange rate and excluding marketing margins show a jump from 0.33 in 1985 to 0.96 in 1986 (Table 3.8). The actual NPCs in 1986 and 1987 were thus very close to parity. Farmers' actual market revenues would have increased, due to this price effect, to the extent that they were not already receiving equally high prices on unofficial markets before adjustment.

The primary rationale for increasing producer prices is to increase production. Unfortunately, the actual production impact of price adjustment in Guinea is especially difficult to ascertain given the low quality of production data. In the case of rice, what little data do exist show an increase in production subsequent to liberalization of official producer prices. The 1989 agricultural census (GOG 1990c) revealed a rice (paddy) production level of 596,000 metric tons. This stood in contrast to a mean production level of 506,000 metric tons estimated for the three-year period 1985–87 (Table 3.3). The Filière Riz surveys in Guinée Maritime (GOG, undated a) and Haute Guinée (GOG, undated b), which questioned farmers on land use, also indicate increases in hectarage cultivated with rice between 1985 and 1987.[16]

In the case of export crops, coffee has appeared to react most dramatically to changes since the liberalization of producer prices and marketing, consistent with reports that coffee is the crop in which Guinea holds the largest comparative advantage (SCETAGRI 1986). Official exports increased forty-six-fold between 1985 and 1986 and have increased at an average rate of 22 percent annually between 1986 and 1989 (Table 3.4). Whereas total official exports were approximately 100 metric tons in 1985, they had increased to 4,600 metric tons by 1986 and to 8,100 metric tons by 1989.[17] Palm kernel exports have also increased, as have exports of pineapples and mangoes.

---

[16] In Haute Guinée, the mean area farmers planted in rice increased from 2.39 hectares in 1986 to 3.23 hectares in 1987. In Guinée Maritime, 61 percent of farmers claimed to have cultivated more land with rice in 1986 than in 1985, and 70 percent were going to increase their 1987 land allocation to rice beyond their 1986 levels. In Guinée Maritime, the increased land planted to rice was due primarily to the clearing of new land and reduction of fallow time rather than to the reallocation of land from the production of other crops such as groundnuts or fonio (GOG, undated a and b).

[17] Some of this large increase reflects the rechanneling of informal sector exports back into the formal sector because of improved incentives. To the extent that coffee sales have simply been diverted from informal sales to official sales, producer welfare may not have increased as dramatically as government export tax revenues. The continued rapid increase in exports (e.g., of 42 percent in 1989), however, is likely also to signify the increased allocation of resources to coffee production since 1985. The 1989 agriculture census data indicate actual hectarage in 1989 to be close to double that estimated in 1985.

Despite these early production responses, however, several factors are acting to limit continued gains in agricultural production and marketing in Guinea. In the case of rice, the first relates to producer prices themselves.[18] After the rapid adjustment in 1986, producer price increases have not kept pace with inflation. Between 1986 and 1989, real producer prices decreased at an annual average rate of 13 percent. Even in nominal terms, producer prices have appeared to stabilize since 1988 and have not risen in pace with the devaluation of the exchange rate. Thus, despite a fall in the world price over the period 1988–90, NPCs for rice have deteriorated from the levels attained immediately after producer price liberalization. They do, however, remain above pre-liberalization levels.[19]

In any case, the current constraints to production of both food and export crops lie beyond prices. Technology has shown no major progress since before adjustment because of years of neglect of local research. There has been a conspicuous lack of research on seed improvement, water management, and farming systems in the case of rice (AIRD 1989). Only 9 percent of interviewed farmers in Upper Guinea used "improved variety" seed (GOG, undated a). Fertilizer, pesticides, and phytosanitary products are still not used by most farmers. Low utilization levels of these inputs are partially a result of the lack of credit and finance among farmers, partially a consequence of the lack of extension, and partially due to the fact that inputs such as fertilizer remain largely unavailable in most rural areas (USAID 1987; GOG, undated b).[20] Transport also constitutes a big bottleneck in Guinea. Trucking capacity is limited and roads are bad. There are approximately 4,000 trucks in Guinea—most old, dilapidated, and small. Scarce credit, inefficient marketing links, bad roads, and limited transportation capability, in combination, are responsible for the high cost of agricultural marketing in Guinea. Indeed, it costs three times as much to transport a kilogram of rice from Boké to Conakry as it does to ship a kilogram from Bangkok to Conakry (Arulpragasam and Sahn, forthcoming).

---

[18] It should be noted that no systematic collection of producer prices has taken place in Guinea since the liberalization of fixed prices. Even when producer prices are recorded at the local level, there is no system whereby this information is transmitted, recorded, or made available to the central government in Conakry. Consequently, the producer price estimates used here are drawn from several available studies and from different parts of the country.

[19] Even if producer prices did not increase in 1991, they would still reach 60 percent of world prices given a fall in the latter.

[20] Only 2–3 percent of interviewed farmers in Guinée Maritime and Haute Guinée used fertilizer and only 1 percent used other plant products (GOG, undated a). The lack of rural supply is in turn attributable to the private sector's negligible involvement in the marketing of agricultural inputs. Moreover, there is anecdotal evidence that the lack of demand is due in part to farmers finding it unprofitable to use unsubsidized fertilizer for rice production, even given current producer prices for rice (AIRD 1989).

Finally, with respect to export crops in particular, there are also heavy bureaucratic costs associated with a multitude of export procedures.[21] Reducing paperwork will lower the time costs of processing exports, thereby eliminating one of the factors that also diminishes the timeliness and reliability of Guinean exporters in the eyes of international partners. To this end, there has been discussion of establishing a *guichet unique* at which all procedures can be undertaken at one point at the border. Eliminating administrative hoops and consolidating processes presumably will also lower the costs incurred in the form of rents paid to civil servants to get the required paperwork done.

Impact on Food Security.   In the short term at least, the greatest scope for food policy to affect household welfare, in both urban and most rural areas, is likely to be mediated through rice. This is especially the case given the importance of rice as a staple in Guinea. In Conakry, rice represents 15.5 percent of food expenditures and 8.2 percent of total expenditures (CFNPP/ENCOMEC survey). Among the ultrapoor and poor, rice is even more important, constituting 23.9 percent and 18.7 percent, respectively, of food expenditures and 15.6 and 11.3 percent, respectively, of total expenditures (del Ninno and Sahn 1990). Despite the importance of other commodities such as bread, fish, and oil, no single other commodity or commodity group is more important in the diet or expenditure bundle than rice.

In discussing rice availability in Guinea, it is important to make the distinction between three general rice types. The differentiation is based not only on source but also on perceived quality, and hence on price. In particular, the two general imported rice varieties in Guinea are cheaper than the local rice variety. Rice commercially imported from Vietnam, Taiwan, Pakistan, and Thailand (*riz Asiatique*) is generally not parboiled, of low quality (usually 35 percent broken), and one-third cheaper on the Conakry retail market than locally produced rice. Rice imported as food aid (largely in the form of PL 480 Title I and II rice from the United States, *riz Caroline*) is preferred less than the local variety but more than the commercially imported *riz Asiatique* and is consequently priced between the two.

Rice availability in Guinea has increased markedly with reform, due primarily to the increase in rice imports after liberalization. In the five years from 1984 to 1989, rice imports (commercial and food aid) doubled,

---

[21] Exporters have been required to submit several forms and certificates. They were also subject to a 2 percent export tax until 1989 and a *taxe de conditionnement* until 1991. For a more detailed description of export procedures, refer to Arulpragasam and Sahn, forthcoming.

increasing by close to 100,000 metric tons (Table 3.9). Whereas in the late 1970s and early 1980s imported rice accounted for approximately 20 percent of total rice availability, by 1990 it accounted for 30–40 percent of total availability. Whereas per capita rice availability nationally averaged close to 60 kilograms per year between 1980 and 1984, it had risen to average about 70 kilograms per capita per year between 1989 and 1990 (Table 3.9). The exact extent of the increase in rice availability is, however, clearly sensitive to the estimate of domestic production used.[22]

Beyond quantities, reform has also had an important impact on the price of rice to the consumer. Under the First Republic, official rice prices were subsidized. In 1980 the official price was 31 percent of the parallel market price, implying a subsidy of 69 percent (Nellum and Associates 1980).[23] The magnitude of the income transfer inherent in the ration program is best measured as the difference between the official and parallel market prices (see, e.g., Alderman, Sahn, and Arulpragasam 1991). In 1984, given official prices of GS 20 per kilogram early that year and parallel market prices of GS 70 per kilogram, the average income transfer through the ration would have been approximately GS 300 per month per person, or GS 3,390 per household (see Arulpragasam and Sahn, forthcoming).[24] This corresponded to approximately 11 percent of mean household expenditures in 1984.

As compared to pre-reform official, subsidized prices, the price of rice to consumers has increased since 1986 with liberalization. Deflating official consumer prices before adjustment and imported rice prices after adjustment by the c.i.f. price reveals a fall in consumer welfare since 1985, to the extent that consumers were actually accessing official retail prices before that date. Whereas the official retail price averaged 53 percent of the c.i.f. price between 1981 and 1984, the actual retail price of imported rice averaged 139 percent of the c.i.f. price between 1988 and 1990 (Tables

[22] Estimates of domestic rice production vary widely. The figure used here utilizes the midpoint of the highest and lowest estimates for domestic production from GOG 1990c (570,000 metric tons) and is lower than the estimate of 596,000 metric tons from the agricultural census. Assuming an estimate of 510,000 metric tons of production in 1988 and 1989 (i.e., zero growth since 1984), per capita consumption between 1989 and 1990 is still much higher than previously, averaging 68 kilograms. If a lowest bound estimate of 425,000 metric tons of domestic production is assumed, per capita consumption between 1989 and 1990 averages 62 kilograms. (If the low estimate is in fact correct, production estimates for previous years would also have to be revised downward.)

[23] This measure of subsidy captures more than just the distortion inherent in rice pricing, however, since it also incorporates distortions in the parallel markets (which are also spillovers from government policy).

[24] Due to incomplete coverage of ration cards and unavailability of rationed rice at the shops, the average per capita monthly uptake of rice at the official price was actually 6 kilograms rather than 8 kilograms. The average household size utilized for these calculations is 11.3.

Table 3.9. Rice production, imports, and availability, Guinea (thousand metric tons)

	Paddy production[a]	Commercial imports	Food aid imports	Total imports	Total national rice availability	Food aid as % of total imports	Total imports as % of total rice availability	Per capita rice availability[a] (kg/yr)
1970	400.0	—	—	35.0	255.0	—	13.73	—
1971	375.0	—	—	35.0	241.3	—	14.50	—
1972	375.0	—	—	45.0	251.3	—	17.91	—
1973	362.0	—	—	40.0	239.0	—	16.74	—
1974	391.0	—	—	35.0	250.0	—	14.00	—
1975	367.0	—	—	42.7	244.7	—	17.45	47.10
1976	432.0	—	—	34.7	272.3	—	12.74	50.61
1977	414.0	36.7	14.4	51.1	278.8	28.18	18.33	51.14
1978	418.0	10.0	34.3	44.3	274.3	77.43	16.15	48.83
1979	348.0	53.0	9.5	62.5	253.5	15.20	24.65	44.77
1980	281.0	44.4	17.5	61.9	216.5	28.27	28.59	37.57
1981	485.0	60.0	12.6	72.6	339.4	17.36	21.39	56.52
1982	490.0	62.0	20.8	82.8	352.8	25.12	23.47	57.47
1983	495.0	58.7	16.6	75.3	347.6	22.05	21.67	63.95
1984	500.0	67.0	30.4	97.4	372.4	31.21	26.15	67.29
1985	505.0	86.5	21.4	107.9	385.7	55.61	27.98	68.05
1986	504.0	79.0	60.0	139.0	416.2	42.66	33.40	72.25
1987	509.0	58.3	59.3	117.1	397.8	26.66	29.62	66.64
1988	—	162.0	31.4	193.4	—	10.86	—	—
1989	570.0	174.4	21.0	195.4	508.9	0.84	38.40	82.16
1990	570.0	110.5	1.7	112.2	425.7	0.00	26.35	65.20

Sources: Fourbeau and Meneux 1989; Hanrahan and Block 1988; Hirsch 1986a; SCETAGRI 1986; USAID 1989; Lowder-milk 1989; Filippi-Wilhelm 1987, 1988; GOG 1991a. Production estimates of 570 for 1989 is average of high and low estimates from GOG 1991b.

Notes: The estimations allow for a 0.55 transformation coefficient in processing paddy into rice and account for a 0.10 loss/waste/seed coefficient. No carry-over stocks and no reexports are assumed. Reexports are likely to vary greatly from year to year. Estimates of the volume of reexports range from 5 to 20 percent of total imports. Overestimation of rice availability by not accounting for reexports is compensated to some extent by the likely underreporting of official imports as reflected in customs data.

[a]Local milled based on a 0.55 conversion from paddy. Losses, waste and seed assumed to be 10 percent of paddy production.

Table 3.10. Official rice retail prices, Guinea

	GF/kg	At official rate Price (US$/MT)	Ratio to c.i.f.	At parallel rate Price (US$/MT)	Ratio to c.i.f.
1975	20	977.52	2.36	350.39	0.85
1976	20	977.52	2.82	328.79	0.95
1977	20	977.52	3.57	328.30	1.20
1978	20	977.52	2.85	654.02	1.91
1979	20	977.52	2.61	654.02	1.74
1980	20	977.52	2.20	479.62	1.08
1981	20	977.52	1.96	273.67	0.55
1982	20	977.52	2.33	203.40	0.49
1983	25	1,221.90	4.48	195.19	0.72
1984	25	1,221.90	4.82	88.21	0.35
1985	25	1,221.90	5.03	80.47	0.33

Sources: Rice c.i.f. price from import quantity and value data from AIRD 1990. Producer prices and official prices from MICA unpublished data. Other sources for rice data: Filippi-Wilhelm 1987, 1988;; Thenevin 1988; Hirsch 1986a; Hanrahan and Block 1988; USAID 1989; World Bank 1984a; Nellum and Associates 1980; Hanrahan and Block 1988; SCETAGRI 1986; Lowdermilk 1989; Chemonics International 1986.

3.10 and 3.11).[25] In other words, the estimated mean per capita income transfer for those previously accessing the *ravitaillement* system was effectively eliminated after 1985.

The income loss due to the elimination of the official, subsidized ration rice that followed the beginning of adjustment, however, must be weighed against gains due to a fall in the parallel market price relative to c.i.f. In 1980, for example, the parallel market price was 345 percent of the c.i.f. price evaluated at the parallel exchange rate (Tables 3.10 and 3.11).[26] The income loss from purchasing 8 kilograms of rice per month on the parallel market (rather than at the c.i.f. rate) could thus have been as high as GS 8,720. In 1990, in contrast, the free market retail price was only 153 percent of the c.i.f. price. Although the observed retail price of rice has fluctuated as a percentage of the c.i.f. price since 1986, the average margin of 45 percent over the five-year period is not much higher than the estimated marketing margin of 30 percent (Tables 3.10 and 3.11).

The formal abandonment of the ration system coupled with the liberalization of imports has thus meant two countervailing price effects with adjustment. In particular, the welfare implications of changes in rice policy

[25] Presumably the recorded price for imported rice captures an average of both imported rice types—the PL 480 *riz Caroline* and the *riz Asiatique*.
[26] The magnitude of the margin between official retail and c.i.f. prices evaluated at the parallel exchange rate also reflects the high transaction costs, risk premia, and scarcity in the parallel markets due to trade and other restrictions. As such, this ratio captures several distortions inherent in the policies of the First Republic.

Table 3.11. Retail rice prices, Guinea

| | Local rice | | | Imported rice | | | | | |
| | | | | Official price | | | Market price | | |
	GF/kg	(A) Ratio to c.i.f.	(B) Ratio to c.i.f.	GF/kg	(A) Ratio to c.i.f.	(B) Ratio to c.i.f.	GF/kg	(A) Ratio to c.i.f.	(B) Ratio to c.i.f.
1985	149.62	30.11	1.98	—	—	—	100.20	20.17	1.33
1986	219.09	2.82	2.60	100	1.29	1.19	129.53	1.67	1.54
1987	194.42	2.20	2.13	100	1.13	1.10	119.25	1.35	1.31
1988	319.64	2.44	2.22	173	1.32	1.20	212.64	1.62	1.48
1989	319.00	1.72	1.59	210	1.13	1.05	230.00	1.24	1.15
1990	406.87	2.41	2.25	210	1.24	1.16	277.05	1.64	1.53
1991	422.27	2.28	2.16	—	—	—	327.23	1.77	1.67

Sources: Rice c.i.f. price from import quantity and value data from AIRD 1990 and from unpublished data from Bureau Veritas. 1988 and 1989 figures are calculated using 0.85 BKK FOB + freight. Retail price data: 1989 data are derived from early 1990 based on ENCOMEC survey results. 1990 and 1991 (first quarter) prices are from unpublished Ministère du Plan et de la Coopération Internationale data. Other sources for rice data: Filippi-Wilhelm 1987, 1988; Thenevin 1988; Hirsch 1986a; Hanrahan and Block 1988; USAID 1989; World Bank 1985b; Nellum and Associates 1980; Fourbeau and Meneux 1989; SCETAGRI 1986; Lowdermilk 1989; Chemonics International 1986; Caputo 1991.

Notes: (A) prices evaluated at official exchange rate. (B) prices evaluated at parallel exchange rate.

varied according to consumers' relative reliance on the parallel market in fulfilling their rice needs before adjustment. The relative losers were most likely the more influential merchants and higher level *fonctionnaires* who had access to the ration system (GOG 1986). The relative gainers from a policy of cheap and abundant imported rice on the free market (rather than a policy of rationed and subsidized rice on the official market) were those who had always had difficulty accessing the official market. These were mainly the smaller merchants, the part-time workers, and the unemployed (GOG 1986). To the extent that most of the population purchased on the parallel market at the margin by the end of the First Republic, however, most consumers have stood to gain from liberalization.[27] In particular, the policy of cheap imported rice is of greatest relevance to consumers in the capital. Rice consumption levels are higher in Conakry than elsewhere in the country, since large quantities of imported rice are readily available there. Estimates place the consumption of rice in Conakry at 18–20 percent of the rice available nationally (Chemonics International 1986; Filippi-Wilhelm 1987), with 82 percent of all rice consumed in Conakry being imported (Arulpragasam, Ninno, and Sahn 1992).[28]

Relative to urban areas, in rural areas the issues of food security in general and the consumption effects of rice policy in particular are much more complex, and data are much scarcer.[29] On aggregate, rice is the most important food crop produced and consumed even in rural Guinea.[30] Changes in producer and consumer prices of rice have, therefore, certainly affected food security in the interior. Moreover, the policies of adjustment have raised the important question of the impact of rice import liberalization, specifically, on domestic production, farmer incomes, rural consumption patterns, and rural food security.

The price differential between local rice and cheaper imported rice even in rural areas highlights the importance of rice import liberalization for the majority of rural consumers who are net rice purchasers.[31] Indeed, the

---

[27] According to survey data (GOG 1986), ration uptake was highest among civil service (and some private sector) officials and large merchants. Furthermore, it was estimated that Conakry households, on average, were getting only 25 percent of their cereals from official sources, whereas households outside the capital ration cardholders were acquiring only 5 percent of their cereal purchases from official sources (World Bank 1984a).

[28] Conakry thus consumes an estimated 60 percent of Guinea's official rice imports.

[29] Moreover, production and consumption patterns are extremely varied across the country. The importance of rice for food security is not as critical in the Fouta Djallon, for example, as it is in Conakry. It is impossible to get region-specific food balance pictures.

[30] Rice is in fact grown on approximately 43 percent of all cultivated land throughout the country, although there is considerable geographic diversity.

[31] Of the 87 percent of Guinée Maritime farm households that purchase rice, 32 percent consumed both imported and local rice, 32 percent consumed only local rice, and 36 percent consumed only imported rice (GOG, undated a).

Table 3.12. Regional variation in consumer price of local and imported rice, Guinea (GF/kg)

Region	Local rice		Imported rice	
	1986	1987	1986	1987
Guinée Maritime	120–140	180–200	115–120	130–150
Moyenne Guinée				
Fouta Djallon	160–190	250	160–175	200–250
Northwest	150–180	175–200	140–180	160–180
Haute Guinée	200–220	195–250	155–160	175–200
Guinée Forestière	125–135	170–175	115–125	160

Source: Filippi-Wilhelm 1988.

price of imported rice is consistently lower than that of local rice throughout Guinea, although the discount for imported rice is lower in producing regions and in areas farther from Conakry (Table 3.12). The fact that imported rice is cheaper than local rice even in the interior lends credit to the argument that there is a quality premium paid for local rice. The narrower price margin in Guinée Forestière, for example, is also consistent with the fact that that region is a high-production area distant from the capital.[32]

The availability of cheaper imported rice on rural retail markets nevertheless raises a controversial argument. It is argued that the adjustment policy of rice import liberalization constitutes a food security threat to the nation by transmitting a production disincentive to domestic rice farmers (Hirsch 1986a; Filippi-Wilhelm 1987; Thenevin 1988). The doubling of the volume of rice imports from 1984 to 1989 has heightened concerns regarding the heavy dependence on imports. In response to these concerns, the government has commenced measures aimed specifically at reducing rice imports and their sale in the interior. Indeed, even after liberalizing internal trade in 1986, the government continued to require approval (*autorisations de transit*) to move imported rice into areas of the interior. Moreover, whereas between 1985 and 1988 rice was exempted from all customs and tariffs, by 1990 the levy on rice had subsequently been raised to 20 percent of c.i.f. value. Furthermore, in March 1990 the government imposed the requirement that 20 percent of the c.i.f. value of imports of consumption goods on order be deposited in advance at the BCRG in order to access foreign exchange. Given the working capital constraints of smaller importers, this requirement severely restricted demand. The proce-

[32] That import rice is still cheaper in Guinée Forestière may also be explained by reexports from neighboring Liberia or Sierra Leone.

dure for accessing foreign exchange has further limited rice imports in that it requires importers to lock in to the exchange rate that prevails at the time of import declaration, exposing importers to substantial exchange rate risk. Currently, moreover, there is a move for a further policy shift. Under the recommendation of Caputo (1990), the government intends to proceed with the examination and design of a variable import levy on rice.

Despite the fact that importers have innovated several new procedures designed to circumvent these regulations, the new policy orientation has had its intended effect. In 1990 there was a large reduction in rice imports. According to customs data, rice imports fell by a dramatic 42 percent between 1989 and 1990. The impact of these protectionist policies on consumer welfare is reflected in the increase in the retail price relative to the c.i.f. price from 1.15 to 1.67 between 1990 and 1991. Although the impact of the protectionist policies on the retail price of imported rice was restrained in 1990 and 1991 because of the concomitant fall in the world price of rice, the increased cost of distortions faced by traders was nevertheless passed on to consumers, who consequently did not benefit from the fall in the world price of rice.

The new policy orientation with respect to rice raises several pressing questions with respect to food security on both the demand and supply sides. With respect to demand, raising tariff and nontariff barriers will ultimately raise the consumer price of imported rice, both relative to c.i.f. and in absolute terms, if there is no decline in the world price. Given the important role of cheap imported rice for food security among urban households and net rice-purchasing rural households, such policies will likely have a negative impact on the welfare of these groups. A related question is whether increased protection will foster domestic production. For tariff protection to effectively promote the production of domestic rice, increasing the price of imported rice must increase the demand for domestic rice. If domestic rice is not a close substitute for imported rice, this may not follow. Poorer consumers who now eat imported rice may switch to another staple rather than to domestic rice. Moreover, even if consumers do switch to domestic rice consumption, the real income loss due to the price increase will reduce overall demand.

With respect to supply too, important questions should be addressed before raising tariff and nontariff barriers on imported rice. For example, if there is an increase in the demand for domestic rice as a result of protectionist policies, will domestic producers be able to increase supply at the prevailing price? It is not yet clear whether the high cost of domestic rice on the Conakry market is currently due to cost-of-production factors or to other impediments, ranging from a shortage of trucks to labor technology and credit constraints. These latter high costs will likely be better addressed by technological innovation and road improvements, for example, than by the imposition of a tariff. Moreover, if markets are so fragmented by marketing and transport costs, by implication there exists built-in protection for domestic rice in rural markets. Indeed, the argument that imported rice has hurt domestic production is contradicted by estimates that indicate a production increase of 15–20 percent since 1983, despite the doubling of imports.

The success of local farms in capturing a larger share of the domestic rice market, especially in Conakry, would seem to hinge importantly on government policies to further technological efficiency in domestic production and to reduce marketing and transport costs such that producers garner a higher share of the market. These same factors, however, may also induce farmers to switch to higher-value export crops instead of producing rice. This scenario will likely have beneficial food security implications as long as the policy of liberalized international and domestic grain trade is maintained in the context of cheap international rice prices.

Financial Sector Policy Reform

Going beyond agriculture, the sweeping reforms in Guinea have also transformed the financial sector. On December 22, 1985, the president ordered the immediate closing of all six state banks, commencing a dramatic reform of the financial sector. Banking since that date has been open to private sector participation, including foreign interests. In response, five commercial banks have opened in Guinea, most founded with foreign participation and management. Reform, moreover, has meant private sector investment in Guinea in general and within the banking industry in particular.

The positive consequences of these reforms are evident in the movements of monetary aggregates (Table 3.5).[33] First, and perhaps most important, is the shift in domestic credit away from state enterprises in the process of being liquidated and toward the private sector. Whereas the share of private sector credit in total domestic credit was on the order of 5 percent in 1980, by 1986 it had increased to slightly over 30 percent, and by 1990 to 73 percent. Second, as a result of the liberalization of the exchange regime, a new investment code, and the influx of donor credit, Guinea's net foreign asset position has turned positive since 1986.

The limits to reform, however, are also evident: the financial sector

---

[33] Given the large-scale changes in data methods, too much weight ought not be placed on the magnitude of changes in monetary aggregates between 1980 and 1986 in Table 3.5. Nevertheless, several observations regarding the direction of changes are noteworthy.

remains shallow (Table 3.5). The stock of money (M1) as a proportion of GDP remains relatively unchanged at approximately 7 percent.[34] It was estimated that the private sector's local currency deposit base at the banking system was as low as 2.1 percent of GDP in 1989 (World Bank 1990e). This low base underlines, first and foremost, the inadequacy of interest rate reform to date. Even after nominal increases, interest rates remained negative in real terms. In 1990, even with a fall in the inflation rate, savings and sight deposits carried a −3 percent real rate of interest.[35] Cheap deposits have combined with subsidized interest rates made available to commercial banks to exacerbate distortions and limit the efficient expansion of the banking system in Guinea. Although policy reform has led to the elimination, in 1989, of interest rate ceilings for regular short- and medium-term loans by commercial banks, the availability of funds at cheap, subsidized rates through the central bank continues to restrain banks from assuming higher-risk loans or promoting interest-paying deposits.

Problems persist not only on the deposit side but also on the credit side. Commercial banks face continuing disincentives to domestic lending, these constraints centering on the institutional and legal void left in Guinea after the Sekou Touré years. In addition to the obvious lack of credit records for clients, the absence of an enforced legal framework reduces information and increases risk for the lender. These factors have also limited credit for short-term loans. Moreover, most of this credit has been channeled to financing commodity imports by large, well-known traders. Credit practices have thus reinforced the consumption goods orientation of the foreign exchange auction.

It is critical for Guinea's long-term growth prospects that domestic savings in general and personal savings in particular be mobilized and directed toward productive investment. To that end, however, efficient financial intermediation still remains a challenge. Gross domestic savings as a percentage of GDP are no higher than in 1986. Although gross fixed investment has increased by 31 percent since 1986, this has been primarily due to a 54 percent increase in largely externally financed public investment. In fact, after an initial improvement in 1987, private sector investment has declined since 1988. It will clearly take time to regain the public's confidence in the banking sector after the Touré years. Moreover, raising savings is also going to require, at the least, that real interest rates be consistently positive.

---

[34] The money supply to GDP ratios in the neighboring CFA countries of Mali and Côte d'Ivoire are 23 and 32 percent, respectively, and 14.5 percent in neighboring non-CFA Sierra Leone.

[35] In 1990 the fall in the inflation rate resulted in a positive interest rate (4 percent) on term deposits for the first time.

Public Sector and Fiscal Policy Reform

The sweeping reforms of Guinea's public sector have been a remarkable testimony to the government's will for change. They were both a response to the fiscal crisis that burdened the country in 1984 and also an integral part of the ideological move away from Touré-style socialism. One thrust of reforms, such as the adoption of the government's first formal budget (Loi de Finances) in 1988, has been to get the fiscal house in order. Another thrust of public sector reform has been to get the state, at least partially, out of the economy. As such, the government has undertaken a series of reform measures within the public sector. These include improved budgetary control, increased efforts at revenue mobilization, privatization and liquidation of parastatals, civil service retrenchment, and the undertaking of a formalized public investment effort. These measures, in combination, have had a discernible impact on the aggregate finances of the state since 1986.

Per the objectives of the reform program, revenues and grants have grown since 1986 (Table 3.13). With adjustment, the share of grants increased rapidly, from 12 to 20 percent of total revenues and grants. Moreover, led by increased tax revenues from goods and services, real nonmining tax revenue increased at an average annual rate of 35 percent over the four-year period between 1986 and 1990.[36] The increase was aided by revenue from a special tax on petroleum products. Despite these increases, however, total real tax revenue fell by 34 percent over the four-year period. This was due in part to the decline in real mining tax revenues associated with a pricing formula contracted with mining companies which discounts cost increases over time (World Bank 1990e). It was also due to continued low tax yields. Guinea's aggregate tax yield (tax revenue to GDP ratio) actually fell from 14 to 12 percent from 1986 to 1989.[37]

Several factors account for the low tax yields. The government is unable to properly identify and track taxpayers because of shortcomings in management, evaluation skills, and information technology. More important, because of discretionary exemptions, corruption, and other loopholes, the current tax base continues to be exploited inefficiently. The low rate of effective tax collection is also a consequence of not-so-transparent budgetary cross-subsidization that has benefited government officials and agencies. The military and several parastatals have been subsidized in this manner.[38] Efficient revenue mobilization and budgetary management is

---

[36] The apparent drop in the share of nonmining revenue between 1985 and 1986 may be a function of differing estimation methods for the two series, which are joined at those years.

[37] This compares to tax yield levels of 24 and 32 percent for Zambia and Gabon, respectively (World Bank 1990e).

[38] For example, until its privatization in 1991, the petroleum distribution parastatal ONAH was unable to remit petroleum tax revenue it collected back to the treasury since it needed this revenue to absorb its losses.

Table 3.13. Financial operations of the central government, Guinea (billion 1986 real GF)

	1986	1987	1988	1989	1990
Revenues/grants	103.40	130.70	120.51	132.33	127.72
Mining rev.	71.10	73.26	56.72	63.73	65.82
Nonmining rev.	11.20	17.12	24.70	29.24	36.52
Taxes: income/profits	1.10	1.30	2.16	2.33	—
Taxes: domestic goods/services	4.10	8.44	15.09	18.86	—
Taxes: international trade	6.00	7.30	7.45	8.40	—
Nontax rev./miscell. taxes	8.30	12.09	14.83	10.57	—
Grants	12.80	28.23	24.26	28.74	25.38
Expenditure	136.20	155.93	172.71	166.23	171.18
Current expenditure	85.20	87.05	94.16	87.16	83.70
Wages/salaries	18.20	17.28	28.97	30.23	32.64
Other goods/services	45.00	42.02	30.56	28.59	27.48
Subsidies/transfers	9.50	12.98	14.58	12.81	7.14
Interest payments	12.50	14.77	19.99	15.59	16.44
Capital expenditure	51.00	68.88	78.62	79.02	87.48
Deficit (−) (commitment basis)	−32.90	−25.23	−52.20	−33.90	−43.46
Net changes in expend. arrears (reduction −)	3.20	−1.30	8.40	−3.72	10.98
Cash deficit (−)	−29.70	−26.53	−43.46	−37.62	−32.49
Financing	29.70	26.61	43.86	37.62	32.49
Current budgetary surplus					
Net external financing	36.70	31.07	41.63	44.47	34.09
Drawings	59.00	53.87	69.77	60.90	—
Amortization	−41.30	−42.35	−47.62	−37.62	—
Changes in amort. arrears (reduction −)	−7.90	10.79	6.49	−22.09	—
Debt relief	26.90	8.776	12.99	43.33	—
Net domestic financing	−7.10	−4.46	2.16	−6.85	−3.70

Sources: 1986–89 data from World Bank (1990e). 1990 data from GOG 1991a.

thus inextricably linked to the elimination of hidden subsidies, patronage, and corruption which remain as legacies of Guinea's First Republic.

The concurrent decline in the share of nontax revenue in nonmining revenue over the reform period has been primarily due to the shutdown of parastatals. By the end of 1988, 25 of 131 original enterprises had been privatized and 68 had been liquidated.[39] And though the speed of reform stands testimony to the government's impressive adherence to reform guidelines, the expected financial benefits to the government from the sale of parastatals have been far from met. Enterprises were often sold at low prices, payable over several years, at zero interest. Assets were also sold at extremely low prices. Revenue considerations were also overlooked: some

[39] In 1990 further progress was made with the liquidation of five commercial enterprises. Most important, the state petroleum company ONAH and the electric utility company Enelgui were the object of restructuring. In addition, six other enterprises were also privatized in 1990.

privatized companies were given significant tax breaks as well as temporary monopoly rights, often beyond the provisions of the new investment code (World Bank 1990e).

Total revenue growth has been slow, but the government has succeeded in keeping expenditures in check over the adjustment period. In real terms, average annual expenditure growth was contained to 6 percent between 1986 and 1990, a rate that about equals that of revenues and grants over the same period (Table 3.13). The growth in expenditures has been primarily due to the public investment program component of reform and to the ensuing influx of capital. Indeed, capital expenditures increased on average by over 15 percent annually in the four-year period between 1986 and 1990. The share of capital expenditures in total expenditures grew from 16 percent in 1984 to 51 percent in 1990 (Table 3.13).

The sectoral allocation of the development budget reflects the government's immediate investment priorities. Of expenditures planned for the 1989–91 period, 62 percent were allocated to infrastructure (almost all related to roads and bridges), 25 percent to rural development, and 15 percent to the social services (World Bank 1989f). In addition, the government has been successful in securing foreign financing commitment for the Public Investment Program (PIP). Eighty-five percent of projects in the 1989–91 PIP were to be financed from external sources. Most of the financing has been at extremely concessionary terms.

Despite these accomplishments, the implementation of the PIP has encountered some difficulties. The government has fallen into arrears (GF 8.0 billion in 1990) in the payment of the local currency component of various projects that constitute the investment program.[40] Policy reform has also run into binding institutional constraints that go beyond the good intentions of both government and donors. In putting together and choosing projects for the PIP, given the limited capacity and experience at the responsible technical ministerial planning units to do such work, little economic or technical project appraisal is undertaken (World Bank 1989f). Moreover, little attention is placed on the recurrent cost implications of projects.

Indeed, in contrast to capital expenditures, real recurrent expenditures grew by less than 2 percent during the period 1986–90 and in fact declined between 1988 and 1990 (Table 3.13). This lack of growth was due to a real decline in expenditures on goods and services and on subsidies and transfers. The numbers reflect an imbalance between nonwage recurrent expenditures and capital expenditures.

From a welfare standpoint, this imbalance is especially troubling at the

---

[40] This slippage has required a revision of the PIP budget, resulting in a net increased commitment of funds to the locally financed component of the investment program (the *budget national de développement*) as well as the elimination of some projects that were initially to be completely locally funded.

sectoral level among the social services. The low levels of government financing budgeted for the educational sector as well as the poor allocation of funds within the sector have been important factors in explaining a persistently weak educational system.[41] In 1989 only 12.9 percent of the government's slim recurrent budget was allocated to education, in contrast to the sub-Saharan average of 20 percent (World Bank 1990e). Although funding for education has increased markedly with a 20 percent increase in the allocation to the subsector in 1991 (complemented by external funding from the first installment of a sectoral adjustment loan), special attention is required to assure adequate funding for operating expenditures and expenditures on materials and goods and services. Real operating expenditures are particularly vulnerable given the reluctance to cut salary expenses (which now account for 80 percent of education sector recurrent expenses) and the low scope of recovering education costs by charging fees.

Likewise, despite some improvements in the financing of the health system through cost recovery (UNICEF 1990), the health sector continues to receive inadequate funding to perform effectively. The share of the total government budget allocated to health, although increasing in 1985, has decreased below its 1984 share of 3.4 percent to a level of 2.2 percent in 1989 (Arulpragasam and Sahn, forthcoming). Here again sectoral improvements in recurrent expenditure lag externally financed improvements in capital expenditures. Although the health sector's share in the investment budget grew from 4.8 to 6.8 percent between 1989 and 1990, the unavailability of recurrent funds is evident in shortages of medical equipment and essential drugs at hospitals. This may also prove to be a constraint on the proper functioning of new health centers.

The decline in real expenditures on goods and services with reform has at the same time been associated with an increase in real expenditures on wages and salaries within the public sector. A primary tenet of the economic reform program has been to reduce the ranks of the civil service while concurrently rectifying the problems of low pay that had made it necessary for most public sector employees to take on additional jobs to make ends meet during the First Republic. Although the process has been slower than planned, the government has nevertheless been fairly effective in taking people off its payroll through a combination of obligatory and early retirement, a voluntary departure program, and testing for skills. Civil service staff had been reduced to 71,000 by the end of 1989, down from its level of 104,000 at the end of 1985 (World Bank 1990e).[42]

[41] Enrollment in schools has actually declined since the start of the Second Republic (GOG 1988; World Bank 1990e). The primary school gross enrollment ratio has shown no improvement with reform, having remained at approximately the same level (28 percent) throughout the 1980s (GOG 1988, 1991b). The illiteracy rate is still high at 74 percent.

[42] The ambitious, initial structural adjustment credit document had actually targeted a reduction of public service employment by 20,000 people by the end of 1986.

Table 3.14. Salary structure of the public sector, Guinea

	1986	1987	1988	1989	1990[a]	1991[b]
Wage bill (billion GF)[c]	18.2	21.3	44.7	57.2	78.5	97.5
Staffing (000)						
Civilian	90	80	70	52	52	52
Military[d]	12	12	13	15	15	16
TOTAL	102	92	83	67	67	68
Average wage						
000 GF/yr	178	232	539	854	1,172	1,434
000 GF/mo	15	19	45	71	98	119
Conakry CPI[e]	100	137	174	223	259	—
Real wage (000 GF/mo)	15	14	26	32	38	—
Rice (import.) equiv. wage (kg/day)	3.86	5.31	7.05	10.29	11.79	12.12

Sources: World Bank 1990e; GOG 1991a; and author's calculations.
[a]1990 wage bill figure from GOG 1991a.
[b]Provisional, July 1991.
[c]Budget item 2101.
[d]Military staff levels are estimated. Military are assumed to be paid at same level as civilian staff.
[e]Period average basis.

Even though civil service employment has been cut back, the public sector still continues to be an important employer in Conakry. Microlevel data based on the preliminary results of the first six months of the CFNPP household survey show that, even in 1990, 27 percent of all workers and 35 percent of all household heads in Conakry still had jobs in the public sector (del Ninno and Sahn 1990). The extent of public sector employment stresses the continued importance of public sector wages for household welfare in Conakry.

Public sector salaries have in fact increased, as planned, since the reform program was instituted (Table 3.14). Yet the average public sector wage still represents a fraction of monthly living costs, especially given the escalation of prices with the post-1986 currency devaluations. Based on preliminary data from the CFNPP (ENCOMEC) Conakry household welfare survey, the average monthly public sector wage of GF 98,000 represented approximately 45 percent of actual average monthly household expenditure in 1990 (del Ninno and Sahn 1990).[43] Moreover, it was even less than the total expenditure level (of approximately GF 110,000 per month) among households that constitute the lowest expenditure decile. Households of public sector workers continue to have to derive a significant part of household income from sources other than the public sector.

---

[43] The average monthly public sector wage rate represented approximately 180 percent of actual average monthly per capita expenditure.

An outstanding issue with respect to the reform program is whether liberalization has enabled those displaced from public sector employment to secure alternative employment and incomes in a new economy bereft of official prices. Preliminary results based on the first six months of the CFNPP data reveal, in fact, that many of those laid off from the public sector are having a hard time finding employment. The unemployment rate in Conakry was estimated to be 12.6 percent in 1989 and may actually be higher.[44] Of those seeking employment, 50 percent had lost their previous job. Of this number, 65 percent had lost a job in the public sector (del Ninno and Sahn 1990).

The questions affiliated with post-reform, public sector employment and wages are clearly ones requiring further attention and study. As elsewhere in Africa, in Guinea the civil service retrenchment program has been one of the most politically sensitive tenets of adjustment. In harming the urban party cadres, civil service retrenchment undermines one of the government's most critical constituencies. Moreover, from a purely welfare angle, retrenchment affects the large segment of the urban population that continues to rely on the wages and benefits of public sector employment.

Conclusions

The experience from Guinea drives home several points. First, it demonstrates the importance and potency of misguided economic policy in driving the decline in Guinea's economy over the twenty-five years since independence. Through agricultural and exchange policies, production fell. Foreign exchange reserves also dropped as a consequence of trade policies, and official imports all but ceased. The public sector was effectively bankrupt. Economic policies made exchange prohibitive, and the formal, monetized economy disintegrated as economic actors turned to barter.

The disintegration of the Guinean economy was as much due to the lack of policy as to bad policy. Although it seems contradictory, the Sekou Touré years represent just as much a condemnation of dirigisme as they do of anarchy. Under the dictatorship, no consistent body of law was developed or enforced. Individual government officials were largely free to interpret decrees and policies as they wished. Moreover, the government did virtually nothing to promote rural credit, nor did it invest in agricultural

---

[44] It should be noted that the unemployment rate does not include the large number of people who are of working age, do not work, but are also not looking for a job. Only about 15 percent of nonparticipants were searching for a job. This rate was 24.5 percent among men (del Ninno and Sahn 1990). Although many of the nonparticipating nonsearchers listed themselves as housewives, students, and apprentices, it is also possible that they elected to enter these activities simply because they gave up on the search for a job.

research. The state hindered the development of private trade and markets, at times prohibiting them, but never assisting in their development.

Finally, the First Republic showed that economic distortions, coupled with an unaccountable state and a lack of law, lead to the institutionalization of distortions. Parallel markets became pervasive as an expensive means of circumventing the official economy. Corruption and patronage became a way of life, providing extra incomes for the privileged and representing the only access to jobs, goods, and other services among the general population.

The breadth and speed of reforms undertaken by the Second Republic, since the death of Touré, have been truly remarkable. In contrast to other countries discussed in this volume, reform in Guinea unquestionably represents a true break from the past with respect to both political ideology and economic policy. Within a year the exchange rate was liberalized; international and domestic marketing were freed up; commercial banking and parastatals were privatized; and government budgeting procedures were made more rigorous.

Another important set of lessons to be learned from the Guinea experience relates not to the speed of reform but to the limitations of the traditional adjustment package. On liberalization, the increase in agricultural exports and foreign capital coupled with the sudden supply of both imported and local goods in the market place were at first the source of much happiness among Guineans and donors alike. The subsequent stall in reform and absence of further observable successes, however, has quelled the initial euphoria. This says as much about unrealistic expectations as it does about the short-term limits to reform. Simply put, despite mandated policy change with respect to prices and markets, Guinea's continued growth is being critically hindered by other factors. Agricultural production is stunted by the lack of rural credit markets and rural roads; banking and credit in turn are constrained by the lack of functioning legal structures that assign and enforce creditor rights; the public investment program is undermined by the lack of investment planning expertise and management skills; tax yields remain low due to weaknesses in the information, evaluation, and enforcement capabilities of the civil service.

Guinea's new economic policy framework is a necessary condition for growth, but such a framework alone is not sufficient. Growth will require more than simply adjusting economic price-determined incentives. Growth and development will require building roads and ports and promoting and enabling the provision of educational and health services. Furthermore, a body of law to establish a new way of doing business, one that moves away from the corruption and patronage that has become an institutionalized modus operandi in Guinea, will have to be developed. The legal system should be capable of protecting the rights of new growth-enabling constituencies against vested interests that may favor the status quo. Private sector development and foreign investment furthermore *require* an institutionalized and accountable legal system. Growth will also depend on institutionalized credit systems, which in turn will depend on rebuilding confidence in the monetary economy as well as in the law. Agricultural growth will depend critically on improved agricultural research and extension. Marketing and export will require time for traders to gain contacts, credibility, credit, and capital. The administration and management of foreign capital and of the country itself will further require a cadre of trained professionals. Implied herein, then, is a redefinition of the role of government in Guinea, not the elimination of government. The government's undertaking of these tasks, moreover, will also point to an obvious role for donor technical and project assistance in the future. Indeed, given its history, there will have to be much more to adjusting for the past in Guinea than mere privatization and price adjustment.

4

Missed Opportunity for Adjustment in a Rent-Seeking Society: The Case of Zaire

wa Bilenga Tshishimbi, with the collaboration of Peter Glick and Erik Thorbecke

A large and sparsely populated country, Zaire is endowed with immense natural resources, including rare and precious minerals.[1] This, combined with its strategically important location in the center of Africa, made the country both a large beneficiary of foreign assistance and the focus of East-West confrontation during the cold war era. At independence in June 30, 1960, Zaire inherited from the Belgian administration a fairly vigorous economy, consisting of well-developed education[2] and health systems and good communication and transport networks, all of which supported a dynamic agriculture and growing industry. Yet, by all measures, Zaire was in 1991 one of the poorest countries on the African continent and in the world, with a per capita income of only US$240 per annum. In spite of almost a decade of stabilization and structured adjustment policies, Zaire's economy has not grown and, even worse, is now experiencing a deep recession and economic anarchy.

The structure of the Zairean economy has not changed dramatically since independence. The mining sector is still important and is dominated

---

[1] The data coverage in this chapter does not extend beyond 1990 and in some instances no later than 1989 or even 1988. This lag is rather typical of Zaire. It reflects the inherent difficulties of data gathering and their publication, not unusual in developing countries in general. The Zaire case, however, has an added dimension: the overwhelming size of the country and the complexity of the research environment there.

[2] Although education did not meet all the nation's needs and was limited to primary and secondary levels to serve mainly the clerical and other needs of the colonial administration, the literacy rate was among the highest among African countries.

by Gécamines,[3] the copper- and cobalt-producing parastatal, followed by industrial diamond-producing Miba. Other major resources include iron, petroleum (although prospects for vast reserves of oil seem limited, unlike in neighboring Angola and Congo), and hydroelectric power. A promising source for future development in Zaire, is agriculture. The potential in this area is largely unexploited. Because of chronic neglect of the transport sector and agricultural services, most production continues to be for own consumption. This is the case even in Bandundu, which is the major supplier of food to Kinshasa: a recent random sample of rural households in Bandundu indicates that fully 90 percent of food production is for own consumption (Ben-Senia 1991).

In addition, Zaire is a dual economy, consisting of a formal but by and large sluggish economy and a thriving informal or "unrecorded" economy. The latter comprises production activities, principally but not exclusively in the informal manufacturing and services sectors, as well as illicit activities including smuggling, transfers in the form of tributes and bribes, and simple misappropriation of public property. The growth of the informal or underground economy has coincided with the decline of the official economy and has no doubt been caused by it. Individuals have turned to the underground economy for their livelihood in the face of growing inefficiencies and distortions and the stagnation of the formal sector output and incomes. As a result, it is increasingly recognized that official figures substantially understate the true extent of economic activity.

Historic Developments and Institutional Environment

The seeds of Zaire's economic troubles were sown in the early years of the country's accession to sovereignty. After independence, in 1960, Zaire, then the Congo, quickly entered a protracted period of turmoil marked by political conflict and secession attempts from major segments of the country. In retrospect, the crisis was inevitable in view of the fragility of the institutions hastily set up by the departing colonial power and the new leadership's political inexperience and lack of vision.

The uncertainty generated by the crisis led to a massive migration of the population from rural to urban areas and a flight of skilled expatriates from the country. As a result, both the infrastructure and the productive

[3] Gécamines stands for La Générale des Carrières et des Mines, a state-owned mining company that extracts ore and refines it into metals such as copper, cobalt, zinc, manganese, and several rare metals. Gécamines is the largest provider of government revenue and the country's foreign exchange.

capacity of the country disintegrated rapidly, driving down production and exports. As exports fell, the country lost a substantial amount of foreign exchange, depriving the government of much needed resources for day-to-day operation and for development.

Almost oblivious to these developments, the new governing elite, in their zest to substitute themselves for the old colonial establishment, adopted a conspicuous and expensive lifestyle inconsistent with their actual contribution to national income. In other terms, the new elite quickly learned how to redistribute income to their advantage without a concurrent ability to generate new wealth. Accordingly, the government sustained a high rate of spending during the period and accumulated large budgetary deficits. Faced with very limited domestic savings and undeveloped financial markets, the government relied increasingly on advances from the central bank, that is, on money creation, to finance the growing budgetary deficits. This reliance, in turn, kindled pressures on the economy in response to which the government adopted deflationist remedies—in particular the devaluation of the currency and the control of money supply. The first of such devaluations (over 40 percent) was implemented in November 1961. It produced no discernible effects. The second devaluation (averaging 50 percent) was implemented in November 1963 and produced only short-lived relief.

Five years after independence, the economy was in a relatively advanced state of disintegration, marked by a breakdown of infrastructure and the collapse of production and exports. At the same time, money supply had expanded up to five times its level at independence, inflation had accelerated, and the balance-of-payments deficit had reached an alarming level. By mid-1967, the central bank's holdings of gold and foreign exchange, an index of its ability to finance purchases of goods and services from abroad, amounted to no more than the equivalent of five weeks' imports (Bank of Zaire 1969/70).

By this time, the army had taken over the government, with General Mobutu as head of state, and calm had been restored to the country. The new government proceeded to reorganize the country politically and economically. President Mobutu vowed to introduce a democratic government and a market-oriented economy. Instead, the new government turned out to be rather autocratic. Described variously as a kleptocracy (Körner 1991) or a partrimonialist (Willame 1972) regime, Mobutu's government has appropriated public offices as its prime source of status, prestige, and reward; ensured political and territorial fragmentation through the development of relationships based on personal loyalties; and built its authority on a partisan, almost private, army or militia, all this disguised under a vast, overreaching single political party, the Popular Movement of the Revolution (MPR).

Zaire's economy has been beset by a particularly pervasive rent-seeking behavior. This applies to cases such as rents associated with import quotas, privileged access to foreign exchange, and monopolies controlled by the government which give rise to various rents, political and administrative corruption, and outright embezzlement of public property. Rent seeking in Zaire has essentially stemmed from and relied on a combination of three factors: the existence of a patrimonial and coercive political and institutional system, which encouraged corruption; the economic difficulties experienced by the country, which created scarcities, particularly in the foreign exchange markets; and the large-scale mismanagement of public resources, made possible by the availability of vast resources and the lack of accountability of the political system in place, a lack of accountability that in turn led to a skewed income distribution favoring the ruling class.

Rent seeking certainly is not uncommon in Africa or, for that matter, in the developing world in general. The case of Zaire, however, stands out by its scale and the impunity with which it is carried out. Zaire also stands out by the size and availability of the resources to which the country, or, more properly, the elite, has had access. As McGaffey (1991) puts it, "Zaire is set apart in Africa to some degree by the limitless pursuit of wealth by the powerful clique that runs the country." Moreover, for a country notorious for its inefficient administration, it nevertheless has functioned most effectively to further the interests of the controlling clique. In fact, there exists what Gould (1980) has called a cleverly "organized system within an overall disorganized regime," designed to maximize the interests of the ruling clique. In this context, Young and Turner (1985) write that by 1975 President Mobutu had become one of the world's wealthiest men through large kickbacks on government contracts and outright embezzlement. Moreover, the president had acquired a vast agricultural empire within the country as a result of the Zaireanization measures. It is worth noting that the unspoken rule for attribution of ownership of enterprises during the Zaireanization operation was that the higher one's actual or ascribed position in the country's political and administrative bureaucracy, the larger one's acquisitions.

Another type of rent-seeking behavior resulted from the centralization of salaries of the civil servants in the early 1970s (Gould and Amaro-Reyes 1983). As part of the recentralization process, crucial to President Mobutu's ruling strategy, salary lists were established in Kinshasa by computer for all 300,000 government civil servants serving everywhere in the republic. At the time, this was considered a normal improvement of management. This centralization, however, opened the door to wide-scale, computerized corruption, with practices such as the introduction of thousands of fictitious names for pay pocketed by the corrupt employees, illegal backpays, and bribery to get normal services involved with this facility (e.g., legal backpays).

Price interventionism was another mechanism through which rent-seeking behavior was manifested, in both industry and agriculture. Concerning the former, an important basis of President Mobutu's power has been his capacity to make appointments to state-regulated industries. Efforts at economic reform repeatedly foundered on his determination to manipulate these industries in order to generate privileges for himself and his followers and to reward those whose support he needed to remain in power (Bates 1981, 97). Concerning the latter, rent seeking contributed to the neglect of agriculture in favor of industry. Price and marketing controls in agriculture contributed, in addition to numerous other factors, to the decline of agricultural output throughout the 1970s. Moreover, the structure of mining and industry, which made development of these sectors attractive to rent-seeking elites desiring control of state agencies and lucrative commissions and contracts ensured that public expenditures and investment would be concentrated in these areas. Agriculture, the development of which required sustained long-term investment, was less immediately lucrative for these groups and was neglected. In this way the interests of the elite coincided conveniently with the pro-industry development philosophy prevailing in the 1960s and 1970s. Accordingly, the agricultural sector declined through the entire post-independence period.

Rent-seeking behavior was also an inhibitive factor in the implementation of exchange rate reform. In fact, writes W. J. Leslie (1987), "Mobutu and his clique have had free access to the state's resources. This group has a neo-mercantilist bullion fixation, a preoccupation with foreign exchange—how to obtain it quickly and in large amounts for their own personal needs."

With regard to Zaire's mobilization of resources, the country has had a wider access to resources, from both domestic and external sources, than the average Sub-Saharan country. On the external front, Zaire's strategic weight in East-West relations helped it muster attention and, with it, a substantial foreign financial assistance from bilateral and multilateral sources through the three decades 1960–90. In addition to donations and soft loans from bilateral and multilateral donor institutions, Zaire took advantage of its good credit rating to borrow heavily on commercial terms from the late 1960s through the 1980s, until the rise in the level of commitments led to Zaire being unable to meet its debt service obligations. On the domestic front, Zaire also mobilized a substantial amount of resources from domestic sources. But, although the actual resource absorption in the economy was relatively high, the allocation of public resources was less than satisfactory. Large shares of the development expenditures covered projects of dubious profitability, the so-called white elephants, carried out mainly for political consideration. Similarly, current expenditures were largely allocated in an irrational and wasteful way. Thus, it comes as no surprise that, with economic growth naturalized, Zaire, with its pervasive rent-seeking behavior, remains in economic turmoil today. The combination of rent seeking, corruption, and economic inefficiency explains to a great extent Zaire's past economic problems as well as its inability to fully implement and sustain economic reforms.

Emergence of Imbalances and Early Adjustment Failures

Emergence of Imbalances, 1967–75

Economic conditions deteriorated quickly between 1960 and 1967, by which time all indicators—money supply, the government's budget, balance of payments, and inflation—had reached alarming levels. The country was on the brink of bankruptcy. In June 1967, Zaire requested and obtained a standby arrangement from the IMF in the amount of US$27 million to finance a stabilization program. The program consisted of monetary reforms, including a major depreciation of the currency and the introduction of a new monetary unit, the zaire.[4] The program also contained demand-cutting measures, including a reduction of the rates of public spending and of monetary expansion as well as a freeze on wage and salary increases.

Performance of the economy during the period covered by this IMF-supported program appeared excellent. This success was, however, primarily the result of external factors rather than of any effective domestic effort; true adjustment did not occur. The economy benefited from the extremely favorable conditions that prevailed in the world's markets for Zaire's main exports in the period 1967–69. Copper export earnings in particular recorded large increases during the period under which the program was to be implemented. The effect of these increases was to induce a substantial rise in government revenue and thus a considerable improvement in the budgetary and balance-of-payment performance. At the same time, the other objectives of the government's economic and financial policies covered by the program were also attained. Inflation was reduced substantially

---

[4] The old currency unit, the franc, was worth one-third of the Belgian franc until June 1967. Following the devaluation, the zaire was worth 1,000 Congolese francs but only 100 Belgian francs, which implied a de facto devaluation of 70 percent in terms of Belgium francs.

through 1971. Real GDP grew vigorously, averaging 4–5 percent annually during the period 1967–74. All sectors except agriculture contributed to this growth. The stagnation or, for some crops, decline of agriculture was due to the lack of public and private investments and to the deterioration of infrastructure, particularly the road network.

As noted above, the program's objectives were achieved mainly through unexpected revenue increases rather than through expenditure cuts or revenue policies contained in the standby arrangement. As a result, no drawing was realized under the standby arrangement. The exceptionally favorable external environment had in this case been a mixed blessing. It had released the government from the need to muster fiscal and monetary discipline in order to meet the program targets. An opportunity was thus lost to steer the country into building a tradition of good financial management, which could have proved helpful in the less favorable circumstances of the next decade.

In retrospect, it seems that this missed opportunity, together with a twenty-five-year authoritarian rule under Gen. Mobutu Sese Seko's government, contributed to building up the wrong customs and institutions, which later led to an unprecedented mismanagement of the economy and to a large-scale misallocation of Zaire's immense resources. The most evident illustration of this mismanagement is that, by the eve of the implementation of the 1983 policy reforms, Zaire ranked among the lowest in per capita income and literacy rate and its malnutrition and mortality were among the highest in Africa and in the world.

The euphoric conditions that prevailed from 1967 through 1974, generated by the world's economic boom of the late 1960s and enhanced by the relatively easy access of Zaire to the world's financial markets, particularly to the Euromarkets, encouraged the Zairean government to embark on an ambitious program of public investment financed by foreign borrowing. This program emphasized large projects in infrastructure and energy, most of which had low returns. The most significant of these was the construction of a hydroelectric dam at Inga coupled with an electric power transmission line from Inga to Kolwezi in the Shaba region.[5]

In the wake of Zaire's independence, the government's view was that the Inga potential could be tapped only in the context of an energy-intensive development environment around Kinshasa, which was part of an overall three-pole development strategy of the Zairean economy formulated by the government in the early 1970s. The other two poles were Lubumbashi in

---

[5] The Inga project was an old dream from the earliest colonial era. With a 100-meter water drop within a dozen kilometers and a debit of about 42 thousand cubic meters per second, the Inga site, on the Zaire River, has an enormous hydroelectric power potential, about 30,000,000 kilowatts, second only to that of the Amazon.

the southeast and Kisangani in the northeast. The Lubumbashi pole was to be steered toward the development of mining activities, of which the Lubumbashi-Kolwezi axis was the moving force. The Kisangani pole was to be steered toward agricultural production and agroindustries, of which Kisangani was the center.

In the Kinshasa-Inga pole complex, planned investments included a steel plant at Maluku and nitrogenous fertilizer and aluminum-smelting plants.[6] As it turned out, some of the industrial operations that were expected to utilize the hydroelectric potential of Inga did not materialize. The smelting plant, which was designed to convert bauxite into aluminum, using imported bauxite from Guinea through a large-scale Guinea–Zaire joint venture, was put off indefinitely.[7] An aluminum-smelting complex of this nature would have been one of the largest electric power-using industries in the region. The steel plant at Maluku was abandoned after a costly investment in equipment and human resources and a very short period of activity. In addition, exports of hydroelectric power to neighboring countries never materialized. Although the government established an industrial-free area, Zone Franche d'Inga (Zofi), to attract foreign investment with low-cost electric power and generous tax concessions, few new investment projects have occurred and a considerable excess capacity has emerged at Inga. In retrospect, the Inga dam and the power line were not then and are still not to be considered good investments.[8] But these views did not prevail in the face of the nationalistic determination of Mobutu's government and the readily available contractor-financed turnkey projects in the euphoric period of the mid-1970s.

The government's efforts to put Inga's potential to good use were frustrated in part by Zaire's external debt problems, which themselves resulted from the massive borrowing for the construction of the Inga complex in the first place. After incurring debt to finance the first phase of the project, the country experienced difficulties obtaining additional external financing. The external public debt (outstanding and disbursed), which was merely US$311 million in 1970, surged to $904 million in 1973 and $1,343

[6] The development of the dam was to be slated into four successive phases of increasing capacity. After construction of the second phase, total installed capacity reached 2,100 megawatts, still only a fraction of Inga's enormous potential.

[7] The postponement was justified, in large part, by delays in the construction of a deep-water port at Banana, which was necessary to ensure routine shipping of the production out of the country.

[8] In 1983 the construction of a 1,100-mile high-tension line from the Inga to Kolwezi in Shaba was completed. This power line provides in excess of 1,700 megawatts to the Shaba region. The construction of the power line was more than an economic decision; it also reflected political considerations, in particular, a desire to link the secession-prone Shaba province to an energy source at close range of the power base in Kinshasa and therefore to make Shaba dependent on a good relationship with the capital.

million in 1974, and it continued to increase through the end of the decade because of sizable new commitments as well as the high cost of borrowing. The ratio of debt to GNP, which was less than 10 percent in 1970, rose to 42 percent by 1980 and would continue to climb. As a result, the country's debt service obligations quickly rose to unsustainable levels. During 1970–76, the average interest rate on Zaire's external public debt was 6.9 percent, with a maturity of 15 years and a grace period of 4.4 years. These terms, reflecting the commercial nature of the debt, were considerably harder than those facing other low-income countries during the same period: for this latter group, the average interest rate was 2.6 percent, maturity was 31 years, and the grace period was 7.8 years. The grant element for the group was almost 60 percent, about three times that for Zaire (World Bank 1979).

The era of economic growth came to an abrupt end in 1975, when, in the midst of a worldwide recession, the price of copper in world markets fell by almost 40 percent in a single year. This drop precipitated a sustained deterioration in Zaire's terms of trade (Figure 4.1) and caused serious damage to the economy. By the end of 1975, the country was experiencing a severe balance-of-payment crisis and was in recession. Real GDP, which had increased by some 7 percent in 1974, stagnated in 1974 and fell by 5 percent in 1975 and again in 1976.

Although the payments crisis was to some extent precipitated by an external factor—the fall in copper prices—the economy suffered mostly from underlying structural and policy weaknesses that had gradually become apparent throughout the period 1973–74 and persist, for the most part, today. On the external front, the economy was vulnerable to fluctuations in copper prices. The reliance on copper, in turn, reflected the lack of diversification of the country's exports. On the domestic side, the government carried out inappropriate public investments, such as the Inga project, which were funded essentially through the accumulation of external debt; it financed excessive public sector deficits by money creation; it mismanaged the parastatal sector, which generated poor financial and economic returns and was a constant drain on the public sector's resources; and it created the misguided Zaireanization policy.

Ostensibly the Zaireanization was designed to diversify the ownership of the country's economic base and thus reduce its reliance on foreign capital and skills. In practice, however, it accelerated the total or partial transfer of ownership of corporate holdings from foreigners to the government or to Zairean nationals. The beneficiaries of these transfers, who were essentially family members or political allies of President Mobutu and his government, performed rather poorly because of a lack of business experience and managerial skills. Major enterprises and key industries were thus al-

Figure 4.1. Evolution of terms of trade, Zaire (1980=100)

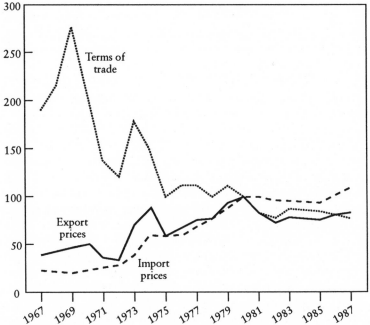

Source: World Bank 1991e.

lowed to break down and were quickly abandoned. As a result, employment and the production and distribution of goods and services were disrupted throughout the country. These policies also led to the transfer abroad of a substantial amount of financial capital, worsening a balance of payments already aggravated by the fall in the terms of trade. To counteract these massive transfers, the central bank, Bank of Zaire, imposed restrictions on capital transfers and reinforced exchange controls, to no avail. At the same time, in order to maintain domestic production activity, the central bank expanded domestic credit to meet the import needs of the newly Zaireanized enterprises. These actions accelerated inflation and thus the depreciation of the zaire in the parallel markets.

Consideration of policy actions regarding agriculture is noticeably absent from this discussion. In spite of all government assertions to the contrary, agriculture was neglected, or worse, penalized by government policies. The neglect of the agricultural sector was consistent with the development thinking of the 1950s and of the early 1960s, when Zaire

Table 4.1. Contribution of agriculture to growth in GDP, Zaire (percentage)

	1965–73	1974–80	1981–88
Real rate of growth of GDP	3.9	−2.0	1.6
Real rate of growth of agriculture	−1.7	1.1	3.4
GDP growth caused by agriculture[a]	—	—	42.0

Source: Tshishimbi and Glick, forthcoming, courtesy of Cornell Food and Nutrition Policy Program.
[a]Equals rate of growth of agriculture multiplied by share of agriculture in GDP divided by rate of growth of GDP.

became independent. Despite repeated pronouncements by President Mobutu's government on the importance of agriculture to national development, such as "l'agriculture est désormais la priorité des priorités," budgetary allocations to the agricultural sector never matched the government's declared intentions. This too contributed to the agricultural sector's decline through the entire post-independence period. In the period 1965–73, agriculture declined in real terms by an average 1.7 percent annually (Table 4.1), even as GDP overall grew by almost 4 percent. During the 1974–80 period of macroeconomic crisis, when GDP declined on average by 2 percent annually, agricultural output was stagnant, growing a mere 1 percent per year, well behind the estimated population growth rate of 3 percent. The poor performance of agriculture through 1980 was characteristic of both the food crop and the export/industrial crop subsectors, as Tables 4.2 and 4.3 reveal. It was only in the 1980s, in the context of the government's reform efforts, that agricultural growth resumed, albeit weakly, and then only for food crops.

Table 4.2. Production of major food crops, Zaire (thousand metric tons)

	Maize	Rice	Manioc	Sweet potato	Beans	Peanuts, unshelled	Plantain	Wheat
1960	330	124	6,045	374	68	175	125	—
1965	232	49	7,785	192	76	137	447	—
1970	428	179	10,346	425	115	267	1,215	3.4
1979	536	223	12,566	324	160	334	1,378	3.6
1980	562	234	13,087	333	162	339	1,408	3.7
1981	639	245	13,172	343	104	347	1,438	3.8
1982	666	251	14,184	353	111	349	1,467	12.1
1983	673	271	14,601	363	156	366	1,496	16.1
1984	704	286	15,038	373	164	375	1,526	9.8
1985	721	297	15,493	382	166	424	1,795	6.0
1986	729	308	—	—	—	443	1,834	—

Source: Tshishimbi and Glick, forthcoming, courtesy of Cornell Food and Nutrition Policy Program.

ble 4.3. Production of industrial and export crops, Zaire (thousand metric tons)

	1970	1971	1979	1980	1981	1982	1983	1984	1985	1986	1987[a]
ffee[b]	68.8	75.0	87.0	89.0	93.0	93.0	84.0	89.0	90.0	95.0	—
gar	43.0	44.0	48.0	48.0	47.0	52.0	52.0	61.0	58.0	61.0	—
coa	4.5	6.0	3.5	4.2	4.6	4.5	4.2	4.4	4.5	6.3	6.0
	7.3	7.3	4.8	4.4	4.8	4.5	4.7	5.0	4.8	4.7	4.2
bacco	—	—	1.6	1.9	2.9	2.8	1.5	1.9	2.4	3.2	4.0
tton, seed	49.0	58.0	19.0	29.0	21.0	24.0	27.0	21.0	22.0	19.0	19.0
bber	35.0	41.0	22.0	21.0	18.0	17.0	16.0	14.0	13.0	13.0	—
m oil[b]	170.0	178.0	98.0	93.0	106.0	94.0	85.0	93.0	89.0	88.0	85.0
inine[c]	—	—	6.0	5.9	6.0	5.9	5.2	4.7	4.8	7.0	—
nber, logs[d]	297.4	288.3	350.0	325.0	350.0	375.0	401.0	410.0	415.0	418.2	420.8
nber, sawn[d]	150.0	149.0	88.0	68.1	61.3	73.5	112.0	115.3	117.6	120.3	127.2

Sources: Tshishimbi and Glick, forthcoming; World Bank 1988a; Bank of Zaire, various years; U.S. Departnt of Agriculture 1988, *Zaire Agricultural Annual.*
Department of Agriculture estimate.
Excluding village production.
Tons of bark.
Thousand cubic meters.

Aside from the political troubles in the period of 1960–67, the stagnation of agriculture before the reforms of 1983 was attributable to a number of inappropriate policies: (1) an overvalued exchange rate; (2) high export duties and burdensome administrative requirements for exporters; (3) price controls on agricultural products at levels below competitive market and world prices; (4) subsidized (through imports) consumer food prices in urban areas; (5) local government control and taxation of production and marketing; and (6) nationalization of foreign interests beginning in 1973. These policies reflected the authorities' emphasis on developing the industrial sector under state control and their general disregard for agricultural development. For example, the government's intervention in the exchange rate market resulted in an overvalued exchange rate, which kept the costs of imported inputs cheap for industry and government but penalized export agriculture. Similarly, the government's heavy reliance on taxation of exports as a major source of budgetary revenue kept farmers' income well below what the world market prices could provide. Pervasive government taxes and control over pricing and marketing led to widespread smuggling and sales on the parallel markets.

The low budgetary allocation to agriculture (less than 3 percent of the total government budget) of current and investment expenditures between 1970 and the mid-1980s also illustrates the government's lack of concern for agriculture (ZTE/COGEPAR 1987). In particular, agricultural services such as research and extension were extremely underfunded. Furthermore, the network of rural roads was allowed to deteriorate, so that transporting

products to urban markets and for export, as well as distributing agricultural inputs, became increasingly costly. Given that Zaire is a large and
landlocked country, these activities depend crucially on a functional road
system; the poor conditions of the transport network remain a major
barrier to agricultural expansion.

Finally, the imposition of nationalization measures in 1973 and 1974
contributed greatly to the decline of output in the modern plantation sector. The total number of holdings dropped from about 1,200 in the 1970s
to 900 in the 1980s after the departure of trained (largely Belgian) management personnel and the abandonment of many plantations (World Bank
1988a). Although the measures were partly rescinded in 1975, the legacy of
Zaireanization was the main obstacle to the government's attempts to
revitalize industrial and export crop production during the 1970s and
1980s.

In the domestic food crops subsector, the policy of setting official prices,
which was supposed to help farmers by ensuring minimum prices, in practice often benefited traders, who treated them as maximum prices to be
paid to farmers. In periods of short supply, this practice tended to raise
traders' margins rather than farmgate prices. Also, because the growth of
food crop production was less than that of the population and lagged far
behind the growth in demand from Kinshasa and other urban areas, the
authorities imported massive amounts of food to meet demand. Tables 4.4
and 4.5 show, respectively, import levels and the import dependency ratios
for several major foods. In 1980, for example, imports of maize totaled

Table 4.4. Imports of major food items, Zaire (thousand tons)

	Maize	Rice	Wheat	Sugar	Meat	Fish
1960	0.02	2.5	—	13.0	4.3	—
1970	64.0	—	—	15.0	—	—
1975	115.0	35.0	90.0	9.0	13.0	11.0
1976	140.0	20.0	125.0	24.5	17.0	6.0
1977	132.0	14.0	120.0	9.0	3.0	2.4
1978	160.0	30.0	104.0	0.5	17.7	NA
1979	141.0	16.0	120.0	3.5	10.4	6.4
1980	147.0	10.0	103.0	17.0	8.7	3.8
1981	119.0	12.0	157.0	8.0	6.1	6.5
1982	69.0	4.0	146.0	9.0	12.0	6.5
1983	51.0	9.0	152.0	25.0	12.0	57.0
1984	38.0	33.0	137.0	23.0	23.0	52.0
1985	35.0	36.0	157.0	23.0	31.0	105.0
1986	58.0	60.0	127.0	—	43.0	107.0

Sources: Tshishimbi and Glick, forthcoming; World Bank 1988a.
Note: NA, not available.

Table 4.5. Dependence on food imports, Zaire (percentage)

	1980	1981	1982	1983	1984	1985
Maize	44	14	16	6	9	11
Rice	3	9	2	16	15	17
Meat	15	10	7	19	33	21
Fish	—	—	—	31	31	53
Sugar	26	32	30	37	29	32

Sources: Tshishimbi and Glick, forthcoming; ZTE/COGEPAR 1987.
Note: Percentage share of imports over total domestic marketed consumption.

147,000 metric tons, about 44 percent of total domestic consumption (excluding consumption of nonmarketed output); imports of wheat, which was not produced domestically, totaled 103,000 metric tons; meat and fish imports each accounted for about a third of domestic consumption. Together with the low mandated prices, the government's policy of providing cheap imports to keep retail prices low in urban markets depressed food crop prices and discouraged domestic production.

Initial Attempts at Stabilization and Reform, 1976–82

The first real attempt at stabilization took place in 1976. To resolve the balance-of-payment crisis after the fall of copper revenues of 1975, the government of Zaire introduced, and the IMF approved (in March 1976), a request for financial assistance for US$150 million. This was the first of four programs, including an extended financing arrangement for more than $1 billion, which were negotiated with the IMF (in 1976, 1977, 1979, and 1981) and resulted in total drawings of SDR 339 million.

The packages included in the stabilization programs, although emphasizing one particular policy or another according to the severity of each crisis, generally combined demand management and supply incentive measures as well as actions designed to rehabilitate infrastructure in the agricultural, transport, and energy sectors. Because of the near continuity of the IMF-Zaire relationship, the effects of one program package are hardly distinguishable from those of another. As it turned out, however, little progress was made during the period reviewed in bringing about domestic and external stability or in restoring economic growth. During the period, real GDP declined by about 1 percent per year, largely as a consequence of the stagnation in Zaire's terms of trade. Nevertheless, domestic policies were also significant. Budget deficits increased, financed by credit expansion. Money supply grew an average of 40 percent annually from 1975 to

1983, while inflation averaged almost 60 percent. By 1980 the government was allocating some 40 percent of the budget to service the debt inherited from the previous years of heavy borrowing.

The failure of the IMF-supported programs to restore budgetary and external balance and reduce inflationary pressures, during this first real attempt at adjustment, to a large extent reflected a lack of commitment on the part of the Zairean authorities to an effective implementation of the prescribed policies. In several instances, programs became inoperative either because intended policies were insufficiently (or not at all) implemented or because their timing diverged substantially from the planned timetable.

The 1977 standby and the 1981 extended fund facility were both cancelled because of the country's lack of compliance. Most of the funds made available by the IMF remained undrawn (SDR 737 million out of a total SDR 912 million in the case of the extended fund facility). In addition, private and foreign investment did not expand as quickly as anticipated in spite of the lifting of nationalization measures in the late 1970s. The absence of investment response followed from the many unresolved issues of compensation and a lingering lack of confidence by the private sector in the government. It was only during the 1983 program, when the government seemed willing to take serious steps to achieve stabilization and to implement economic and financial reforms that would remove distortions and provide suitable incentives to production, that investment began to show signs of resumption.

Reform Policies and Performance

Zaire's most significant adjustment effort to date came with the stabilization package adopted in September 1983 and implemented beginning in December of that year. Earlier steps toward this adjustment effort began in 1982. For about a year before the adoption of the 1983 program, the government took several actions, particularly with regard to the reduction of the public sector's deficit, as a precondition for IMF support. This so-called shadow program was designed to demonstrate the good faith of the government of Zaire and its willingness to undertake rigorous reform measures to redress the entrenched imbalances and distortions in the economy.

The 1983 stabilization program set out to restore macroeconomic stability and improve efficiency through the tightening of fiscal and monetary policy, the reform of the exchange rate and the exchange and trade systems, and the liberalization of all domestic prices (including interest rates). Subsequent programs were formulated to address the longer-term, structural problems facing the economy with additional assistance coming from the World Bank.

The overall 1983–90 adjustment effort can therefore be analytically divided into two periods: that from 1983 to 1986, during which efforts were focused on—but not restricted to—exchange rate price and trade liberalization under the 1983 IMF-supported stabilization program; and that from 1987 to 1990, during which efforts at structural transformation were undertaken with the support of both the IMF and the World Bank.

Adjustment Policies, 1983–86

The cornerstone of the 1983 program package was the adjustment of the exchange rate and the reform of the exchange rate and trade systems. Other important policies included an adjustment and a liberalization of domestic prices, including interest rates, agricultural producer prices, and the prices of petroleum products; a tight control of government spending, through a restraint on wage increases and reductions in public sector employment; a large-scale reform of public enterprises aimed at lowering their funding requirements, thus alleviating their burden on the government's budget; and a revision of the investment code with a view to attracting private and foreign investment.

Exchange Rate Reform.   For most of the 1970s and the early 1980s, the zaire was grossly overvalued and the exchange system was characterized by a multiple-rate regime. In August 1983, one month before the implementation of the reform measures, the average unofficial rate was five times as high as the official rate, reflecting an annual appreciation of 17 percent with respect to 1982.

To bring the Zairean currency to a more realistic level and reduce the incentives for economic agents to engage in parallel market and informal activities, the monetary authorities devalued the zaire by 77.5 percent in September 1983. With a nominal devaluation of this size, the real devaluation also turned out to be substantial. The following year the depreciation of the zaire continued, recording an annual rate of 60 percent (Table 4.6). The government also introduced a significant degree of flexibility in the determination of the exchange rate to ensure that it fully reflected the interaction of market forces.

Before the 1983 reforms, the economy operated under several exchange rates: one legal and official rate, several semiofficial rates, as well as a multitude of unauthorized rates. The latter were related to transactions taking place outside the official banking system. The official rate was used for all official foreign transactions except certain mining transactions. The

Table 4.6. Evolution of exchange rates, Zaire

	Z/US$[a]	Z/SDR[a]	Real effective exchange rate (1980=100)
1970	0.5	0.5	—
1971	0.5	0.543	—
1972	0.5	0.543	—
1973	0.5	0.603	—
1974	0.5	0.612	—
1975	0.5	0.585	—
1976	0.861	1.0	62.7
1977	0.831	1.0	107.0
1978	1.007	1.313	141.7
1979	2.025	2.667	123.5
1980	2.985	3.81	100.0
1981	5.465	6.349	92.6
1982	5.746	6.349	97.7
1983	30.12	31.534	114.3
1984	40.45	39.649	45.6
1985	55.793	61.284	41.3
1986	71.1	86.969	41.2
1987	131.5	186.554	35.6
1988	274.0	368.722	32.3

Source: IMF, various years (b).
[a]End-of-period market exchange rate.

semiofficial rates applied to purchases of gold and diamonds by specialized marketing agencies. Both the official and semiofficial rates were determined and regularly published by the central bank.

Also, as part of the reform package of September 1983, the government unified the rates in the official and semiofficial markets and gradually eliminated the spread between the official rate and the parallel markets. These actions were designed to eradicate the rents provided by these markets and to provide incentives to economic operators to shift transactions from the parallel to the official markets. The first step toward the unification of the rates was the introduction of a dual exchange arrangement, which consisted of an official market restricted to the transactions of the central bank and a free market involving all the other transactions of the banking system. The exchange rate applicable to the free market was determined at a weekly interbank session between the central bank and all other participating financial institutions and reflected the rates applied to the parallel markets transactions during the previous week. The official exchange rate was then determined in reference to the free market rate as described below.

The overall objective was to contain the margin between the official and free market rates. Initially the margin was set at 10 percent; it was gradually reduced to 5 and then to 2.5 percent before being eliminated altogether. The complete unification of the two rates was achieved on March 1, 1984, within a period of less than two years from the inception of the reform, as planned. Subsequently, the unified rate was to be determined in much the same way as the free market was during the transitional period, that is, reflecting the interaction of market forces.

Reform of the Exchange and Trade System.   Beyond the reform of the exchange rate regime, the government decided to rationalize the entire trade system. In particular, it abolished a provision that required commercial banks to surrender 30 percent of their foreign exchange receipts to the central bank within a period of three months following the export of goods or services; gradually liberalized the system of allocation of the foreign exchange applicable to the commercial banks; eliminated the system of retention of exports proceeds, except those specified under international credit arrangements and for the state company, Gécamines; removed, in July 1984, the prohibition against transfers of dividends by companies with foreign participation; and, in general, liberalized the overall import licensing procedures. This last measure was mostly a formalization of what had been going on informally for some time. For most imports, indeed, the license was no longer subject to prior approval of the central bank. Instead, a simple import declaration submitted to an authorized commercial bank was sufficient. In this context, the old system of *licences sans achat de devises,* that is, licenses concerning imports financed without recourse to the country's official foreign exchange reserves, was abolished.

Furthermore, the government simplified the tariff structure. In 1983, as part of the package, it reduced the number of taxes from four to two, and the rates on raw materials and essential food and nonfood products were lowered from a range of 10–20 percent to 3 percent. At the same time, tax rates on luxuries were increased to up to 200 percent.

In some areas (explained below), the new tariff regime had unanticipated effects. In the manufacturing sector, for example, since raw materials and intermediate inputs hardly required any duties and consumer and luxury imports paid very high rates, the effective rate of protection on some industrial goods such as home appliances, motorcycles, and bicycles, which are considered luxuries and contribute little to the domestic economy, averaged some 300 percent (World Bank 1986f); this was due to the high reliance of these industries on imported rather than locally produced inputs. In agriculture, products such as sugar and rice faced greater competition since, under the more liberal price policies, these goods could be imported from neighboring countries, where production was in many cases

subsidized by the government (World Bank 1986f). To resolve these unde-
sired effects, further revision of the tariff structure in April 1984 substantially lowered tariffs on most items with a view toward reducing incentives
for evasion and encouraging manufacturing activity.

Monetary and Credit Policies. Because of a long-standing lack of adequate monetary and financial institutions, the government tended to rely
almost exclusively on domestic banking credit to finance its budget. This,
combined with the scarcity of domestic savings as well as with the high cost
of Zaire's international borrowing in the mid-1970s, contributed to inflationary pressure, which adversely affected savings and investment and
therefore reduced domestic production. One of the major goals of the 1983
reforms was to improve the credit structure and money and financial markets through the creation of institutions such as an effective interbank
market and a market for treasury bills. At the same time, interest rates were
liberalized to encourage savings while discouraging consumption and capital flight.

Whereas Zaire had followed a passive policy regarding interest rates
during the 1960s and 1970s, in the early 1980s the Bank of Zaire began to
use, actively and increasingly, interest rates to control monetary expansion
and the allocation of credit to the economy. Initially, credit was regulated
exclusively through quanitative limitations. Gradually, credit allocation
relied partly on ceilings and increasingly on mandatory reserve requirements and interest rate policy. The introduction of treasury bills, a valuable
development in its own right, opened up the possibility, for the first time in
many years, of interest payments on reserves held by the Bank of Zaire.
Beginning with the 1983 reforms, interest has been paid on that portion of
the reserve requirement of the commercial banks that is held in the form of
treasury bills.

To improve resource allocation, commercial banks were allowed, in
1981, to set lending rates freely, except for interest rates on loans to non-
coffee agriculture. The 1983 reform measures liberalized most of the remaining interest rates.[9] At the same time, the authorities created a market
for short-term treasury bills, designed to provide incentives for holding
zaires instead of foreign exchange. In the first two years of implementation
of treasury bills, the sale target was successfully met and real interest rates
became positive for the first time.

Over the years, the wisdom of implementing an active interest rate policy
became debatable. Unless a parallel effort is made to keep inflation low
enough, achieving positive real interest rates may require unreasonably

---

[9] The rates applying to agricultural credit were liberalized only in 1987.

high nominal rates. Fiscal and monetary policies served to restrain inflation in the period immediately after the 1983 reforms. More recently, however, the financial environment in Zaire was excessively inflationary. In 1987–88 the inflation rate rose above 100 percent. In 1990 it was about 300 percent. In 1991 it broke all records, 1,000 percent in the first quarter and 2,000 percent in the second, with over 3,000 percent expected for the year. Under such conditions of hyperinflation, only prohibitive nominal interest rates can bring about positive real interest rates.

Public Finance and Public Enterprise Reform.   Under the 1983 stabilization program, the authorities were committed to expenditure cuts through reductions in government and public sector employment and through a comprehensive reform of the public enterprises, including the closing of unprofitable units and large-scale privatization. At the same time, the new program envisaged a broadening of the tax base, not only through tax increases or new impositions but also through better tax collection and a rationalization of the regime of exemptions. The new program also included a plan to undertake public investment in a few key sectors that needed to be revamped, such as in education and health.

With regard to the reform of parastatals, the most significant changes involved Gécamines. Gécamines's difficulties stemmed in large part from changes in world copper prices. In 1983 the government introduced important changes in the company's tax regime to allow the latter to be more responsive to fluctuations in revenues. In 1984 the government adopted a reorganization recommended by the World Bank and designed to reduce the operating costs of the copper-mining industrial complex and to ensure greater transparency.

Price and Marketing Reform in Agriculture.   Reforms in the agricultural sector began in June 1982, as part of the shadow program mentioned above. The government began implementation of a three-year agricultural recovery plan designed to correct the distortions and government neglect of the agricultural sector of previous years. Actions considered essential to agricultural recovery included price, marketing, and exchange rate reforms to restore incentives to agricultural production; strengthening of agricultural training, research, and extension services; increasing the availability of credit for crop marketing; restoring incentives to agricultural investment by the private sector; and institutional reorganization to improve planning and programming capacities. Some of these measures were, to a large extent, implemented. Most dramatically, price controls on most agricultural products were removed by May 1982, and the right of local authorities to intervene in marketing was abolished. In 1983, as part of the adjustment reform package, price controls on all remaining crops were removed. Furthermore, the massive exchange rate devaluation of that year was expected to give a boost to the export crop sector.

Performance under the Stabilization Program, 1983–86

In 1984, the first year of implementation, adjustment appeared to be taking effect. One area of success was the reduction of inflation. As seen in Table 4.7, the Kinshasa CPI, which had risen over 100 percent in 1983, increased less than 20 percent in 1984. This decline in inflation, together with the nominal devaluation introduced in September 1983, brought about a substantial depreciation of the real effective exchange rate for 1984 as a whole. Despite the depressed export prices for coffee, total export earnings increased substantially due to higher than anticipated exports of diamonds, crude oil, cobalt, and coffee. As a result, the balance of payments improved markedly and Zaire met its external debt service commitments through the year. On the domestic front, the public sector's deficit was restrained within the limits laid down by the adjustment program.

With this notable improvement in economic and financial performance, real GDP grew by some 3 percent for the year (Table 4.7). Although surely a modest increase and no greater than the growth of population, this GDP

Table 4.7. Basic data, Zaire

	1980	1981	1982	1983	1984	1985	1986	1987	1988	1!
Annual rate of growth										
GDP (1970 prices)	2.4	2.9	−3.0	1.3	2.7	2.5	2.7	2.6	2.2	−
Consumer price index[a]	46.7	53.0	41.0	101.0	14.5	39.0	38.3	106.5	93.7	10
Money supply[b]	61.5	37.9	73.5	73.8	84.2	27.3	58.8	96.7	127.3	7
Population	3.0	3.0	3.0	3.0	3.0	3.1	3.1	3.2	3.2	
Terms of trade	−8.1	−16.5	−5.5	5.7	1.3	−2.4	−2.3	−8.2	19.6	−
Percentage of GDP										
Consumption	86.0	92.3	91.3	91.8	82.6	81.9	87.8	90.8	91.5	9
Gross domestic investment	15.0	15.0	14.4	10.9	13.9	13.5	11.2	11.9	11.0	1
Resource balance (gap −)	−1.0	−7.3	−5.7	−2.7	3.5	4.6	1.0	−2.7	−2.5	−
Gross domestic savings	16.2	15.8	11.3	13.2	18.8	16.9	16.4	10.0	9.8	1
Budgetary deficit (−)	−2.7	−7.2	−10.5	−4.0	−3.8	−1.5	−5.2	−9.7	−21.2	−
Long term public debt	41.5	46.3	45.8	58.0	101.6	120.4	111.1	137.2	118.0	12
Debt service[c]	22.5	21.9	13.2	13.7	20.8	25.7	22.0	22.7	15.9	1

Source: Tshishimbi and Glick, forthcoming, courtesy of Cornell Food and Nutrition Policy Program.
[a]Kinshasa CPI (end of period).
[b]Broad money.
[c]Debt service (after rescheduling) as percentage of exports of goods and services. Includes payments to I and on short-term debt.

growth represented a turnaround from the contraction experienced a year earlier. Growth in 1984 was led by the mining sector, which expanded by almost 7 percent. Commerce, services, and manufacturing also picked up.

Agricultural Performance, 1983–86

Data collected by the Institute Nationale de la Statistique on real producer prices of food crops are shown in Table 4.8. Farmgate prices of maize, rice, cassava, and peanuts decreased steeply in real terms from 1975 through 1981. The data suggest that the effect of the removal of price and marketing controls in May 1982 was strongly positive: real producer prices increased substantially over 1981 levels for all four crops and almost tripled for rice and cassava. Real prices for all crops but cassava fell off somewhat in 1983–84 but continued to be well above pre-reform levels (except for maize in 1986).

Unfortunately, the price liberalization measures did not bring forth a major supply response, as the output data for various food crops in Table 4.2 indicate. Overall, the growth in aggregate food crop production did increase slightly after 1982, but was still less than population growth. Part of the explanation of this less than encouraging response is that the effective realization of the government's price and marketing liberalization policy was geographically uneven, with little improvement in some regions. The attitudes of regional or local authorities was important in determining whether the reform policies were successfully implemented. Officials had to be willing, first, to refrain from intervening between farmers and traders

Table 4.8. Real producer prices of food crops, Zaire (1974 Z per kilogram)

	Maize	Rice	Cassava	Peanuts
1975	7.0	8.0	2.0	8.0
1976	6.3	6.3	5.3	7.9
1977	3.9	3.9	3.2	4.9
1978	4.9	3.5	4.9	5.6
1979	3.8	5.5	3.8	5.5
1980	3.7	5.2	3.7	5.2
1981	3.5	4.4	3.5	3.9
1982	10.5	10.1	10.5	7.3
1983	9.5	9.8	7.2	7.1
1984	6.1	7.6	10.7	—
1985	5.0	11.0	8.0	—
1986	3.0	7.0	6.0	—

Sources: World Bank 1988a; Institut National de la Statistique and Ministry of Agriculture data.

to set de facto maximum prices and, second, to encourage competition in marketing and transport (Sines et al. 1987). The latter was crucial, since competition would make it difficult for traders to continue to pay official minimum prices to farmers. If these conditions were not met, prices received by farmers were not likely to change greatly from the official prices.

The uneven implementation of agricultural reforms at the regional and local levels is undoubtedly part of the explanation for the rather weak supply response seen at the aggregate level. Output of food crops increased at an average annual rate of 3 percent in 1982–86, a slight improvement over the rate of the previous four years but still no faster than population growth. It is encouraging, however, that where price incentives were real, small farmers indicated that they were responding by raising output. Where price controls were lifted on one crop but not on another (e.g., maize and cotton in Shaba), farmers adjusted the pattern of production accordingly (Sines et al. 1987).[10]

Chronic transport problems were another major factor in the poor aggregate supply response of agriculture. These problems, including the deteriorated state of the rural road network, the high cost of fuel, and difficulty in getting spare parts for vehicles, add up to very high transport costs which depress farmgate prices. For example, over 40 percent of the retail value of certain food commodities from Bandundu that are sold in Kinshasa goes to the wholesale margin, some 90 percent of which is accounted for by transport costs (Jabara 1990b).

With regard to urban food prices, a negative potential welfare impact of food price increases is the reduction in the real incomes of the urban poor which may result from increases in prices of staples in urban retail markets. Figure 4.2 shows prices of food items in Kinshasa retail markets relative to the general retail price index for the years 1970–88. Before liberalization, food prices followed an uneven course for varying reasons. Through the mid-1970s, imports of cereals served to keep prices low for cereals and, through substitution in consumption, for other foods as well. In the period 1975–78, relative prices rose sharply for most staple foods as a result of disruptions caused by nationalization policies. During the period 1979–82, prices declined, most likely because of reductions in the purchasing power of Kinshasa residents (World Bank 1988a) and increased imports of cereals (e.g., bread, rice).

Against this background, price liberalization does not appear to have

---

[10] Most of the growth that has occurred has come through the expansion of cultivated area rather than through the increased use of cash inputs such as fertilizer or better seed. Such a pattern of extensive rather than intensive growth is to be expected given the general availability of land in Zaire as well as the inadequacy of agricultural research and extension services, which would promote the development and use of new inputs.

Figure 4.2. Relative retail prices of staple foods, Kinshasa

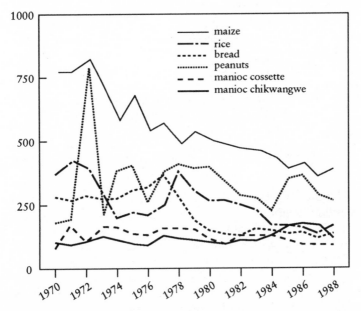

Note: Prices are shown relative to the general retail price index published by the Economic and Social Research Institute, University of Kinshasa.
Source: Republic of Zaire 1989b.

had a major sustained impact on food prices in urban Kinshasa markets. As seen in Figure 4.2, prices for some nontraded (local) products (e.g., cassava and peanuts) did rise faster than the general price index after 1982, but these increases followed previous declines in relative prices and were not fully sustained. Indeed, for nontraded foods taken as a whole, prices were as low at the end of the 1980s as at the beginning of the decade. Prices of tradables (importables) generally fell through the mid-1980s in response to increased imports, as was noted above for cereals. Since 1986 relative prices have risen for some foods (e.g., meat and fish) for which imports have fallen. Nevertheless, relative prices of imported foods as a group were still lower at the end than at the beginning of the 1980s (Republic of Zaire 1989b). The lack of any upward trend in retail food prices relative to other prices in urban markets, despite rapid urban growth and stagnant agricultural supply, reflects the importance of imports in meeting urban demand for food.

With regard to industrial and export crops, price liberalization com-

Table 4.9. Real producer prices of industrial and export crops (deflated to 1981 prices), Zaire (Z per kilogram)

	1981	1982	1983	1984	1985	1986
Arabica coffee	5.0	4.8	4.6	6.3	15.8	9.7
Robusta coffee	1.5	1.5	1.7	4.1	3.5	3.6
Cocoa	3.0	3.7	4.8	5.2	—	—
Seed cotton						
1st quality	1.2	1.3	1.5	2.5	2.2	1.7
2d quality	0.9	1.0	0.8	1.2	—	—
Tea	0.15	0.2	0.3	0.3	—	—
Palm nuts	0.1	0.1	0.1	0.3	—	—

Source: "Synthèse de la situation actuelle de l'agriculture zaïroise," July 1987.

bined with the 1983 exchange rate devaluation resulted in significant gains in real producer prices of export and industrial crops for which data are available (Table 4.9).[11] Table 4.3 indicates, however, that in spite of improvements in prices after 1982, production of industrial and export crops continued to stagnate or decline. Reflecting this, the trend in export volumes has been flat or negative for palm oil, rubber, tea, cocoa, and coffee. The share of agricultural exports in total export receipts has remained at about the same level—slightly over 10 percent—since 1982, with the exception of 1986, when a jump in international coffee prices and the relaxation of export quotas caused the share to more than double.

As with food crops, the deterioration of transport infrastructures has been a major barrier to industrial and export supply response. Another factor inhibiting output increases of export and industrial crops has been the continuing burden of regulation and taxation on the plantation subsector, which is easier for the authorities to monitor than are the much more numerous small farms. At the local level, problems of interference by regional administrations persist, and exporters are also burdened by export taxes and control procedures. Recently, in measures accompanying the 1987 structural adjustment program, administrative requirements have been somewhat simplified and export taxes eliminated on all agricultural products except coffee and wood, but the regulatory burden on exporters remains substantial.

With the implementation of price reform and devaluation, most observers of the situation in Zaire conclude that the structural and policy-related constraints on the supply side constitute the main barrier to expansion of industrial and export crops. Demand is not considered a limiting

---

[11] As with food crops, the effects were probably not uniform. Thus Bandundu and Shaba farmers in the 1989/90 survey discussed above were virtually unanimous in the view that, whereas food crop prices did rise, coffee and cotton prices did not rise as a result of price liberalization.

factor. For both industrial and export crops, exchange rate reform has given Zaire the potential to be competitive at current world prices. A 1986 study by ZTE/SOCFINCO (cited in World Bank 1988a) showed that the costs of domestic resources used to produce rubber, tea, chinchona, coffee, and cocoa for export and palm oil and cotton for domestic use are less than the world prices, indicating that the country has a comparative advantage in the production of these crops; only sugar production appears to be uncompetitive.[12] This is the case even though productivity remains low in both the smallholder and plantation sectors.[13] Zaire appears, therefore, to have significant potential for expansion of production of these crops at current yield levels, if barriers on the supply side are reduced.

These mixed results of the adjustment measures reflect, first, uneven enforcement at the local level of the new policies, so that the actual impact on producer prices and controls was limited in some regions. The burden of local regulation and taxation remained a major constraint, particularly in the plantation sector. Second, the capacity of the agricultural sector to respond to price incentives was constrained by the deficiencies in the transport system, particularly road transport.

Need for Further Actions

At the end of 1984, in spite of early favorable developments, Zaire's overall financial situation remained vulnerable. In particular, the financial position of state enterprises was so weak that they continued to be a drain on the budget. In addition, the industrial sector, in particular manufacturing, was hurt by increased competition from imports, many of which were subsidized in their countries of origin (e.g., cement, sugar).

To restore a viable and long-term balance-of-payment position and attract foreign investment, which were needed to assure resumption of growth, Zaire, with IMF encouragement, concluded that a sustained adjustment effort over a longer period was required. Accordingly, a new twelve-month adjustment program was signed in April 1985. The new program provided for an increase in the ceiling for credit to the private sector and a reduction in the bank financing of the government's budget in order to encourage domestic production, particularly in the manufacturing sector. At the same time, the government continued its efforts at restructuring and reforming parastatals. In more than one case, the reforms brought

---

[12] Domestic resource cost measures the cost in domestic resources of earning a unit or saving a unit of foreign exchange. In the present context, the former applies to export crops such as coffee and rubber and the latter to import-competing crops such as cotton and palm oil used as inputs in domestic industry. If the cost of domestic resources to produce a product is less than its value in terms of foreign exchange added or saved, the country has a comparative advantage in the production of the product (ZTE/SOCFINCO, 1987).

[13] Wages are low enough to offset low productivity of labor.

about positive results. The reorganization and changes in the tax regime of Gécamines, together with the 1983 devaluation, resulted in a substantial improvement of the company's financial position in 1984 and 1985. The net loss of Z 1.5 billion recorded in 1983 was transformed into a net profit of Z 1.5 billion in 1984 and Z 2.2 billion in 1985. The government also commissioned external audits of other major state enterprises. The obvious goal of these audits was to restructure the management of the enterprises involved, reduce their workforce, adjust tariffs, and eliminate their non-economic activities. As a result of these efforts, the management of some public enterprises started to improve; the financial management of others, however, such as the Société National de l'Electricité (SNEL), remained weak and continued to be a drain to the budget.

In 1986, program implementation encountered a number of difficulties. Some of these were external, but most stemmed from domestic developments. The government granted increases in wages and salaries of civil servants and did not sufficiently restrain its other recurrent expenditures on goods and services. As a result, the government budget deficit registered a sharp increase that was financed through government's increased recourse to domestic bank financing. This, in turn, induced a sharp expansion of money supply and an acceleration of inflation. Meanwhile, in the interbank market, the depreciation of the zaire slowed down and the spread between the official and the parallel market rates widened, further contributing to the depletion of the country's foreign exchange reserves. In response to these developments, the government suspended all external debt payments except to a few countries and to the IMF and the World Bank.

Despite the relative success of the initial policy measures, then, by 1986 the economy was once again under considerable pressure. As is not surprising, the trouble stemmed in part from fluctuations in the international economic markets but largely from the inadequate adjustment of 1985. Moreover, distortions and structural weaknesses brought about by years of inappropriate economic policies remained entrenched in many segments of the economy. The gradual liberalization had, as intended, contributed to the elimination of major price-related distortions. But inflation remained high, debt service was still unsustainable, and private investment had not yet responded to the reforms as anticipated. Furthermore, the level and structure of public expenditure were still inconsistent with the program's objectives.

Structural Adjustment Policies, 1987–90

To address these deep-seated problems and thus provide strong incentives to production and growth, the government began implementation in

1986 of several structural adjustment measures with the support of an industrial sector adjustment credit of US$80 million, consisting of a World Bank credit of $20 million and an IMF special facility for Africa credit of $60 million. In 1987 the government adopted a structural adjustment program, financed partly by a World Bank structural adjustment credit of US$165 million and by an IMF structural adjustment facility of US$75 million. The structural adjustment credit consisted of a World Bank credit of $55 million, an IMF special facility credit of $94.3 million, and a special Japanese joint financing of $15.7 million evenly divided into grant and loan components. The aim of the new policies was to lay the basis for long-term growth and a sustainable external financial position through improvements in macroeconomic management, reform of agricultural and transport sector policies, and enhancement of incentives for the private sector. The major production increase was to come from the private sector, particularly nontraditional tradables, which were viewed as a means of eventually achieving external balance. The industrial sector credit was designed to provide additional foreign exchange to finance imports of raw materials and spare parts needed in the manufacturing sector.

Further changes in the tariff regime were also enacted with the aim of restoring incentives to the production of tradables, in particular nontraditional tradables In July 1986 export taxes were abolished on all manufacturing exports. The only remaining export taxes were on mining products, wood, and coffee, with rates for these items ranging from 3 and 4 percent (arabica and robusta coffee, respectively) to 40 percent (copper). The government also took steps to make export procedures less cumbersome, including the elimination of the requirement of approval from the Bank of Zaire and, for most exports, the need for approval from the Department of National Economy.

The tariff schedule was also rationalized in 1986, narrowing the range of rates to a minimum of 10 percent and a maximum of 60 percent. Exceptions were permitted at both ends of the range: 5 percent tariffs were required on some basic consumer goods (e.g., sardines, smoked and salted fish) and agricultural inputs (e.g., tractors, fertilizer, animal feed), but luxury cars, tobacco, textile products, and alcohol tariffs remained above 60 percent. These exemptions were eliminated progressively starting in September 1987, and the range of tariff rates narrowed, although at a slower pace than initially planned.

With regard to the public sector, the major adjustment measures consisted of rationalizing public sector expenditures and rehabilitating infrastructure to support private sector investment. The 1986–89 public investment program (PIP), revised in 1987 in consultation with the World Bank, emphasized infrastructure rehabilitation, improvement of basic services to agriculture and industry, and improved productivity of major parastatals. Reforms of the civil service were also implemented, resulting in the removal from the payroll of some 25,000 fictitious workers since 1987. Also included in the PIP was an emergency rural roads rehabilitation program, financed in part through revenues received by the Office des Routes from increased fuel taxes, to alleviate constraints on the provision of agricultural inputs and evacuation of products. Producers of import-competing agricultural commodities were also expected to benefit from the reforms in the tariff structure, as noted above. Finally, financial reforms such as the removal of commercial banks' interest ceilings on agricultural credit were implemented to improve the mechanism of funding agricultural investment and marketing activity.

Agricultural Reforms, 1987–90

In the context of the 1986–90 development plan as well as the 1987 structural adjustment program, additional measures were taken to benefit the agricultural sector, including a revision of the tariff structure to eliminate disincentives to the production of agricultural tradables; reduction of duties on agricultural exports except coffee and wood, and simplification of export procedures; and measures to remove institutional and infrastructural constraints regarding marketing credit, public investment projects, transport, and agricultural research services. In this regard, the Bank of Zaire removed commercial banks' interest ceilings on agricultural credit. Moreover, as part of the structural adjustment program, the PIP allocated 10 percent of the funds (Z 9.7 billion) to agriculture, a substantial improvement over previous programs; in recognition of the barriers to marketing imposed by the poor condition of the rural roads network, an emergency rural roads improvement program was created with a budget of Z 400 million, with an additional Z 300 million anticipated from a rural roads tax; finally the restructuring of the national agricultural research institute, the Institut National pour l'Etude et la Recherche Agronomique, was undertaken in an effort to reverse that institution's long decline.

Performance under Structural Adjustment, 1987–90

The performance of the economy after implementation of the structural adjustment measures was disappointing in more than one respect. In 1987, the first year of implementation, real GDP grew by only 2.6 percent (Table 4.7), short not only of the program's target but also of the population growth rate. The primary and readily advanced reason for this poor performance was, as always, stagnation in production of the mining sector. This once again highlighted the disadvantages of an excessive reliance on a single major export product. The principal sources of difficulties included Gécamines's implementation of its 1986–90 rehabilitation program, which deemphasized output expansion while striving to reduce production costs, and the delay in the implementation of investment geared toward increasing the capacity of the Office des Mines d'Or de Kilo-Moto (IMF 1988e). In contrast, the manufacturing sector, helped by domestic credit policies and disbursements of foreign exchange under the industrial sector credit, recorded substantial growth in 1987. For agriculture unfortunately, there are no comprehensive, economy-wide data to evaluate the impact of the adjustment measures.

Despite small positive effects of the recent efforts at reform, the overall performance of the economy has been poor. The major obstacle to the successful implementation of the overall structural adjustment program remained the irresponsible conduct of fiscal policy. At the same time that revenue declined slightly, expenditure targets were widely exceeded in most categories. Revenue decline followed from poor tax collection and a low elasticity of revenue with respect to inflation and nominal GDP. In particular, the revenue from direct and indirect taxes in 1987 (excluding mining and petroleum export taxes and royalties) increased 64 percent, below the more than 100 percent rate of inflation that year. Moreover, the anticipated increase in revenues resulting from changes in the tariff failed to materialize. This failure was also attributable to collection inefficiency and excessive investment code exemptions. Similarly, the increase in revenues from import duties was limited to 60 percent, even though imports in local currency terms rose by over 80 percent that same year (IMF 1988e). As a result of these developments, the central government deficit approached 10 percent of GDP in 1987, nearly doubling from the previous year (Table 4.7). Monetary financing of the deficit added greatly to inflation, which exceeded 100 percent. Reflecting this instability, private investment ventures, which the structural policies had been designed to encourage, came to a halt.

Program performance weakened further in early 1988, particularly with respect to the budget and financial management. As a result, only the first tranche of the World Bank structural adjustment loan (SAL) (US$82.5 million) and the first installment of the structural adjustment facility (US$58.2 million) were released. Unable to control public expenditure, the government again resorted to money creation to finance the budgetary deficit. This contributed to maintaining a high rate of inflation, which remained in the same range as in 1987. Moreover, according to revised national accounts data (Republic of Zaire 1990), output fell by 1 percent in 1988, in contrast to the modest 2.6 percent growth achieved in 1987.

Given the conduct of fiscal policy and monetary policy, the standby arrangement with the IMF lapsed and Zaire became ineligible for further purchases. The World Bank responded in a similar manner, withholding the release of the SAL's second tranche as well as the processing of an energy sector adjustment credit, and other bilateral and multilateral donors adopted a wait-and-see attitude. The net effect of these developments was a sharp decline in the balance-of-payment assistance in 1988, which was to some extent offset by higher copper export revenue. Zaire responded to the reduction in balance-of-payment support by suspending its debt service and accumulating a substantial amount of payment arrears. During most of 1988, the standby arrangement remained off track.

In May 1989, Zaire and the IMF initiated discussions leading to a new agreement for a twelve-month standby arrangement in June 1989. After a ten-month period of attempts to implement the new program, various new difficulties emerged. In early 1990, after a less-than-successful first review, the program registered serious slippages once again with respect to public spending and the PIP. The minimum investment expenditures targets with respect to transport, health, and education infrastructures, the backbone of the structural program supported by the World Bank, did not receive sufficient attention. Meanwhile, expenditures were allocated to unplanned objectives, particularly to public consumption and wages and salaries. Monetary objectives were set at levels too high to keep demand down, particularly public demand. A further sticking point, from the World Bank perspective, was the excessively high level of production costs of Gécamines, the main source of revenue for the government.

As a result of these developments, both the IMF and the World Bank suspended disbursement of their principal operations in Zaire in early 1990. At that time, the adjustment effort ceased completely. The economic data indicate a daily deterioration of the economic and financial situation of the country. Inflation soared to more than 1,000 percent in late 1990 as money supply growth escalated to unprecedented levels. Reflecting as well as adding to inflation, the parallel market exchange rate rose from Z 500 per U.S. dollar in April 1990 to Z 2,500 per dollar in December 1990. Inflation abated somewhat in January/February 1991 before rising again to new records in March/April, more than Z 5,000 per dollar. From April to the third quarter of 1991, the economy continued to deteriorate and inflation was estimated to run as high as 3,000 percent a year.

Conclusions

Eight years after the inception of major policy reform in 1983, it appears that long-term internal and external balances have clearly not been restored in Zaire, and the prospects for improvement are dim. The question therefore is not whether the adjustment policies were successful—for they clearly were not—but why adjustment in Zaire has brought such little concrete results or, put differently, why economic performance has been so poor.

A straightforward answer is that economic developments in Zaire were the outcome of inappropriate macroeconomic policies implemented by the government in the course of three decades, irrespective of whether these policies were introduced in the absence or the presence of any adjustment program. A more elaborate response has to explore further whether the adjustment policies proclaimed by the government—presumably appropriate and adopted in compliance with the Bretton Woods institutions' recommendations—were implemented. If they were, why did they bring about such mediocre results? If they were not, why not?

Except for a short while in the 1983–84 period, adjustment appears to have been implemented inadequately or not at all. In 1983–84, when stabilization measures were for the most part appropriately implemented, the relevant macroeconomic variables attained the programmed targets, in particular the budgetary and balance-of-payment deficits and inflation. It is not, however, clear whether the gains derived from the achievement of such macroeconomic goals trickled down to all areas of the economy to benefit all socioeconomic groups.

When measures were not implemented or were implemented only incompletely, macroeconomic program targets were missed, programs were cancelled, and new programs were agreed on. The conventional view was that, in the post-independence era, Zaire has been managed under a joint tutelage of the IMF and the World Bank. In practice, these institutions' attempts to help implement standby arrangements and structural adjustment programs were often thwarted, and the sporadic and inconsistent implementation achieved at best only mixed results. In fact, it can be argued that geopolitical considerations forced the IMF and the World Bank to demonstrate extreme lenience in evaluating Zaire's performance under the adjustment programs examined above.[14] Zaire was repeatedly cited as a model pupil of the Bretton Woods institutions and as such received huge loans through a series of financial programs without adequate monitoring of either program implementation or resource allocation. Taking advantage of this expression of good faith, the Zairean government skillfully managed to avoid implementing painful stabilization and structural adjustment policies.

[14] Zaire was generally considered one of the staunchest allies of the Western powers in the East-West confrontation during the cold war era.

During the three decades of authoritarian rule, President Mobutu enjoyed considerable influence in Washington, Paris, and Brussels. He manipulated this influence to mobilize the support he needed to get IMF and World Bank staffs to devise reform programs to which his government easily adhered, on paper. As soon as it received the financing made possible by the promise of carrying out these reforms, the government then just as quickly ignored the reform commitments or reversed some of the reforms actually implemented. In a too well known pattern, the government then renegotiated new reform programs, beginning the process again. This was an easy process, because Mobutu had no trouble convincing his single party's loyalists and the nominal parliament. The general population was never part of the consultation process.

In reality, it was not until the mid-1980s that the international financial community began to realize that IMF facilities and World Bank credits were gradually becoming self-preserving devices, as much needed by the lending institutions and donor countries as by Zaire itself. On the lending side, these IMF and World Bank–sponsored arrangements became essential for Zaire to meet its debt service toward the IMF and World Bank. At the same time, on Zaire's side, the macroeconomic reforms made it possible for the country to benefit from debt relief arrangements provided by the Paris and London clubs.[15] In turn, the debt relief arrangements were indispensable to open opportunities for new credit lines, including those from private lenders, which the country needed to finance its imports of current consumption as well as imports of intermediate and capital goods required to sustain production and development. While this superficial "adjustment effort" was going on, the management of the country's vast resources was largely inadequate, opportunities were missed, and as a result the country's performance with regard to the realization of objectives in the areas of basic infrastructure, economic targets, health, and education proved very poor.

The latest estimate of Zaire's per capita income is US$260 (1989) (World Bank 1991e).[16] This figure, which has been revised upward to account for the parallel economy, is still low not only in relation to the rest of the developing world, but also relative to the potential income of a country as richly endowed with resources as Zaire. Poverty is an extremely worrisome problem in Zaire, particularly in the urban areas and even more specifically

[15] The Paris Club is a group of OECD country creditors whose representatives meet in Paris. The London club is similar but involves the private banking community.

[16] This is an amended figure, following a recent comprehensive revision of Zaire's national and economic accounts. The revision takes into account, inter alia, the very important unrecorded activity that thrives in the Zairean parallel economy. The corresponding figure published previously was US$150.

in Kinshasa, where there is general agreement that, in view of the acceleration of inflation in the past few years, per capita income has been falling rapidly. The results of two household surveys of Kinshasa taken in 1969 and 1986 suggest that real per capita expenditures dropped about 20 percent during the interim period (Houyoux 1986). The extent to which this drop reduced available caloric or nutritional levels is unclear (Tshishimbi and Glick, forthcoming; Tabatabai 1990), but anecdotal evidence certainly indicates a deterioration in food consumption, especially in recent years. It is now common in Kinshasa for families of modest means to eat only one meal a day or even one every two days.

The impact of inflation on real official remunerations, which are already so low as to be well below levels needed to meet a family's basic needs, is particularly pernicious. Trends in nominal and real wages for private sector and public sector employees are shown in Table 4.10. Public sector salaries in 1989 were just one-fourth their 1975 levels. In response to this decline, workers in the formal wage sector, particularly the public sector, have apparently turned to second and even third jobs in the urban informal sector and to activity in the parallel economy. This pattern has undoubtedly contributed to the expansion of informal and parallel economies in recent years and provided further incentives toward rent-seeking activities.

After the interruption of the financial assistance programs, the economic

Table 4.10. Salary indices of private and public sectors, Zaire (1975 = 100)

		Private sector		Public administration		Minimum legal salary	
	CPI[a]	Nominal	Real	Nominal	Real	Nominal	Real
1974	74.1	—	—	88.7	119.7	85.8	115.8
1975	100.0	100.0	100.0	100.0	100.0	100.0	100.0
1976	170.9	131.4	76.9	127.9	74.8	131.4	76.9
1977	278.9	159.5	57.2	133.4	47.8	137.1	49.2
1978	428.8	206.9	48.2	155.7	36.3	157.1	36.6
1979	965.0	277.3	28.7	320.3	33.2	209.9	21.8
1980	1,420.4	453.6	31.9	371.5	26.2	222.8	15.7
1981	1,977.5	721.6	36.5	468.9	23.7	229.6	11.6
1982	2,727.4	1,172.1	43.0	593.1	21.8	261.4	9.6
1983	4,688.5	2,510.2	53.5	684.5	14.6	665.1	14.2
1984	7,437.4	5,613.3	75.5	1,189.1	16.0	1,801.8	24.2
1985	9,209.3	6,637.8	72.1	2,142.6	23.3	1,801.8	19.6
1986	13,511.1	9,863.8	73.0	3,430.6	25.4	1,801.8	13.3
1987	25,724.4	16,482.4	64.1	5,759.8	22.4	1,801.8	7.0
1988	42,689.5	34,764.7	81.4	9,549.7	22.4	1,801.8	4.2
1989	67,662.9	—	—	16,826.7	24.9	1,801.8	2.7

Source: Bank of Zaire, annual reports, in Tshishimbi and Glick, forthcoming, courtesy of Cornell Food and Nutrition Policy Program.
[a]Weighted average of price indices for market- and store-purchased items.

situation continued to deteriorate in 1990 and 1991. Inflation, after abating somewhat in 1989 owing to a reduction of money supply growth, was estimated at over 300 percent in 1990 and at over 3,000 percent in 1991. The instability of the economic situation was also reflected in the soaring parallel market exchange rate, which, as noted earlier, rose from Z 2,500 per U.S. dollar to Z 65,000 per dollar between December 1990 and December 1991. Clearly, over eight years following the initial adoption of major policy reforms, internal and external balance has not been brought about. With interruption of the IMF and World Bank programs, Zaire has abandoned any pretense of conducting a coherent policy.

The economic picture is further clouded by the prospect of imminent changes on the political scene. On April 24, 1990, President Mobutu proclaimed the liberalization of political institutions and recognition of the need for freedom of expression and association, thus responding to numerous calls by the population to dismantle the single party, reform the constitution, and adopt more democratic political institutions. The process of political reform, however, quickly stalled.

At the end of 1991, Zaire was on the verge of bankruptcy. Food, pharmaceuticals, and other necessities were in short supply. The civil service and employees of public enterprises and major private concerns were on a lingering strike. The country needs a resolution of the political crisis before its economic problems can be addressed. This will undoubtedly require a fundamental change from an authoritarian and army-led government to a responsible and accountable government. A new system of government, based on accountability to the population, will make it possible for economic policy, including stabilization and structural adjustment policies, to be conducted in a more rational manner and will make possible a resumption of economic growth.

5

Oil Boom and Bust: The Harsh Realities of Adjustment in Cameroon

David Blandford, Deborah Friedman, Sarah Lynch, Natasha Mukherjee, and David E. Sahn

Often held up as one of the few countries in Africa to have achieved constant and steady economic growth, Cameroon currently finds itself in an unprecedented economic crisis. At first glance this situation is puzzling, given all the factors working in the country's favor. Cameroon has a particularly rich and diverse natural resource base and an underexploited agricultural potential. It is considered a middle-income economy by the World Bank, having attained an average per capita GNP of US$1,010 in 1989—one of the highest in sub-Saharan Africa and nearly two and one-half times greater than Guinea, which has the second highest GNP per capita of the countries included in this book. In addition, Cameroon has enjoyed considerable political stability since gaining independence in 1960.

While most sub-Saharan countries were being buffeted by the series of shocks that disrupted their economies in the early 1980s, the discovery of oil in 1978 and the substantial earnings that came with it enabled Cameroon to emerge as a rare success story. Healthy revenues from high prices for the country's exports seemed to produce the expectation that the favorable economic trend would continue notwithstanding structural weaknesses. Even more important, its government has earned a reputation for relatively responsible fiscal policy. Cameroon's prudence in spending oil revenues, which became increasingly significant beginning in 1977/78,[1] compares favorably with other oil-exporting developing countries. The contrast is especially clear with respect to neighboring Nigeria's oil-

---

[1] Split-year notation refers to the Cameroonian fiscal year, which runs from July to June.

financed consumption and deficit spending. In fact, until recently Cameroon has made relatively minimal use of foreign resources for its development and therefore has incurred a low level of debt to national income.

Whatever appreciation there was of the vulnerability inherent in, and inefficiencies fostered by, Cameroon's export-led growth, it was not sufficient to prepare the country for the decline in export prices that was to occur. Specifically, in 1985/86, Cameroon was hit with the simultaneous decline of the price of its oil and its principal export crops, notably coffee, cocoa, and rubber. This shock exposed significant structural problems in the economy which the infusion of oil revenues had masked. Abundant oil revenues had absorbed the losses of public enterprises, artificially boosted growth in manufacturing, and helped create what was to become an insolvent banking sector. Furthermore, oil revenues had compensated for weak performance in the agricultural sector, which had experienced almost no productivity increases since the mid-1970s despite its rich natural resource base. Human resources also remain underdeveloped, with investments in human capital favoring sophisticated rather than basic services, particularly in education and health. Life expectancy remains relatively low, and child and infant mortality rates are still high compared to other middle-income countries. These observations temper the customarily optimistic view of Cameroon and point to the failure of government policy to use the oil windfall to steer the productive and social sectors of the economy appropriately.

Extensive reforms are required to cure Cameroon's economic crisis. The deteriorating situation prompted the country to develop its own austerity program in 1987/88, to seek an IMF standby credit in 1988, and, finally, to adopt a structural adjustment program in 1989 around loans from the World Bank and the African Development Bank. In this chapter we focus on the policies and events that led Cameroon's success story to crisis. To the extent possible as of this writing, we also review the structural adjustment process under way, which, unlike those in the other countries included in the volume, did not really begin until the closing months of the 1980s.

The Boom Years: 1978–85

Rate and Structure of Economic Growth

The discovery and subsequent extraction of oil in the late 1970s, at a time of relatively high world prices for the resource, yielded a revenue windfall for Cameroon. Rigorous analysis of these years is difficult not only because of many deficiencies in the coverage and quality of statistical *e 5.1.* Annual growth of GDP, Cameroon

	Current CFAF				Constant 1987 CFAF		
GDP[a] (billion CFAF)	GDP growth (%)	GDP capita (thousand CFAF)	GDP/capita growth (%)	GDP (billion CFAF)	GDP growth (%)	GDP capita (thousand CFAF)	GDP/capita growth (%)
318	11.2	49	8.6	1,320	2.9	203	0.4
341	7.0	51	4.4	1,370	3.8	206	1.3
378	10.9	55	8.1	1,402	2.3	205	0.2
424	12.2	60	9.3	1,479	5.5	210	2.7
519	22.5	72	19.1	1,637	10.7	227	7.7
612	17.9	82	14.5	1,625	0.1	218	−3.6
693	13.1	90	9.8	1,695	4.3	221	1.2
834	20.4	105	16.7	1,839	8.5	232	5.1
1,052	26.2	129	22.2	2,109	14.7	258	11.1
1,259	19.7	149	15.9	2,389	13.3	283	9.7
1,569	24.6	180	20.7	2,762	15.6	318	12.0
1,980	26.2	221	22.3	3,119	12.9	347	9.4
2,303	16.3	249	12.8	3,201	2.7	346	−0.5
2,785	20.9	291	17.2	3,450	7.8	361	4.5
3,273	17.5	332	13.9	3,651	5.8	371	2.6
3,839	17.3	378	13.7	3,931	7.7	387	4.4
4,166	8.5	397	5.2	4,246	8.0	405	4.7
3,969	−4.7	367	−7.7	3,969	−6.5	367	−9.4
3,695	−6.9	330	−9.9	3,662	−7.7	327	−10.6
3,495	−5.4	303	−8.4	3,538	−3.4	306	−6.5
3,288	−5.9	278	−8.3	3,397	−4.0	287	−6.2
3,207	−2.5	264	−5.0	—	—	—	—

urces: World Bank 1991e; IMF 1991a.
At market prices except for the years 1990 and 1991, for which GDP figures are provisional and apparently calculated at r cost.
Denotes fiscal years ending in June, e.g., 1970 = 1969/70.

data in general but, especially, because of the secrecy surrounding critical oil sector data. These data are only partly revealed in the national accounts, government accounts, and balance-of-payment and official reserve records; estimates are used here to reflect orders of magnitude but should be treated with caution.

From independence in 1960 until the discovery of oil, the Cameroonian economy, fueled largely by the agriculture sector, grew steadily at an average rate of 5 percent per year (Table 5.1). With the discovery of oil,[2] the rate of growth of the economy accelerated sharply, averaging over 14 percent between 1977/78 and 1980/81. There was a subsequent slowdown, as oil earnings began to stabilize, but real growth continued to average

[2] Under present conditions of world market price, production costs, and revenue-sharing agreements between the Cameroonian government and the oil companies, the volume of economically recoverable oil reserves is limited; current reserves are likely to be depleted by the mid-1990s.

around 7.5 percent annually from 1982/83 until 1985/86. Because of rapid population growth, averaging 3.2 percent annually, per capita GDP growth was lower, but it still averaged over 6 percent per year between 1978/79 and 1984/85. Much of the increase in public and private resources derived from oil was channeled into capital formation; the share of GDI investment in national expenditures rose from 21 percent in 1979/80 to almost 33 percent in 1985/86. These developments, favored by rapidly increasing national savings, resulted in an estimated increase in GDP per capita in current prices from US$485 in 1978 to $915 by 1985.

Agriculture accounted, on average, for 36 percent of GDP, 84 percent of employment, and 87 percent of export earnings for several years preceding the oil boom. In the early 1980s, however, the newly developed oil sector increased its share of GDP from 4 percent in 1979/80 to 20 percent in 1984/85 (Table 5.2). Growth in the oil sector allowed for the expansion of investment and domestic demand, which in turn induced the rapid growth of the non-oil sectors, particularly manufacturing, utilities, and construc-

Table 5.2. Composition of GDP, Cameroon

	Sector product as (%) of total GDP			
	Agriculture	Non-oil industry	Oil	Services
1970[a]	37.2	16.3	—	46.6
1971	36.3	17.6	—	45.9
1972	37.5	16.6	—	46.0
1973	37.0	16.4	—	46.7
1974	35.0	15.6	—	49.4
1975	35.7	16.3	—	48.0
1976	34.7	16.8	—	48.5
1977	32.4	18.4	—	49.3
1978	29.3	16.7	1.0	53.0
1979	29.7	20.4	2.0	47.9
1980	27.9	21.9	4.0	46.3
1981	27.9	22.0	8.0	42.1
1982	28.2	21.9	12.0	37.8
1983	24.2	23.0	15.0	37.7
1984	24.9	21.5	16.0	37.6
1985	20.5	12.3	20.0	47.1
1986	21.8	22.4	10.5	45.3
1987	24.6	22.3	6.3	46.8
1988	25.8	22.3	6.3	45.6
1989	26.9	20.8	7.7	46.1
1990	26.6	20.2	7.9	—
1991	—	—	8.4	—

Sources: World Bank 1991e; IMF 1991a.
[a]Denotes fiscal years ending in June, e.g., 1970 = 1969/70.

tion; the share of these in GDP increased from 17 to 22 percent in the period 1978/80–84/85. Public administration maintained its share of GDP at 7 percent. The slow growth in agriculture at this time contributed to the decline in its share of GDP from over 30 percent to 20 percent (Table 5.2).

The imbalance in the rate of growth of the various sectors of the economy since 1980 provide some, though limited, evidence of "oil syndrome" or "Dutch disease."[3] During the oil boom, construction and services, including government, and to a lesser degree domestically consumed food crops registered the highest growth rates. Growth of manufacturing was relatively high, but distortions through protectionist policies resulted in an incentives system that encouraged the development of highly capital-intensive industries geared to the domestic market. Cash crop production remained stagnant as a consequence (aside from the sharp decline through drought in 1983) of price and commercial policies as well as an institutional support system that did little to encourage farmers (World Bank 1990b). Although Cameroon seemed to experience only a slight acceleration of inflation during the period 1979–85 as measured by official indices, domestic prices nonetheless rose faster than export prices. Consequently, there was a differential between the growth of prices of tradable and nontradable goods which contributed to the appreciation of the real effective exchange rage and to the further stagnation of export-oriented agriculture.

Other major structural changes, with effects went far beyond those of price policy and micro management, were occurring during this period. Prominent among these was that, throughout the boom years and partially in response to the growth of work opportunities in construction and services, a rapid demographic shift was taking place. The urban population increased at the rate of 7 percent per year, while the rural population grew at just over 1 percent. Recent estimates indicate that at least 40 percent of all Cameroonians reside in towns, up from 14 percent in 1960 and 28 percent in 1976, placing the country among the highest levels and rates of urbanization in Africa. This rapid movement of population has produced serious problems of urban unemployment, overcrowding, and overburdened social infrastructure, despite the high level of investment during the years of the boom.

---

[3] "Oil syndrome" is classically manifested in inflationary pressures and the skewing of production toward nontradable goods and services. "Dutch disease" refers to the tendency for a country suddenly infused with a new source of revenues—usually from a newly exploited natural resource—to experience a decline in domestic industries not associated with the new resource and a sharp increase in imports, particularly for consumption. The name for this phenomenon stems from the Dutch experience with an influx of revenues from oil and natural gas in the 1970s.

Table 5.3. Export performance: annual growth, contribution to GDP, and relative composition, Cameroon

	Real[a] annual export growth (%)	Exports as % of GDP	Composition of exports (% of total)		
			Nonfuel primary products	Fuels	Manufactures
1970[b]	2.2	19.9	91.6	0.0	8.4
1971	−9.8	17.3	90.5	0.0	9.4
1972	−6.2	15.8	90.2	0.2	9.6
1973	36.6	20.6	90.8	0.2	9.1
1974	17.3	21.7	91.1	0.2	8.7
1975	−22.2	17.1	89.7	0.3	10.0
1976	6.2	17.4	90.4	0.4	9.3
1977	30.5	20.9	93.9	1.9	4.2
1978	0.2	18.3	93.1	3.1	3.8
1979	25.0	20.1	71.5	22.8	5.8
1980	10.8	19.3	61.7	34.8	3.5
1981	22.6	21.0	35.1	60.0	4.9
1982	2.0	20.8	29.8	65.0	5.2
1983	20.6	23.3	30.7	64.7	4.7
1984	18.2	26.0	31.1	64.6	4.3
1985	17.8	28.5	31.2	64.1	4.7
1986	−29.8	18.5	37.8	54.3	7.9
1987	−30.5	13.8	44.3	46.8	8.9
1988	−13.1	13.0	41.2	47.1	11.7
1989	22.9	16.5	46.9	44.1	9.0

Source: Data from World Bank 1991e.
[a]"Real" exports = exports in millions of current U.S. dollars (f.o.b. prices) times the annual average exchange rate adjusted by the GDP deflator.
[b]Denotes fiscal years ending in June, e.g., 1970 = 1969/70.

Trade

The oil boom caused a tremendous change in the magnitude and relative composition of Cameroon's foreign trade.[4] The real value of exports grew at an average rate of 17 percent per annum between 1978/79 and 1984/85 and increased from 20 to almost 30 percent of GDP over the period (Table 5.3). Although export earnings from each of the exporting sectors grew absolutely during this period, non-oil export growth was sluggish—averaging around 3 percent a year in volume—owing to the quasi-stagnation of traditional agricultural exports. In contrast, between 1978

[4] The government placed substantial sums of oil export revenue in separate bank accounts—often held outside the country, usually in the United States. These monies were *not* included in the official national accounts. Therefore, because the data analyzed here can take account only of that portion of this revenue that was transferred into the government expenditure stream—for which fairly accurate estimates do exist—they most likely undercount oil export revenue, especially on a year-to-year basis (see section on public finance below).

Table 5.4. Agricultural exports and agricultural share of total exports, Cameroon

	% of agricultural exports					Total agricultural products as % of total exports
	Cocoa/cocoa products	Robusta/ arabica coffee	Logs/wood products	Cotton	Other	
1978a	42	33	14	3	8	87
1979	40	33	15	4	8	76
1980	32	37	18	6	7	62
1981	32	36	17	9	6	40
1982	30	31	17	12	10	30
1983	31	39	14	10	7	28
1984	37	39	9	8	7	28
1985	37	38	12	4	8	27
1986	34	40	12	5	8	36
1987	38	36	11	5	11	42
1988	37	35	11	4	13	48

Source: Data from World Bank 1989c.
aDenotes fiscal years ending in June, e.g., 1970 = 1969/70.

and 1985 the share of industrial products in non-oil exports rose from 10 to 20 percent (World Bank 1987). Yet it was the oil sector that was overwhelmingly responsible for the impressive export performance. This is evidenced by the fact that the share of petroleum and oil products in exports increased from 3 percent in 1977/78 to 65 percent in 1984/85, whereas the share of agricultural products in exports fell from 87 to 26 percent in the same period (Tables 5.3 and 5.4).

The rapid growth in exports was apparently not accompanied by a steady growth of imports (Table 5.5).[5] Though in the very early years of the oil boom recorded imports registered an average annual real growth of almost 12 percent, from 1979/80 to 1981/82 the real value of imports actually decreased almost 10 percent. From 1982/83 through 1986/87, the real value of imports increased again, on average 6 percent annually. Imports as a percentage of GDP fell, however, from almost 23 percent at the onset of the oil boom to 14 percent as it ended (Table 5.5). During this period the share of consumer goods in total imports increased from 17 to 22 percent, with luxury food and drink figuring prominently in consumption imports. The share of intermediate goods increased somewhat, from an average of 39 percent in the first half of the oil boom period to an average of 43 percent from 1981/82 through 1984/85, a growing proportion of which consisted of semiprocessed goods for an essentially import-

[5] But many sources indicate that imports may have been significantly underreported during the boom years.

Table 5.5. Imports: annual growth, percentage of GDP, and composition, Cameroon

	Real[a] annual import growth (%)	Imports as % of GDP	Composition of imports (% of total)				
			Total consumer goods	Petroleum products	Raw materials	Total intermediate goods	Tc ca goc
1970[b]	20.2	20.9	25.0	4.5	4.9	38.8	2(
1971	1.0	20.3	24.2	4.7	4.8	38.6	2
1972	5.9	21.0	23.9	5.2	4.2	41.2	2
1973	−4.7	19.0	21.6	5.0	4.7	36.8	3
1974	12.9	19.4	23.2	6.7	5.1	40.1	2
1975	11.2	21.7	19.7	9.3	5.3	40.1	2
1976	−7.3	19.3	18.7	7.2	5.5	40.1	2!
1977	24.5	22.2	17.1	9.7	5.2	36.9	3
1978	18.3	22.9	16.6	10.3	3.9	37.8	3
1979	8.3	21.9	17.0	8.8	3.8	39.0	3
1980	8.5	20.5	16.4	12.7	3.7	41.0	2(
1981	−7.1	16.9	15.0	14.8	3.7	37.6	2!
1982	−2.6	16.0	16.2	4.9	6.2	44.4	2!
1983	7.1	15.9	17.7	1.4	6.0	45.5	2!
1984	0.9	15.2	16.0	1.5	6.9	43.6	3.
1985	1.2	14.3	22.2	0.8	5.1	37.1	3.
1986	20.7	15.9	24.7	0.7	4.7	39.2	3(
1987	2.1	17.4	26.4	0.5	4.3	40.5	2!
1988	−26.0	14.0	—	—	—	—	-
1989	7.5	15.5	—	—	—	—	-

Sources: World Bank 1980a, 1987a, 1991e; Blandford and Lynch 1990.
[a]"Real" imports = imports in millions of current U.S. dollars (c.i.f. prices) times the annual average exch: rate, adjusted by the GDP deflator.
[b]Denotes fiscal years ending in June, e.g., 1970 = 1969/70.

substituting industrial sector. The share of capital goods in imports grew slightly over the period.

The expansion of oil exports, therefore, produced a substantial surplus in Cameroon's merchandise trade account. In 1984/85 exports exceeded imports by roughly US$1.2 billion. Net exports of goods and nonfactor services as a proportion of GDP increased steadily from −2.5 percent in 1977/78 to 11 percent in 1984/85. The current account balance, which had typically been negative, moved into surplus in 1983/84 and was equivalent to roughly 4 percent of GDP in 1984/85.

Public Finance

Cameroon's public finances have traditionally been in good shape. Despite a relatively low tax ratio, between 1971 and 1979 the government's operating budget was regularly in surplus, budget deficits never exceeded 2 percent of GDP, and external borrowing remained limited (World Bank 1990b). In fact, it was during this pre–oil boom period of responsible macroeconomic management that Cameroon earned its reputation for fiscal prudence, in contrast to the irresponsible behavior of the state in many of the other countries discussed in this volume. Even though the surge in earnings during the oil boom induced dramatic changes in revenue and expenditure patterns, the government of Cameroon continued to pursue what were generally regarded as prudent fiscal policies. This characterization may be overgenerous, however, in light of the experience since 1990.

Revenues.   Only oil revenues stemming from the royalties and profit taxes of the oil companies operating in Cameroon were included in the regular government budget. Indications are, however, that the greater part of Cameroon's oil earnings—revenues accruing directly to the government from its production-sharing agreements with the oil companies—were for the most part held in overseas accounts, until they were repatriated, at which time they were transferred to extrabudget accounts called *comptes hors budget* (CHB).[6] The extent of overseas holdings was a carefully guarded secret at least partly to dispel public expectations on government expenditures.[7]

A substantial portion of the oil revenues accrued by the government, then, entered into the expenditure stream through the CHB. In effect, the CHB was a transit account outside the regular budget into which oil revenues were paid and then used to finance certain types of expenditure, mainly investments. The CHB were directly managed by the president, who transferred revenue in as he deemed necessary. The share of the receipts the presidency decided to allocate to the extrabudget accounts each year was not known in advance, even by the technical ministries concerned. During much of the oil boom these transfers represented on average one-fourth of total government receipts.

As the oil boom got under way, total government revenues[8] from both the oil and non-oil sectors increased sharply—from CFAF 222 billion in 1978/79 to CFAF 877 billion (over US$2.2 billion) in 1985/86, represent-

---

[6] The government participates in oil production in conjunction with foreign firms through the parastatal Société Nationale des Hydrocarbures (SNH), created in 1980. SNH receives up to 70 percent of production under production-sharing agreements. In joint ventures, companies are expected to bear the entire cost of exploration, 50 percent of which is reimbursed by SNH only if new reserves are discovered. The CHB existed before the oil period, but during the 1980s oil earnings were by far the major source of the CHB.

[7] And thus this was, to some extent, an attempt to avoid "Dutch disease." Furthermore, financial operations through such accounts may have violated the agreement governing the Central African Monetary Area.

[8] Estimates of oil revenue transferred in from the CHB are incorporated into the numbers given here.

ing a fourfold increase (Table 5.6).[9] In particular, whereas revenue as a share of GDP averaged 16 percent in the late 1970s, it increased steadily during the boom, amounting to more than 25 percent of a much larger GDP in 1982/83. The absolute level of revenues steadily increased through 1985/86, though revenue as a share of GDP decreased to an average of 21 percent in the last years of the boom. From 1981/82 through 1985/86, the oil sector contributed on average over 40 percent of total government revenues. The expansion in economic activity generated by the oil sector increased tax inflows from the rest of the economy, particularly through taxes on income and profits.[10]

Expenditures.   The increase in government revenues generated a parallel increase in government expenditures (Table 5.7). Total annual government expenditure increased dramatically from CFAF 173 billion in 1977/78 to CFAF 1,229 billion in 1986/87. Whereas expenditures averaged 15 percent of GDP in the late 1970s, they averaged 22 percent of GDP between 1980/81 and 1985/86 and peaked at 31 percent of GDP in 1986/87. Thus the expenditure data, like the revenue information presented above, indicate a significant increase in the size and share of the government sector during the boom years.

For the most part, revenues exceeded expenditures during the boom years. From 1977/78 to 1984/85, government generally ran a budget surplus, averaging approximately 5 percent of total annual receipts; only in 1983/84 did the government register a small deficit, amounting to half a percentage point of receipts in that year. But in 1985/86 and especially in 1986/87, as the oil boom came to an end, expenditures increased sharply, resulting in deficits equivalent to 6 percent and 71 percent of revenue, respectively—and 13 percent of GDP in 1986/87—implying that commitments were made based on expectations that the revenue boom would be long-lasting. It is clear that the precipitous downward turn of international prices in these years was unanticipated.

Current expenditure generally rose in the same proportion as government receipts during the boom years, but there was a shift in the composition of expenditures. A particularly sharp rise in "subsidies and transfers"

[9] The CFA franc, the Cameroonian currency, is fixed against the French franc at 50 to 1. In the 1980s the value of the CFA franc has fluctuated at CFAF 270–450 to the U.S. dollar (IMF 1990b).
[10] The percentage of current revenue accruing from taxes on income, profits, and capital gains (excluding oil) was, however, lower during the boom years (roughly 18 percent) than the average in other lower middle-income countries (about 38 percent), whereas taxes on international trade accounted for a higher percentage of current revenue (26 percent) as compared to the average in the same sample of other middle-income countries (18 percent). In recent years this trend has been reversed (World Bank 1984d, 1987a, 1988i, 1991b).

Table 5.6. Government revenue, Cameroon

	Nominal revenue (billion CFAF)	Oil sector % of total[a]	Annual real[b] revenue growth (%)	Revenue as % of GDP
1971[c]	52.1	—	—	15.3
1972	56.5	—	0.4	15.0
1973	59.3	—	−1.6	14.0
1974	67.9	—	3.7	13.1
1975	86.8	—	7.5	14.2
1976	104.3	—	10.8	15.1
1977	128.2	—	10.7	15.4
1978	178.8	—	26.9	17.0
1979	222.4	—	17.8	17.7
1980	230.7	7.0	−3.8	14.7
1981	447.0	38.1	73.3	22.6
1982	535.1	41.8	5.7	23.2
1983	705.9	49.9	17.5	25.4
1984	716.9	37.0	−8.5	21.9
1985	813.9	39.8	4.1	21.2
1986	877.2	43.6	7.3	21.1
1987	720.5	35.0	−19.4	18.2
1988	598.7	34.6	−17.6	16.2
1989	563.0	33.0	−4.0	16.1
1990	468.0	28.9	—	14.2
1991	467.0	40.3	—	14.6

Sources: IMF 1987b, 1990c, 1991a; World Bank 1980a, 1987a.
[a]Includes both budgetary oil revenues (i.e., royalties and taxes of oil companies) and transfers from CHB.
[b]Adjusted by GDP deflator.
[c]Denotes fiscal years ending in June, e.g., 1971 = 1970/71.

occurred, although this seems to be somewhat understated in the official expenditure data shown in Table 5.7. In fact, one estimate indicates that in 1984 total subsidies to public enterprises alone were CFAF 150 billion (US$366 million), or 50 percent of government oil receipts in that year and 18 percent of total expenditure (World Bank 1990b). These transfers were necessitated by the continued weak financial situation of many large public enterprises, the rising operating costs of public institutions in social services, and subsidy programs in the agricultural sector (World Bank 1990b).

Oil revenues were also used to increase public consumption, mainly through wages. Whereas the share of wages and salaries in current expenditure fell during the boom years, outpaced by the rapid increase in subsidies and transfers, the significant absolute increase in wages and salaries during this period reflects both the proliferation in the number of government jobs and an array of nonwage payments and benefits for which civil servants became eligible. The public sector became an increasingly attractive place

Table 5.7. Government expenditure, Cameroon

	Expenditure			Current expenditure (% of total)	Allocation		Capital expenditure (% of total)	Financed by CHB (% of Capital)
	Nominal (billion CFAF)	Real growth (%)[a]	% of GDP		Wages/ salaries (% of Current)	Subsidies/ transfers (% of Current)		
1971[b]	49.1	—	14.4	90.2	52	10	9.8	—
1972	56.0	5.6	14.8	88.2	52	10	11.8	—
1973	60.2	0.8	14.2	84.0	56	11	16.0	—
1974	68.8	3.5	13.2	85.2	57	10	14.8	—
1975	78.7	-3.8	12.9	90.1	52	24	10.6	—
1976	94.4	10.6	13.6	83.6	57	18	15.7	—
1977	133.5	27.4	16.0	82.5	51	13	18.6	—
1978	173.2	18.0	16.5	71.8	49	14	28.2	—
1979	189.0	3.3	15.0	70.9	50	15	29.1	—
1980	224.9	10.4	14.3	67.5	47	17	32.5	33
1981	443.7	76.5	22.4	49.0	43	16	51.0	50
1982	494.3	-1.6	21.5	52.7	41	17	47.3	57
1983	606.2	9.3	21.8	59.1	38	26	40.9	56
1984	720.4	7.0	22.0	61.9	39	26	38.1	42
1985	813.7	3.6	21.2	62.3	40	25	37.8	48
1986	926.4	13.4	22.2	49.2	49	13	50.8	30
1987	1228.8	30.1	31.0	43.4	52	11	56.6	43
1988	813.0	-34.4	22.0	65.2	50	9	34.8	26
1989	700.0	-12.1	20.0	75.4	49	8	24.6	—
1990	716.0	—	21.8	74.3	47	7	25.7	—
1991	674.0	—	21.0	82.3	48	—	17.7	—

Sources: IMF 1987b, 1990b,c, 1991a; World Bank 1980a, 1987a.
[a]Adjusted by GDP deflator.
[b]Denotes fiscal years ending in June, e.g., 1971 = 1970/71.

to work, with benefit packages including low-cost housing loans, the free use of government vehicles, and free public utilities. Government employment rose from roughly 124,000 in 1982 to 174,300 in 1986, an increase from 27 to 30 percent of total formal sector employment. Between 1980 and 1984, the private and parapublic sectors generated about 19,000 new jobs a year, whereas public sector employment rose by about 9,000 a year.

Public Investment.   A particularly important feature of the oil boom was public investment; the share of central government in total investment increased from one-fifth in 1979/80 to over one-third by 1984/85, or 6.7 percent of GDP.[11] The expanding public enterprise sector accounted for another 12.5 percent of gross fixed capital formation in 1984/85, or 2.5 percent of GDP (World Bank 1990a).

For several years preceding the oil boom, capital expenditures represented on average 15 percent of official government expenditures. Between 1977/78 and 1980/81, capital expenditure grew almost fourfold in real terms, but then did not grow again in real terms until 1985/86, apparently because the authorities recognized the constraints on their capacity to absorb additional investments efficiently (World Bank 1987a). Capital expenditures did not displace current expenditures in the early boom years, they simply grew much faster (Table 5.7). From 1980/81 through 1985/86, capital expenditure generally accounted for between 40 and 50 percent of total expenditure and, including extrabudgetary outlays, peaked at 57 percent of total expenditure in 1986/87.

Although the details of actual investment spending are difficult to determine, one revealing fact is the divergence between planned and realized spending. In particular, the Fifth Five Year Development Plan (1982–86), adopted by the government in 1981, has been noted for its realistic approach both to the temporary nature of the oil boom and to the constraints on investment absorptive capacity. Levels of investment were planned to increase only moderately relative to savings and GDP. Furthermore, planned investment heavily favored support to agriculture, rural development, and the social sectors. But the actual sectoral allocations of investment departed considerably from planned targets, at least as evidenced by the partial data available (Table 5.8). Agriculture, rural development, and the social sectors, in particular, received a much lower share of actual investment than planned.

[11] Fixed investment was an important determinant of the rapid growth of the Cameroonian economy, especially in the early boom period. Early on—from 1979 to 1981—the rapid growth of fixed investment was largely due to the gross capital formation of enterprises, reflecting investment for oil development, the completion of the oil refinery, and the execution of several large-scale public enterprise projects in industry. Though the government financed a large part of these investments, only from 1983 did government direct investment begin to increase as a share of total investment and in relation to GDP (World Bank 1987a).

Table 5.8. Distribution of planned and actual investments, Cameroon

	Fourth Plan, 1977–81 (%)		Fifth Plan, 1982–86 (%)	
	Planned	Actual	Planned	Actual[a]
Agriculture/rural development	17.3	13.7	23.7	14.6
Manufacturing/mining/energy	30.9	44.7	16.4	25.4
Commerce/transportation	1.8	1.4	5.5	6.0
Tourism	4.9	6.5	2.2	2.5
Infrastructure	21.6	22.1	21.1	18.5
Urban, housing	12.3	2.6	11.0	7.1
Education/training	5.0	2.6	8.8	4.1
Health/social affairs	1.7	1.1	4.0	1.4
Other	4.5	5.3	7.3	20.4

Source: Data from World Bank 1987a.
[a]Actual for 1982–84.

To make matters worse, the bulk of investments in education and health during the boom period went to sophisticated technical schools and hospitals rather than to primary education facilities and public health projects. Even in 1989/90 planned investment in these social sectors concentrated limited funds in inappropriate projects—including the Spanish-financed Yaoundé University project, the Canadian-financed technical schools, and the Yaoundé and Douala hospitals—essentially extending decisions made and projects started during the boom years.

At least three quarters of extrabudgetary (CHB) resources were used to finance capital expenditures in social and economic infrastructure additional to those entered in the budget, as well as investment subsidies to the public enterprises and other public establishments (World Bank 1990b). But the ad hoc nature of CHB management made it almost impossible for decisionmakers to make a reasonable estimate of total resources available to distribute between recurrent and investment expenditures and consequently greatly weakened the planning process. Official planning was further aggravated by the fact that externally financed investments were not programmed together with the domestically financed investment budget. In effect, then, there were three investment budgets: domestically financed (36 percent of the total in 1984), CHB-financed (42 percent of the total in 1984 and managed by the presidency), and externally financed (22 percent and separately programmed and developed by the Ministry of Finance) (World Bank 1986a). The quality of management of all these, and the level of coordination between them, appears to have been weak. Furthermore, there is every indication that the recurrent cost implications of the heavy investment push were not fully taken into account. These factors, in combination, contributed to Cameroon's later economic crisis.

Financial Sector

Cameroon is the largest member of the CFA monetary area in terms of population and among the wealthiest. Membership in the franc zone system has been an important element in the development of the francophone African economies, particularly in maintaining the links of these economies with the rest of the world through the full convertibility of their currencies.[12] Still, the rules of the franc zone—though generally well designed for the least developed economies of the zone states—represent an increasing constraint on the development of the Cameroonian financial system. Consequently, there seems to be a general consensus that the level of financial sophistication of the Cameroonian economy is far behind what it should be given the country's per capital income (World Bank 1986a).

The central bank for the CFA members—the Banque des Etats d'Afrique Centrale (BEAC)—is responsible for regulating the volume of domestic credit, which it does by several means, including setting interest rates and regulating the access of commercial banks to its rediscount facilities. To this end, then, Cameroon and other member governments forfeit autonomy in monetary, credit, and interest rate policies though they retain an important role and vote in policy formulation.

Membership in the monetary area has produced substantial domestic price and currency stability in Cameroon. Given the monetary discipline provided by BEAC, the rate of inflation in Cameroon has been modest in comparison to many other African countries, though it remains high by the standard of the major industrial countries. As the oil boom progressed, the overall rate of inflation increased. The annual rate of change in consumer prices rose from 9.5 percent in 1979/80 to 16.7 percent in 1982/83 and is primarily attributed to the prices of food items, which increased from 5 to 16 percent over the same period.[13]

Though the nominal value of the CFA franc depreciated relative to the U.S. dollar in the early 1980s as the latter gained strength, Cameroon's high rate of domestic inflation in comparison to its trading partners meant that the real effective rate slowly appreciated. The real appreciation of the CFA franc added to the problems of Cameroon's export and import-

---

[12] Although the CFA franc is not traded on world currency markets, it is convertible against the French franc. Member countries are required to transfer to the Bank of France a minimum of 65 percent of their foreign exchange earnings. In exchange, French francs are then deposited into the accounts of the individual countries. These accounts are then drawn on by the Banque des Etats d'Afrique Centrale (BEAC, the central bank for the CFA franc members) so long as it does not have a deficit with the Bank of France. The franc zone has historically been in surplus with its operations account at the French treasury. It was not until the mid-1980s that the BEAC moved into a deficit position.

[13] The rate of inflation increased from the early part of the boom period to the later boom years, but on average the rate of inflation during the entire boom period was not significantly different from the equivalent period immediately preceding the oil boom years.

competing industries as it weakened the competitive position of Cameroon internationally.

Traditionally a net lender to the domestic monetary system, the government of Cameroon expanded this role significantly during the oil boom. Government deposits increased from CFAF 17 billion in 1978 to a peak of CFAF 172 billion in 1985, and during the same period net foreign assets rose from CFAF 16 billion to CFAF 166 billion. The increase in oil revenues added substantial liquidity to the financial system, resulting in a significant expansion of private sector lending. Between 1978 and 1985, claims on the private sector more than tripled, to over CFAF 800 billion.

Credit expanded rapidly in real terms in the first few years of the oil boom. During this time, as the productive capacity of the oil industry was being established, credit to the mining and petroleum sector grew rapidly from a mere 1 percent of total credit in 1977/78 to 13 percent in 1982/83. Commerce, on average, received approximately 26 percent of total credit between 1977/78 and 1984/85 and subsequently was to account for a large portion of the nonperforming loans of the formal sector banking system. The manufacturing/utilities sector's share of credit declined in the same period from roughly 28 to 20 percent in 1984/85 and down to 16 percent for several years after that. But perhaps the greatest indication of the structural problems inherent in, or at least aggravated by, the oil boom was that agriculture received on average only 5 percent of the total volume of credit in the period 1977/78–84/85. The credit provided to agriculture was low and, as was the case with investment in agriculture, was not in accordance with the continuing importance of the sector in the economy. Thus, the government's efforts to avoid "Dutch disease" were only partly successful in that the traditional sectors of the economy, most notably agriculture, did not benefit from the oil income windfall through expanded investment and access to credit.

The above discussion of the formal financial system in Cameroon is enlightening, but it is not complete. In particular, the *domestic financial market* in Cameroon is highly segregated between the formal and informal sectors, the latter being of considerable importance though not adequately taken into account when examining savings and investment activity. Although the government, larger private firms, and parastatals acquire credit through the banks (liquidity permitting), much of the rest of the economy operates outside the banking system in local savings clubs, referred to as "tontines."[14] There is practically no contact between the commercial banks and the tontines or among the tontines themselves.

During the 1980s the commercial banking system was highly concentrated and fragile. Four of a total of ten banks controlled roughly 85

[14] The tontines, social or clan affinity groups averaging 15–20 members, act as informal credit cooperatives.

percent of all lending; this concentration rendered any free-market deter-mination of interest rates impracticable.[15] Even during the height of the oil boom, the commercial banks were for the most part technically insolvent, as a result of several ill-considered loans that led to nonperforming assets some six times greater than the bank's bad debt reserves, representing as much as four times the banks' current capitalization (World Bank 1986a). This situation left the banks more risk-averse with respect to domestic lending and reduced the already limited access of smaller-scale Cameroo-nian investors to capital, leaving the latter even more dependent on the tontines.

The rates of interest offered to savers by the commercial banks were low—less than 8 percent, except for large depositors (over CFAF 25 million)—and given the rate of inflation (approximately 12 percent) yielded a negative pretax real return. BEAC maintained these low interest rates ostensibly to minimize production costs and encourage domestic investment by small- and medium-scale enterprises. But low real interest rates relative to international rates prompted an outflow of bank deposits to foreign accounts.

Thus the boom years exposed the fundamental weaknesses of the formal banking sector and left the informal sector—the tontines—in a very prom-inent role in Cameroon's financial system, supplying the financial require-ments of those not ordinarily served by the formal sector and representing a substantial share of credit to the economy. In fact, it was estimated that in 1985 the tontines held more than a third of all private deposits in the financial system, or CFAF 200–300 billion, as compared to CFAF 700 billion held by commercial banks and the Development Bank of Cameroon (World Bank 1990b). Borrowing from the tontines remained expensive: their interest rates were usually several multiples of those of the formal sector and maturities were short (almost always under one year). Although the continued importance of the tontines, though symptomatic of the un-derdeveloped financial system in Cameroon, has been viewed as an impedi-ment to the further financial deepening of the economy, the fact remains that they meet the needs of a large number of informal and small-scale enterprises. Furthermore, these informal credit institutions have been able to persist and remain viable, before, during, and after the oil boom, and avoid the pitfalls and excesses of the state-run financial sector.

Agricultural Sector Developments

Given the favorable and diverse natural conditions in which a wide range of crops can be grown, Cameroon has a strong comparative advan-

---

[15] All but two of these banks have foreign partners, some in a majority position; the four largest banks are all associated with French banks.

Table 5.9. Contribution to agricultural GDP, Cameroon

	% of agricultural GDP					Total agricultural GDP (billion 1980 CFAF)
	Food crops	Export crops	Livestock	Forestry	Fishing	
1984[a]	47.4	27.1	14.9	9.5	1.2	516
1985	50.0	23.6	14.7	10.9	0.9	495
1986	54.8	20.1	15.3	8.9	1.0	535
1987	58.0	17.9	14.7	8.3	1.0	553
1988	57.7	17.8	14.7	8.7	1.1	556

Source: Data from World Bank 1989c.
[a]Denotes fiscal years ending in June, e.g., 1984 = 1983/84.

tage in agriculture. But, with an average long-term growth rate of about 4.4 percent since the mid-1960s, the sector's performance has been modest relative to its resource base (World Bank 1989b). Particularly worrisome is the fact that, since 1982, the agricultural export subsector has stagnated, and its share of agricultural GDP has declined (Table 5.9). The scope for expanding agricultural production is significant, but it requires overcoming numerous constraints in the policy environment; these constraints, which intensified during the boom years, include inappropriate domestic marketing and pricing policies and inadequate institutional support to farmers.

Although the government of Cameroon demonstrated a commitment to agricultural development—in its willingness, at least, to dispense resources to the rural sector—its actions were not part of a well-defined strategy. As a result, much of public spending on agriculture did not contribute to higher growth. The fact that the food crop subsector, largely ignored by government policy during the oil boom period, has been relatively successful in comparison to the agricultural export subsector seems to emphasize the failure of government policies in agriculture.

The agriculture sector's share of total public investment has remained relatively constant, between 14 and 15 percent, since the early 1960s (World Bank 1989b). The Fifth Five Year Development Plan (1982–86) intended to increase the relative share of public investment in agriculture substantially, yet the sector's share was not appreciably changed (see Table 5.8). Moreover, public investment during the boom years largely focused on the estate subsector, despite the fact that the traditional smallholder has been the most important aspect of the agricultural sector in terms of employment and production—accounting for 90 percent of agricultural output, 80 percent of marketed output, 90 percent of the cultivated area, and over 75 percent of agricultural employment during the 1970s and 1980s (World Bank 1989b).

For the most part, Cameroon's agricultural sector has not experienced increases in productivity; yields have remained stagnant and what little growth has occurred has come about mainly through location effects and increases in acreage. An exception to this has been the significant growth registered by cotton and rice; here parastatals played large supportive roles but have come under fire for the high cost of their operations. A poignant example of stagnation in Cameroon's most important export crop, cocoa, was presented in the agricultural census of 1984: the average age of cocoa growers was fifty-two, almost the average life expectancy. Equally startling was the fact that the average age of all farmers was apparently forty-seven years.[16]

Marketing and Pricing Institutions. The export crop subsector has been rife with government control and regulation. The greatest interference occurred through the price-setting mechanism, referred to as the *barème* system. For most export crop subsectors, the *barème* procedure entailed the setting of prices that agents in the marketing chain could claim for its services. Prices for the major cash crops were set at the beginning of each season—independently of the relevant world market prices—by presidential decree, subsequent to consultations led by the Ministry of the Economy and Planning together with representatives of the parastatal export marketing monopsony, the Office Nationale de Commercialisation des Produits de Base (ONCPB).

As the leading actor in the marketing of export crops, ONCPB's mandate was to stabilize domestic prices for export crops and thus protect producers from fluctuations in world prices. Established in 1976 through the merger of marketing organizations inherited at independence, ONCPB was given financial autonomy but came under the control of the Ministry of Commerce and Industry. In principle, ONCPB had a monopsony on farmgate purchases and exports for all export crops, but in practice it delegated marketing responsibilities to cooperatives, private buyers, and parastatal development agencies, with a different arrangement for every crop and in several cases different arrangements among provinces.

Theoretically the stabilization objective had its merits insofar as it protected the farmer from temporary downswings in world market prices and prevented overextension when prices were peaking, but in practice the distortion of price signals ultimately did not benefit the Cameroonian

---

[16] The government policy of regularly increasing producer prices paid to farmers for cash crops was partially intended to restrain the rural exodus of the younger population (World Bank 1990b). But, whereas in much of the country the rural exodus is due to the relatively low returns to agricultural labor, in the densely populated west it is the result of land tenure customs that force young men to leave the countryside. It cannot be altered by price policy.

farmer. Fixed prices and retention of revenues by the parastatal in times of favorable world markets lowered investment and precluded farmers from increasing production in response to higher world prices. This was especially debilitating in the case of traditional export crops. For example, during the 1970s cocoa farmers in Côte d'Ivoire were the beneficiaries of higher world prices that led to substantial real income gains for producers, who responded with a massive planting effort that effectively tripled Côte d'Ivoire's production capacity and ensured export supplies for the next twenty years (World Bank 1989b). In Cameroon, on the other hand, producers had much less incentive to plant because of the decision to stabilize prices at lower levels during the same period. Thus, at present, Cameroonian cocoa farmers are faced with deteriorating tree stocks in a market with soft prices and substantial competing supplies from such countries as Côte d'Ivoire. When world export prices were high, the Cameroonian incentive structure created low relative prices of export to food crops. This, coupled with the fact that food crops are generally less labor-intensive, resulted in farmers substituting food crops for export crops in their planting decisions. Table 5.10 illustrates the levels at which producer prices were maintained and the degree to which producer prices of the most important export crops diverged from their respective export prices, a degree that until the late 1980s is generally far in excess of the costs of marketing.

During the commodities boom, the *barème* thus provided ONCPB with a large amount of discretionary revenue with which it financed not only activities within the agricultural sector but also its own organizational expansion. Apparently the revenues gained by the ONCPB were also used at times to finance activities outside the agriculture sector; in this way the agricultural sector was taxed heavily relative to other sectors of the economy.

The unwieldiness of the *barème* system also introduced significant delays into the marketing network, inflated costs, and brought other inefficiencies associated with rent-seeking behavior. Various agents in the processing chain were said to have made handsome profits from the *barème,* notably the Union Centrale des Cooperatives Agricoles de l'Ouest. Furthermore, the system of pan-territorial pricing enabled buyers in favorably located areas to prosper because of lower than average transport costs. In other instances, costs were not adequately covered by the *barème*'s price schedule, which often ended up hurting the farmers who found it difficult to market their products at administered prices.

Fertilizer Subsidies.   Cameroon's decision to subsidize fertilizer has its roots in the action and urging of various donors in the late 1960s to compensate farmers for prevailing low producer prices and to encourage

Table 5.10. Producer prices relative to export prices, Cameroon (CFAF per kilogram, f.o.b.)

	Cocoa		Robusta coffee		Arabica coffee		Cotton	
	Producer price	Producer/export price ratio	Producer price	Producer/export price ratio	Producer price	Producer/export price ratio	Producer price	Producer/export price ratio
1970[a]	85	0.59	117	0.59	200	0.72	30	—
1971	85	0.66	125	0.61	176	0.67	30	—
1972	90	0.40	125	0.60	165	0.67	31	—
1973	90	0.40	125	0.60	175	0.62	31	—
1974	100	0.36	130	0.51	200	0.68	38	—
1975	120	0.51	135	0.52	190	0.59	43	0.39
1976	130	0.43	145	0.36	235	0.51	43	0.37
1977	150	0.28	195	0.24	305	0.32	55	0.45
1978	220	0.32	250	0.29	325	0.32	65	0.46
1979	260	0.39	280	0.46	360	0.55	65	0.53
1980	290	0.50	310	0.43	350	0.42	75	0.51
1981	300	0.64	320	0.43	340	0.49	80	0.54
1982	310	0.65	330	0.57	350	0.50	100	0.47
1983	330	0.46	350	0.40	370	0.53	105	0.43
1984	370	0.44	390	0.36	410	0.37	117	0.59
1985	410	0.43	430	0.34	450	0.35	145	1.11
1986	450[b]	0.48	470[b]	0.46	515[c]	0.36	155[d]	0.85
1987	450[b]	0.63	470[b]	0.45	515[c]	0.70	155[d]	0.94
1988	450[b]	0.70	470[b]	0.75	515[c]	0.70	155[d]	—
1989	420	0.95	440	0.74	475	—	140	—
1990	250	0.71	175	0.50	220	—	95	—
1991	200[e]	—	—	—	—	—	87[e]	—

Sources: For 1970–88, World Bank 1989c; for 1989–90, IMF 1990a; for cotton's pr/ex ratio, World Bank 1989a.
[a] Denotes fiscal years, e.g., 1970 = 1969/70. The producer price for cocoa is announced in September, for coffee in January, and for cotton in November.
[b] Producer price + CFAF 30 bonus.
[c] Producer price + CFAF 40 bonus.
[d] Producer price + CFAF 15 bonus (first-quality).
[e] Estimate.

fertilizer use as a means of increasing yields and modernizing the agriculture sector. These subsidies soon became the most expensive item in the Ministry of Agriculture's investment budget (World Bank 1989b). Unfortunately, the cumbersome fertilizer network that evolved was largely unsuccessful, with supplies arriving long after they were needed and often used inappropriately. Profits were made, but not by those originally targeted.

The proportion of fertilizer consumption that was subsidized progressively increased over time, so that by 1987, at the onset of the fertilizer subsidy removal program, roughly 60 percent of all fertilizer was subsidized at 67 percent of its full cost, costing the government roughly CFAF 6 billion annually.[17] The fertilizer subsidy programs traditionally focused on export crops, so much larger proportions of fertilizer were applied to export crops (90 percent of the cotton farms and 55 percent of the coffee farms) than to food crops (24 percent of food crop farms) (World Bank 1989b). Most of the fertilizer used by the traditional farm sector, which was primarily engaged in producing food, was also subsidized.[18]

In practice, the failure of the fertilizer subsidy to provide farmers with fertilizer of appropriate nutrient content, in a timely fashion, and at a reasonable cost was a reflection of the failure of the state-run monopoly, established by the government and jointly managed by the Ministry of Agriculture and FONADER (Fond National de Développement Rural, the rural development credit agency), which were responsible for procurement and distribution of subsidized fertilizer. Although the private sector was involved in several limited, discrete tasks of procurement and distribution, all decisionmaking and discretion remained in the public sector (Truong 1989). The fertilizer procurement and distribution system was complicated and time-consuming and involved a chain of responsibilities and decisions made successively by the coffee cooperatives (who initiated the process by submitting estimates of their fertilizer needs), the Ministry of Agriculture, FONADER, the Ministry of Finance, the presidency, the Ministry of Commerce and Industry, and the Ministry of Computer Services and Public Contracts. The process was designed to take a little less than six months,

[17] Total fertilizer use was estimated at about 105,000 tons. Despite the high level of public expenditure, an active parallel market for fertilizer existed, implying that not all the intended beneficiaries' needs were met; parallel market fertilizer prices were significantly more than the subsidized price: in the western province in 1987, for example, parallel market fertilizer cost CFAF 65 per kilogram as compared to CFAF 45 per kilogram for subsidized fertilizer (Truong 1989).

[18] The crop development parastatals—and in particular SODECOTON, the cotton parastatal—purchased and distributed almost all the nonsubsidized fertilizer (Truong 1989). SODECOTON purchased unsubsidized fertilizer, but this became part of a package of inputs that could be sold at a discounted price to the cotton farmer; the subsidy was recovered by reducing the cotton price paid to farmers (International Fertilizer Development Center 1986).

but this schedule proved entirely unrealistic, with delays building on each other between the numerous tiers of the procedure, resulting in a process that took as long as eighteen months to complete. Various surveys have indicated that most fertilizer deliveries to farmers rarely took place before September or October, instead of the months of maximum agronomic usefulness, April and May.

The quantity of fertilizer actually arriving at the cooperative was usually much less than requested. In the western province, the major coffee-producing region, the fertilizer obtained by the coffee cooperatives was increasingly used by the farmers to produce maize, given its relative profitability over heavily taxed coffee. Several studies suggest that in some parts of the country as much as 90 percent of the coffee fertilizer was used on food crops. Because the fertilizer provided for coffee is not appropriate for maize, there was an obvious inefficiency in fertilizer use, with negative consequences for soil fertility (World Bank 1989b).

The haphazard use of fertilizer was exacerbated by its price structure. The price of all types of subsidized fertilizer was determined by the government without regard for nutrient content or distance to the farm from the point of transport at Douala. As a result, a bag of ammonium sulfate cost the same as a bag of urea, regardless of farm location. The tenders board was predisposed to award a large number of relatively small contracts, often to higher priced—and politically advantageous—bidders. Frequently, the contracts were suboptimally small, given economies of scale in shipping, resulting in high freight costs. Apparently sizable rents were made by several participants in the system of public tenders at the import level and among those awarded inland transportation contracts by the government (USAID 1990b).

The Shock and Its Ramifications

The year 1985/86 marked the beginning of a precipitous fall in the U.S. dollar–denominated prices of Cameroon's major export commodities—oil, coffee, and cocoa. A 40 percent depreciation of the U.S. dollar against the CFA franc between 1985 and 1988 added significantly to the impact of the fall in world prices.

In the two years between 1985 and 1987, the deterioration in Cameroon's terms of trade was massive: Cameroon's export price index fell by 65 percent for oil, 24 percent for cocoa, 11 percent for coffee, and 20 percent for rubber. This led to a 47 percent decline in the country's terms of trade in just two years. Total export earnings fell by 30 percent in *each* of the two years after the price shock. Revenues from oil and petroleum products fell by 41 percent in 1985/86, then by an additional 39 percent in the following year. For agricultural products, the decline was similarly dramatic: in 1985/86 they fell by 22 percent, then by another 50 percent over the next two years. The country's balance-of-payment situation deteriorated drastically, plummeting from a positive 4.4 percent of GDP in 1984/85 to a deficit amounting to 8.8 percent of GDP in 1986/87.

When the shock first hit the economy, the government was not able to respond quickly enough to the changed circumstances by scaling back public expenditures. Government revenues fell by CFAF 157 billion, or 19 percent in real terms, in 1986/87 while government expenditures increased by CFAF 302 billion (US$935 million), or 30 percent in real terms. Wages and salaries, subsidy payments, and other miscellaneous items, all subsumed under current expenditures, increased steadily. The state and parastatal apparatus that had evolved in the oil boom years required, on average, approximately CFAF 500 billion yearly in the period 1985/86–88/89. Total government expenditures jumped 33 percent in 1986/87 but then turned negative in the following year, declining 34.4 percent in real terms in 1987/88 and another 12.1 percent in 1989 (Table 5.7).

The decline in government spending primarily came at the expense of investment. Whereas government-financed capital outlays were increased sharply in 1985/86 and 1986/87, they experienced real reductions of 60 percent in 1987/88 and another 38 percent in 1988/89. By 1991 the continued contraction of investment spending was so severe that its nominal value was only 17 percent of that just five years earlier, as the government attempted to adjust to the harsh realities of reduced revenues from oil. Private investment also took a downward plunge of 40 percent between 1985/86 and 1987/88, as reduced government spending led to a contraction of aggregate demand.

The government initially financed the deficit by a combination of measures, including building up domestic arrears to CFAF 240 billion—mainly to domestic contractors working on investment projects—and drawing on its reserves held domestically and abroad. By 1986/87, Cameroon was also increasing its reliance on foreign borrowing to maintain the level of public spending. Foreign financing of government fiscal operations accelerated rapidly, increasing from CFAF 48 billion (US$148 million) in 1985/86 to CFAF 164 billion ($532 million) in 1988/89. The external public debt situation deteriorated from 26 percent of GDP in 1985/86 to nearly 40 percent of GDP in 1988/89. The country's overall debt service ratio rose from 8 percent of exports of goods and services in 1984/85 to 31 percent in 1989/90.

The banking sector, along with construction, was one of the first to suffer the impact of the economic crisis. By mid-1987, and in several cases many years before this, most of the commercial banks were technically insolvent. The World Bank estimated that in 1989 the aggregate deficit in the banking sector was CFAF 300–375 billion, roughly equivalent to one-third of the banks' portfolio and about 10 percent of GDP (EIU 1991b).

The use of domestically held deposits to continue financing current expenditures resulted in a severe squeeze on the banking sector's liquidity. Traditionally both the government and the parastatal sector had been major depositors in the domestic banking system, and during the oil boom their roles had expanded significantly. In 1987/88 the liquidity crunch worsened when the government, in an effort to reduce current expenditures, drastically curtailed subsidy payments to the parastatals, who were then forced to make large withdrawals from the banking system to cover their operating deficits. To make matters worse, the public enterprise sector had incurred large payments arrears to domestic suppliers, who in turn were unable to service their obligations to the banks, exacerbating their liquidity positions.

A vicious circle was created. In the agricultural sector, for instance, the parastatal monopsony in charge of export crops, ONCPB, was suffering unsustainable losses on its activities. Its growing deficit soon exhausted the liquid reserves ONCPB held in the banking system, as well as its account with the treasury. The end result was that ONCPB accumulated large arrears and was able to pay only a mere fraction of the official producer price, usually with considerable delay, to its producers. Certain large banks had held ONCPB deposits, which meant that the high exposure to crop credits had made them especially vulnerable to the terms-of-trade shock: over the course of the three-year period 1986–89, crop credit extended by commercial banks to ONCPB rose from CFAF 76 billion (US$219 million) to CFAF 230 billion ($720 million).

Overall, the crisis produced a rapid decline in real GDP: in the first year after the shock, real GDP fell by approximately 6.5 percent, and by almost 8 percent in the year thereafter. By 1989/90 real GDP was estimated to have fallen 20 percent from its 1985/86 level (Table 5.1).

Adjustment

Confronted with a rapidly deteriorating economy, the government of Cameroon found itself forced to consider a series of measures to correct the precarious state of its public finances. The first step was taken in 1986, very shortly after the oil boom had ended, with the establishment of the interministerial Parastatal Rehabilitation Commission, whose mandate was to find ways to reduce the heavy financial burden the sector was imposing on the treasury. The next step came when President Biya announced an austerity budget for 1987/88, which outlined comprehensive measure to reduce expenditures and increase revenues. Current expenditures were cut by a third and public investment halved, but revenue continued to fall. In early 1988 the government also initiated the formulations of a medium-term adjustment program; a committee of directors and senior technicians from all major operational ministries was assembled and charged with drawing up the government's "Statement of Development Strategy."[19]

Several months later, in May 1988, the final report of the Parastatal Rehabilitation Commission was submitted to the presidency. With the assistance of foreign experts financed by the World Bank and the United Nations, the five-member technical team proposed that, of the 75 parastatals examined (out of a total of more than 150), 15 be liquidated, 12 privatized, 4 merged, and 38 rehabilitated. Cost estimates for the parastatal overhaul over a four-year period were set at approximately CFAF 616 billion, not including possible costs associated with liquidation and privatization (IMF 1990b).

The magnitude of the crisis, so poignantly exemplified by this one problem, meant that recourse to international institutions became inevitable. At this point, the government approached the IMF for an 18-month standby credit of SDR 69.5 million and an additional credit of SDR 46.4 million under the IMF compensatory financing facility. The loans came into effect in September 1988.

Two months later, in December 1988, a further interministerial committee was established, charged with the task of drafting a proposal for a structural adjustment program to be submitted to the World Bank, formulated around measures outlined in the development strategy. In May 1989 negotiations were concluded with the World Bank on a structural adjustment loan in the amount of US$150 million, which was to be released in three equal tranches; the first installment of $50 million was disbursed in November 1989. The loan has been cofinanced by $125 million from the African Development Bank and designated to support the first phases of the government's adjustment program, which include restructuring of public finances, the public enterprise sector, and the banking sectors; liberal-

---

[19] According to van de Walle (1990), "the IMF and the World Bank had been negotiating with the government on and off since mid-1986, without reaching agreement. At this time, Biya staked the national prestige on refusing the tough austerity programs of those two institutions. Through 1986 and 1987, he insisted that Cameroon would undertake an adjustment on its own, and seek only nonconditional capital from bilateral donors and the private banks . . . . in late 1987 Biya announced the creation of a new 'anti-crisis' ministry, the Ministry for the Stabilization of Public Finances. . . . this example of economic nationalism was applauded at home and by the pan-African media."

ization of agricultural marketing practices; deregulation of internal commerce; improvement of incentives for petroleum exploration and production; reorientation of policies in the forestry, health, and education sectors; and establishment of specific action programs to reduce the adverse social impact of adjustment.

To date, performance under the structural adjustment program has been uneven, with significant reforms in some sectors alongside almost complete lack of progress in others. The second installment of the World Bank's loan was disbursed in April 1991, albeit with some reticence. Cameroon turned again to the IMF at the end of 1991 for a second standby arrangement, in the amount of SDR 28 million over nine months.

Fiscal Policy Reforms

A disappointing performance on the fiscal front constitutes the most serious constraint to progress under the structural adjustment program. The third tranche of the IMF facility, which expired at the end of June 1990, remained undrawn because of recurrent budget deficits and, in particular, failure to meet tax collection targets (EIU 1991b). Non-oil tax revenues have continued to fall far short of projections, in part because of the delayed adoption of discretional tax measures and also because of deteriorating compliance. In contrast to the boom years, revenue as a share of GDP was down to 14 percent in 1991, with capital expenditures accounting for only 18 percent of the much smaller government budget.

The extended contraction in economic activity in general, and especially in government outlays, and the poor performance of public financial and nonfinancial enterprises, has seriously eroded the tax base of the economy. The closure of numerous private enterprises has sharply reduced the corporate profits tax base, while a sharp rise in unemployment—the loss of some 200,000 jobs between 1986 and 1990—has considerably weakened not only the income tax base but the base for consumption-related taxes as well.

Widespread smuggling and evasion have plagued the tax system in recent years and have substantially contributed to the fall in government revenue. The Cameroonian government itself estimates that as much as 60 percent of imported goods evade duty (EIU 1991a). Since 1988/89 the government has instituted several measures to counter these problems, including a general restructuring of the customs services, new tax inspectorates, the placement of patrol vessels to survey the coastline and major smuggling water areas inland, as well as tolls to be levied on some of the major roads. Also, accompanying the tightening up of authorizations for tax exemptions, all major revenue items are being computerized to enhance management and detection of noncompliance with tax laws. Efforts to improve revenue collection were challenged, however, by a civil disobedience campaign launched by the political opposition (EIU 1991a).

The wage bill is another major point of contention with the IMF and the World Bank. Despite official announcements of extensive job cuts in the public sector, official estimates for 1989/90 show an increase in the number of employees in the public sector, from 179,118 in 1987/88 and 185,362 in 1988/89 to 187,863 in 1989/90; similarly, the number of civil servants in 1989/90 was estimated at 78,269, up from 75,315 the year before (EIU 1991b). This phenomenon has been blamed on the automatic right of trainee civil servants to a post on graduation from national training colleges. The restructuring of public enterprises calls for job losses on the order of 12,000, but, fearing the political implications of such measures, the government has proceeded slowly on this front. Civil service salaries have been frozen, but not cut as in countries such as Côte d'Ivoire (EIU 1991a). Since the introduction of austerity measures in July 1987, the government has annually announced a severe crackdown on perquisites in the public sector but has been very slow to implement cuts.

Public Enterprises and the Banking Sector

The need to restructure and rationalize the public, parapublic, and banking sectors was a key area targeted in the structural adjustment program. Restructuring of the public enterprises is taking longer than planned. By the end of 1991, only eight of the twenty-nine priority nonfinancial enterprises designated for rehabilitation had begun the process (IMF 1991a). Privatization had been completed for two of the twelve enterprises to be privatized, and only one of the fifteen enterprises to be liquidated had been eliminated.

Reforms in the banking sector began with courageous measures: the interest rate structure was liberalized and simplified, rates were raised, and several unsound banks were closed. By reducing its share in the capital of banks, the government is trying to limit its role in the banking sector (IMF 1991a). Comprehensive plans for restructuring and recapitalizing the commercial banks have been made for almost all the banks. Nevertheless, a chronic and systematic shortage of liquidity plagues the system (IMF 1990a), and progress in this sector has stalled of late.[20] Part of reform

[20] The formal financial sector remains highly concentrated: in December 1989, the four largest banks held 98.9 percent of the banking system's domestic assets and 81.4 percent of all deposits. The smaller banks concentrate on specific areas of the market, primarily short-term loans to the small-scale private sector and trade financing (IMF 1990a). Costs of restructuring the banking system are estimated at CFAF 400–500 billion (World Bank 1990a).

measures designed to resuscitate the banking system included a new crop financing system with the BEAC area, starting with the 1988/89 crop season. The system has covered only part of the export costs—set below prevailing export prices so that a margin is left to be covered by nonbank resources. "The main objective of the scheme is to prevent the recurrence of crop credit arrears to banks" (IMF 1990a).

The prolonged economic crisis in so many CFA countries has prompted a debate over a possible devaluation of the CFA franc. "The French government and the IMF have both officially dismissed the possibility but, behind the scenes, many officials in both institutions suggest that a devaluation must, in the long term, take place" (EIU 1991b).

Agriculture

In what is considered one of the most decisive moves of the structural adjustment program, the government overhauled the operations of the cash crop sector, significantly reducing producer prices to bring them into line with the world market, limiting the role of the marketing boards (including laying off several thousand employees), suspending export taxes, and rescheduling crop credit arrears with the banks as mentioned above.

At the start of the 1989/90 cocoa season, a presidential decree announced a cut of almost 40 percent for the price of first-grade cocoa from CFAF 420 per kilogram to CFAF 250 per kilogram (Table 5.10). Similarly, producer prices for best grade robusta coffee and cotton were slashed to CFAF 175 per kilogram and CFAF 95 per kilogram, respectively, representing price reductions on the order of 60 and 32 percent. Unfortunately, these cuts are likely to have significantly adverse effects on farmer motivation to use inputs and carry through replanting programs and thus are likely to reduce production and affect quality; lower prices are also likely to have severe social effects, particularly in the cotton-growing north where cotton constitutes the primary livelihood. Moreover, cuts in producer prices of cash crops are encouraging farmers to switch to production of food crops. Thus, although there seems to be little recourse but to adjust export prices downward in the face of failing world prices, the fact that the agriculture sector paid the price of high taxation during the oil boom, and was perhaps hardest hit during the period of adjustment, is a cause for concern.

Along the lines of government policy to liberalize agricultural marketing, the ONCPB has been closed and replaced by the Office National du Café et du Cacao (ONCC), essentially a quality control board that is to supervise the crop collection system but not directly intervene as a trader; movements in world market prices are to be absorbed by producers and private exporters. ONCC is to set producer prices in relation to export prices and in accordance with world market trends, monitor stock levels, and ensure the proper grading of beans. Similar practices are being adopted for rice and cotton (IMF 1991a).

The Fertilizer Subsector Reform Program

In 1987 the government embarked on a fertilizer reform program (FSSRP) aimed at phasing out subsidies over a five-year period. Privatizing the distribution of fertilizer was intended to lower costs to the government, minimize the final cost to small farmers, and improve the timeliness of fertilizer delivery. This program was initiated with a US$20 million grant from USAID[21] and a government contribution equal to the subsidy outlay; these funds have been made available to commercial banks to encourage and support loans to fertilizer importers and distributors.

FSSRP demonstrated significant progress in its first two years. The new system managed to import the same volume of fertilizer as in 1987, approximately 64,000 tons (Abbott 1991). The government realized cumulative savings of US$14 million over this period. The final cost to the farmer proved less than the full cut in the subsidy because of increased competition among, and consequently reduced margins for, importers and distributors.[22] Though unit subsidies were reduced by over 75 percent in 1990 as compared to 1987, costs to the farmers rose only about 30 percent (Abbott 1991). Delivery times for fertilizer were cut in half.

The system has been liberalized, but it is as yet only "quasi-market" given that a subsidy still exists, importers and distributors of fertilizer receive preferential rates of interest on credit (as an incentive to enter the business), the government is managing the FSSRP, and donors are intensively involved in the program (Blane et al. 1990).[23] The future sustainability of the privatized system is still not clear. Furthermore, given the

[21] Of this, $17 million is nonproject assistance and $3 million project assistance.

[22] Participation by private actors was not as great as hoped for, but, nevertheless, efficiency gains were realized even with the limited number of participants. By the second year of the program two importers, each accounting for about half the imported tonnage, took part; nine distributors participated, including five cooperatives and four private enterprises; two commercial banks were involved, one of them financing 83 percent of all shipments in the program (Abbott 1990).

[23] Donors who have traditionally supported the export crop subsector in Cameroon (the EEC and France) believe that the pace of privatization urged by USAID is too rapid. Reasons offered include concern over undeveloped internal distribution channels and the view that the relative success of SODECOTON and SEMRY (Société d'Expansion et de Modernisation de la Riziculture à Yagoua) in credit recovery and production follows the effective integration of input supply with monopoly crop purchases. Others feel that these donors prefer the pace of privatization to be slower because of their established aid programs, which are often tied to fertilizer purchases in their own countries and underwritten by their own banks (Lele et al. 1989).

extent to which producer prices have been slashed, which both limits farmers' purchasing power and encourages them to switch to food crops, the demand for fertilizer has been greatly reduced. In 1990 only 22,000 tons of fertilizer were imported through the program in response to the sharp drop in fertilizer consumption in 1989. Recommendations have been made to maintain the subsidy temporarily at current levels.

Conclusions

The key characteristic of economic adjustment in Cameroon has been the need to reduce aggregate demand to cope with the dramatic decline in export revenues that followed from the fall in oil prices. Adjustment has been primarily expenditure reducing rather than expenditure switching. This is a consequence of the cause of the economic crisis, in particular the reliance on oil revenues, which fell precipitously during the last half of the 1980s. The fact that Cameroon is member of the franc zone eliminated the option of nominal exchange rate devaluation as a policy instrument, thereby making it more difficult to adjust to the severe negative terms-of-trade shock. Although trade policies such as tariffs and subsidies can in theory be used as expenditure-switching instruments, in practice the structure of Cameroon's trade limits the effectiveness of these policy options. Therefore, fiscal policy reforms have been the core of the process of adjustment.

Given the prominent role the government sector assumed in Cameroon's economy, it comes as little surprise that dramatically lower revenues and expenditures resulted in negative growth rates in recent years. If there is any surprise, it is that the fiscal policy reforms aimed at reducing public investments, as well as the level of recurrent expenditures, represented such harsh medicine for what was one of the best-performing and presumably best-managed economies in sub-Saharan Africa. In fact, the Cameroon economy, both before and during the oil boom, was considered an example of one with few economic distortions, relatively little corruption and rent seeking, and a relatively large degree of fiscal prudence. The way the state responded to the discovery of oil, in fact, was with a degree of restraint, as juxtaposed with other oil-exporting countries. It did not fall into the trap of mortgaging its future oil revenues. At the same time, however, Cameroon could not help but become adjusted to life with oil, as observed by the massive investment push, the increase in public consumption, and related changes such as the construction boom, which indeed were difficult to adjust to as oil receipts fell precipitously.

Although Cameroon's experience of adjusting to the oil boom and subsequent bust finds few obvious parallels with the other countries included in this volume, both because prior conditions were so different and because the process of adjustment was so heavily biased to stabilize an economy through contracting demand, there are in fact some important points in common. First and foremost is that many of the economic institutions, particularly those under direct or indirect state control, were extremely weak, in terms of both management and financial stability. The oil revenues masked many of these weaknesses and gave Cameroon a reprieve from restructuring until the late 1980s, unlike other sub-Saharan countries, which began to adjust earlier in the decade. But the poor management of public and quasi-public sector institutions in Cameroon was similar to that found elsewhere in Africa. It made adjustment inevitable, with the only questions being when, and how much pain the process would entail.

Another common point is that there was a general neglect of agriculture, and in particular of export-oriented agriculture, which performed poorly throughout the 1980s. Furthermore, those state agencies that absorbed a large share of agriculture's share of public spending were poorly managed, inefficient in delivering services and inputs to farmers, and inept in their role of marketing output. Thus not only did agriculture receive a disproportionately small share of government expenditures during the boom, but the limited resources allocated to the sector ended up in the hands of inefficient enterprises, often engaged in inappropriate activities. At the same time, no sector was hit harder than export agriculture by the oil bust—induced austerity that accompanied the decline in world prices for agricultural exports. The farmers thereby failed to prosper during the boom and were left exposed to the harsh realities of the negative terms-of-trade shock during the process of adjustment. It is interesting to note that agriculture's burden was not fully shared, as indicated by the protection of civil servant employment. Likewise, the slow pace at which privatization and elimination of parastatals has proceeded is once again testament to the political influence of the urban population, and to the generalization that priorities during adjustment often reflect the same set of interests that helped precipitate the economic crisis.

It is also the case that adjustment in Cameroon, if it is to succeed, will have to entail a variety of reforms to improve competitiveness and diversify exports. The high cost structure of the economy, fueled by oil revenues and aggravated by the dollar's slide against the CFA franc, has made Cameroonian export products uncompetitive. Given that devaluation of the currency is not possible, at least in the short term, productivity-enhancing measures are paramount. Similarly, the lowering of the cost structure of production is essential to promote a sustained shift in the economy toward greater production of tradable goods.

Finally, as the economic crisis that besets Cameroon appears likely to persist for at least the medium term, the possibility of exchange rate devaluation needs further consideration as a strategy for addressing the acute economic imbalances. All official pronouncements suggest that the CFA franc will not be devalued. There is, however, little doubt of the usefulness of such a measure as a means of counteracting the lack of competitiveness of Cameroon. On balance, membership in the franc zone has likely been beneficial to Cameroon, although less so than to some of its neighbors, but continued long-term affiliation without an adjustment in the exchange rate may prove more problematic.

6

Economic Fallout from a Uranium Boom: Structural Adjustment in Niger

Paul Dorosh

Niger's economic future seemed bright in the late 1970s as high world uranium prices and increased export volumes boosted revenues. Massive investment programs in public infrastructure, manufacturing, and irrigation were begun between 1978 and 1982 in an attempt to accelerate income growth in an economy based on semiarid agriculture and livestock. Unfortunately, Niger's development push, like that of many other sub-Saharan countries, led to serious macroeconomic imbalances in the early 1980s as debts arising from the government's investments coincided with adverse movements in the external terms of trade (a decline in world uranium prices) and a slowdown in foreign credit to Niger.

The government adopted stabilization and structural adjustment programs supported by the IMF and the World Bank beginning in 1983 in order to address its budgetary and balance-of-payment problems and restore long-run growth. Three adverse external shocks—drought in 1984, economic recession in Nigeria (the economic giant to the south of Niger), and falling world uranium prices—also limited Niger's economic growth, despite reforms in agricultural and industrial policies and conservative fiscal policies. In addition, although Niger's membership in the Union Monétaire Ouest-Africaine (UMOA) helped to prevent some of the financial imbalances found in other developing countries, it also removed the option of nominal exchange rate devaluation as a policy instrument, making adjustment to adverse external shocks more difficult.

Table 6.1. Population and per capita GDP of Niger and other West African countries

	Population, 1989 (millions)	Population growth rate, 1970–89 (%)	GDP/capita, 1989 (thousand CFAF)	GDP/capita growth rate 1970–89 (%)
Niger	7,479	3.156	87	−2.69
Burkina Faso	8,776	2.36	93	1.61
Chad	5,357	2.04	60	−0.43
Mali	8,212	2.30	81	1.88
Average Sahelian	7,456	2.46	80	0.09
Benin	4,593	2.92	117	−0.48
Cameroon	11,554	3.07	302	2.19
Côte d'Ivoire	11,713	4.05	254	−0.54
Senegal	7,211	2.94	206	−0.57
Togo	3,507	2.95	122	−0.17
Average other	7.716	3.19	200	0.09
Average all	7,600	2.86	147	0.09

Source: Computed from World Bank 1991d.

Structure of the Economy

The fragile economy of Niger reflects its difficult, harsh environment. The country, 600 kilometers from the sea at its closest point, is about 12.5 times the area of France. Most of the 1.27 million square kilometers of land is desert, and 90 percent of the population lives in the southern fringe of the country within 150 kilometers of the border with Nigeria. Annual average rainfall is only 200–600 millimeters, even in the wetter southern part of the country (Horowitz et al. 1983).

Per capita GDP in 1989 was CFAF 87,000 per person (about US$273 per person at the 1989 exchange rate), similar to that of Mali and Burkina Faso, Niger's western neighbors in Sahelian West Africa, or of Nigeria,[1] but less than half the per capita GDP of Côte d'Ivoire or Cameroon (Table 6.1). The population has grown rapidly, increasing by an average of 3.15 percent per year between 1970 and 1989 to 7.5 million people. Rapid increases in population are likely to continue: 44 percent of the population is less than fifteen years old (Ministr̀e du Plan 1990b), and the total fertility rate (the average number of children a woman bears in her lifetime) of 7.1, among the highest in the world, has actually increased slightly since 1970. Meanwhile, the infant mortality rate dropped substantially between 1970 and

[1] The GDP figure for Nigeria is converted to CFA francs using the official exchange rate for 1989.

Table 6.2. Structure of GDP, Niger, 1987

	Value added (billion CFAF)	Share (%)
Agriculture	241.4	35.4
Crops	135.7	19.9
Livestock	83.4	12.2
Forestry and fish	22.3	3.3
Mining	43.9	6.4
Industry	57.9	8.5
Services	311.5	45.7
of which public administration	74.0	10.9
Indirect taxes on commodities	26.7	3.9
TOTAL	681.5	100.0
Formal	204.4	30.0
Informal	450.5	66.1
TOTAL INDUSTRY	57.9	100.0
Formal[a]	25.8	44.6
Informal[b]	32.1	55.4
TOTAL SERVICES	311.5	100.0
Formal	134.6	43.2
Informal	176.9	56.8

Source: Dorosh and Nssah 1991, courtesy of Cornell Food and Nutrition Policy Program.
[a] Formal industry includes informal mining activities.
[b] Informal industry includes formal meat processing activities.

1989, from 170 to 131 (World Bank 1991d). Projected population in 2025 is 24 million, nearly eight times the population at independence in 1960 (World Bank 1990g; IMF 1991b).

Rural activities dominate the economy. Nearly 85 percent of the total population lives in rural areas, and farming or livestock herding is the major occupation of 74 percent of the labor force (Ministère du Plan 1990b; see Dorosh and Nssah 1991, tab. 3.2). Apart from the capital, Niamey (1988 population, 398,000), only two other cities (Zinder and Maradi) have populations in excess of 160,000 (Ministère du Plan 1990b).

The agricultural sector (crops, livestock, fishing, and forestry) accounts for 35.4 percent of value added, mining 6.4 percent, industry 8.5 percent, and services 45.7 percent (Table 6.2). Millet and sorghum, two drought-resistant crops, are the major staples and are typically cultivated on small family plots, often in association with cowpeas.[2] In total, these two grains are planted on about 4.4 million hectares of the approximately 5.0 million

---

[2] In this chapter, data on the structure of the economy in 1987 are taken from the new, detailed set of Niger's national accounts for 1987 (Ministère du Plan, unpublished), which are not yet official. Time-series data on the national accounts are based on the older, official data. The summary of rural activities draws heavily from a more detailed description found in Jabara 1991; see also SEDES 1987 and 1988.

hectares cultivated annually. About 20,000 hectares of rice, a crop that requires much more water, are grown along the Niger River. Roughly half this rice is irrigated and is managed by a parastatal, the Organisation Nigérien de l'Aménagement Hydro Agricole. Cowpeas, groundnuts, vegetables (especially onions and string beans), and cotton are the major cash crops. About 50,000 hectares is planted in vegetables and other crops under some form of irrigation during the dry season.

Livestock (mainly cattle, sheep, and goats) play a key role in Niger's rural economy as a source of income and food (meat and milk) and as a store of wealth. The 1972–73 drought devastated livestock holdings, reducing cattle population by 48 percent, from 4.2 million head in 1972 to 2.2 million head in 1973. By 1982, restocking had increased the herd size to 3.524 million head, 84 percent of the 1972 level. Cattle population declined again after the 1984 drought, however, and estimated cattle population in 1989 was only 1.635 million head, 61 percent below the 1972 population. The drought was especially disastrous for nomadic livestock herders, many of whom were forced to settle in villages or urban centers. In the mid-1970s approximately 15 percent of Niger's population was estimated to be pastoral (Horowitz et al. 1983), but by 1988 the share of nomads fell to only 3.6 percent according to the population census.[3]

Data from the 1988 agricultural census indicate that, before the 1984 drought, 55.9 percent of households owned cattle. The average number of cattle per household was 3.3. Similarly comprehensive data for the post-drought period are not available. Using the Ministry of Agriculture's aggregate estimates and survey data from villages in western Niger based on Hopkins and Reardon (1989), Dorosh and Nssah (1991) estimate that 30.2 percent of rural households owned cattle.

Because of the importance of crops and livestock in Niger's economy, rainfall has a much bigger effect on national income than do other purely economic factors such as exchange rates or the level of government spending. Droughts in 1972–73 and 1984 not only severely diminished cereal production in those years but resulted in the death of large numbers of cattle and other livestock (Figure 6.1). Reductions in the size of the animal herds during the drought reduced the capital stock and thus the national income derived from livestock for subsequent years as well.[4] Moreover, livestock owners who suffered a loss in their stock of wealth would be

[3] The numbers are not strictly comparable since the definition of nomads changed between the 1977 and 1988 population censuses. Previously, nomads were defined as individuals living in nonpermanent housing (*campements*). In 1988 nomads were defined as those households that engaged in nomadic activities (Ministère du Plan 1990b).

[4] Production of the livestock sector in the national accounts includes the value of milk, animals that are slaughtered, and a portion of the increase in total herd size, which is considered investment.

Figure 6.1. Welfare indices: GDP, food production, and cattle, Niger

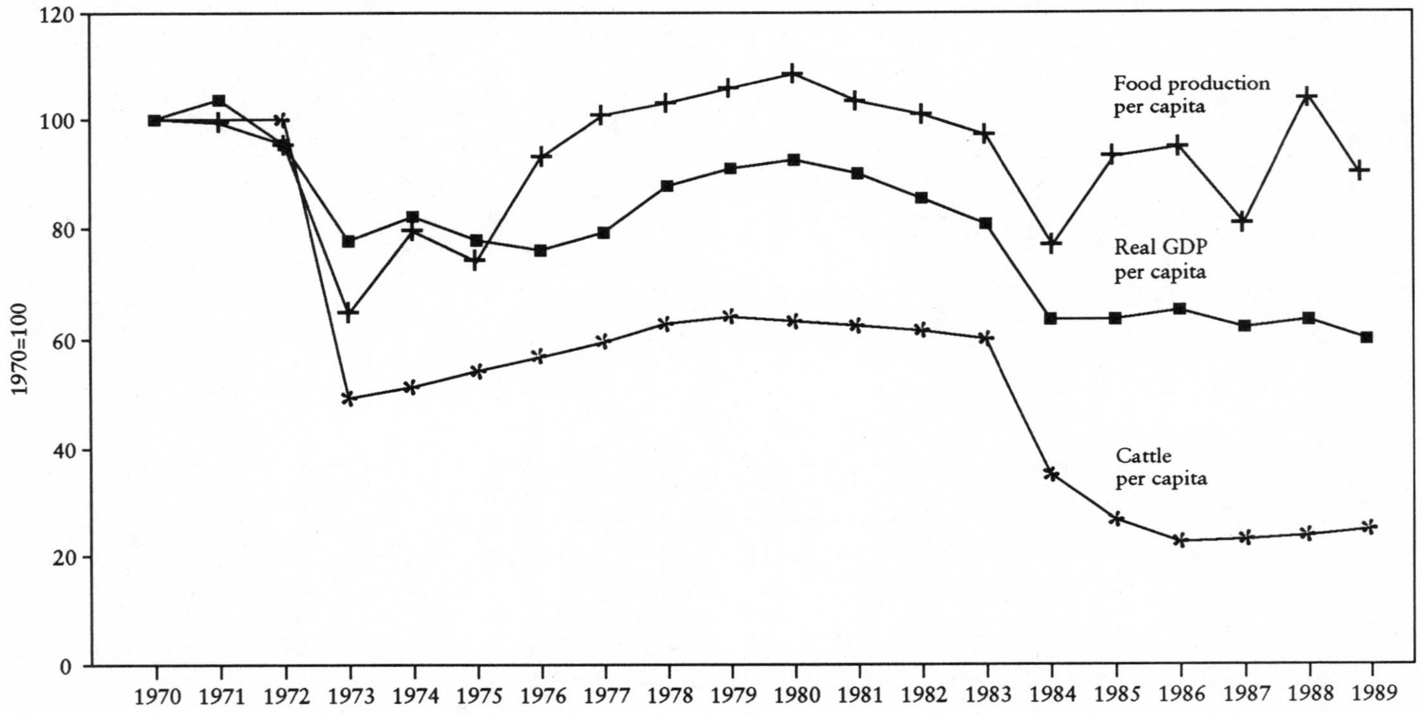

Sources: World Bank 1991d; Ministère du Plan 1991.

likely to reduce their consumption expenditures, weakening aggregate demand and further depressing the national economy. Real GDP per capita fell by 14.3 percent between 1972 and 1974 and by 19.9 percent in 1984.

The uranium mining sector is the country's most important earner of foreign exchange revenues as well as a leading source of tax revenues. Uranium exports rose rapidly in the late 1970s, from CFAF 2.0 billion in 1971 to a peak of CFAF 100.8 billion in 1980, 74 percent of Niger's revenues from exports of goods and nonfactor services. This huge gain in exports was due in part to a more than fourfold increase in world prices from 1973 to 1977, after the world oil crisis of 1973. In addition, Niger's output of uranium more than doubled after the opening of a second mine in 1978.

Before 1976, the contract price, negotiated annually between the Niger government and foreign shareholders of the two major uranium mining companies—Société Minière de l'Air and Compagnie Minière d'Akouta—was less than the world spot price (Jabara 1991). The contract price received by Niger more than doubled between 1975 and 1977, and when uranium prices on the European spot market dropped sharply thereafter the contract price averaged 82.7 percent above the world market price from 1980 to 1989 (Table 6.3). Thus, despite the 57.5 percent drop in the CFA franc price in the world spot market, Niger's export revenues declined

Table 6.3. Uranium exports, prices, and subsidies, Niger

	Uranium exports (metric tons)	Real uranium exports (billion 1980 CFAF)	Niger export price (CFAF/kg)	European spot price (CFAF/kg)	Price subsidy (%)	Subsidy/ merchandise exports (%)
1977	1,466	41.9	21,566	22,857	−5.6	−3.9
1978	2,206	60.8	23,099	21,687	6.5	4.8
1979	3,422	93.5	23,971	19,938	20.2	13.4
1980	3,956	96.3	24,353	16,270	49.7	26.3
1981	4,971	93.4	20,418	16,737	22.0	13.9
1982	3,832	76.0	23,875	17,350	37.6	20.0
1983	3,491	73.2	26,977	19,493	38.4	20.5
1984	3,468	68.8	28,989	18,506	56.6	27.4
1985	3,042	63.3	29,888	14,827	101.6	39.3
1986	3,026	67.6	29,975	13,523	121.7	43.4
1987	2,948	62.3	29,353	11,428	156.9	42.7
1988	2,950	57.4	27,513	10,593	159.7	45.4
1989	2,950	53.6	25,000	6,922	261.2	53.8
1990[a]	2,211	—	20,387	6,086	235.0	NA

Sources: USAID 1986b; Ministère du Plan 1989c, 1990a, 1991; unpublished data, U.S. Department of Energy.

Note: NA, not available.

[a] 1990 export figure and export price are actually production quantity and ex-mine price.

by only 23.4 percent in current CFA francs between 1980 and 1989. More recently, given continued low world prices of uranium, the contract price has declined, although the implicit subsidy continued to increase, reaching 261.2 percent in 1989.

Despite the massive investment program in the late 1970s, Niger's industrial base remains small. The formal sector accounted for only 30.0 percent of GDP in 1987. Apart from uranium mining, other major formal sector industries are textiles, food processing, chemicals, metal works, and paper products, accounting for a combined 4 percent of GDP in 1987. Informal industry (mainly meat processing, clothing, other food processing, and woodworking) produces 55 percent of total industrial value added and 4.7 percent of GDP (Ministère du Plan 1989b).

Altogether the formal sector (including nongovernment formal sector services such as transport and telecommunications) employs only 30 percent of the nonagricultural labor force. Expatriate workers figure prominently in the formal sector labor force: in 1988, 6.8 percent of formal sector workers, earning 28.1 percent of the total wage bill, were expatriates. Niger's public administration is largely funded by revenues from the uranium sector and by foreign grants and loans. In 1980 and 1981, 14.8 percent of government revenues (CFAF 22 billion) were derived from the uranium sector. By 1984 this share had declined to 9.6 percent of nongrant revenues, but it rose again in 1985 and averaged CFAF 10.1 billion per year (13.9 percent of nongrant revenues) from 1985 to 1989. Additional sources of funding are a high priority for Niger's government, leading to much discussion over how to tax wealthy traders and enterprises in the informal sector. Project spending, funded almost entirely from foreign grants and loans, accounted for an average of 46.9 percent of government expenditures during the uranium boom between 1978 and 1982 but only 15.5 percent of government expenditures in 1988.

Data on Niger's foreign trade are highly uncertain because of the large amount of unrecorded trade across Niger's porous southern border with Nigeria. Livestock and cowpeas are exported by Niger in exchange for small manufactured goods, gasoline (petrol), and other consumer goods.[5] By evading customs officials, smugglers avoid cumbersome administrative procedures and delays, export taxes (in place until 1988), and import taxes. Uranium accounted for 70.2 percent of official exports of CFAF 123.8 billion in 1987; illegal trade in livestock and cowpeas was estimated by Dorosh and Nssah (1991) as CFAF 29.4 billion, 20.3 percent of total

---

[5] Cook (1988) estimates that 90 percent of livestock are exported illegally. According to Dorosh and Nssah (1991), about 77 percent of cowpea production is exported outside the legal channels. Changes in the system of reporting export and import customs data in 1987 have further complicated trade estimates. As of mid-1991, the latest year for which official balance-of-payment data were available was 1989.

Table 6.4. Export and import of goods and services, Niger, 1987

	IMF		Social accounting matrix	
	Million CFA	Share (%)	Million CFA	Share (%)
Exports				
Uranium	86,900	70.2	86,991	60.0
Livestock	13,000	10.5	16,667	11.5
Cowpeas	100	0.1	17,861	12.3
Other	23,800	19.2	23,581	16.3
TOTAL	123,800	100.0	145,100	100.0
Imports				
Consumer goods	74,500	49.8	NA	NA
Cereals	7,900	5.3	11,300	5.8
Petroleum products	7,200	4.8	16,590	NA
Other	59,400	39.7	NA	NA
Primary materials/				
capital goods	58,900	39.4	NA	0.0
Services	16,205	10.8	10,600	5.5
TOTAL	149,605	100.0	193,800	100.0

Sources: IMF 1990f; Dorosh and Nssah 1991; Ministère du Plan, unpublished national accounts data.

Note: NA, not available.

export revenues (Table 6.4). Because of the flows of goods, money, and people across the border, Niger's economy is greatly influenced by economic activity in Nigeria (Azam, n.d.; Université de Clermont [CERDI] 1990). Although comparisons of per capita GDP are made difficult by the large fluctuations in the value of the Nigerian naira, Nigeria's population and its economy are approximately fifteen times those of Niger. Domestic market prices for Niger's main non-uranium exports (livestock and cowpeas) are thus largely determined by supply and demand in Nigeria's markets. Similarly, because such large quantities were imported on the parallel market by Niger, Nigeria's subsidies on fertilizer determined market prices of fertilizer in Niger in the mid-1980s (Peterson 1989).

Niger's membership in the UMOA limits the government's control of monetary and exchange rate policy. The Banque Centrale des Etats de l'Afrique de l'Ouest (BCEAO), the central bank for UMOA, issues a common currency, the CFA franc.[6] France guarantees the full convertibility of the CFA franc at the fixed exchange rate of CFAF 50 per French franc, and each member country is normally required to deposit at least 65 percent of its foreign exchange reserves in the operations account (*compte d'operations*) with the French treasury. Given targets for national incomes, price

[6] The other six members of UMOA are Benin, Burkina Faso, Côte d'Ivoire, Mali, Senegal, and Togo. Together with the member countries of the Banque Centrale des Etats de l'Afrique Centrale, they make up the CFA franc zone. CFA was originally an abbreviation for Colonies Francaises d'Afrique but now stands for Communauté Financière Africaine (Pryor 1963).

levels, and the balance of payments, together with levels of foreign exchange reserves and other variables, overall credit ceilings for each country are set annually by the board of directors of the BCEAO, which consists of representatives of France and the seven West African member nations. Other restrictions on monetary policy include a limit on government borrowing from the BCEAO to 20 percent of the government's previous year's fiscal receipts and a limit on BCEAO loans to commercial banks of 45 percent of the commercial bank's total credit (see Guillamont and Guillamont 1984; Bhatia 1985; Plane 1989). Despite these controls, other means for government financing are available, including government borrowing through public enterprises, accumulation of arrears to the private sector, and, most important in the case of Niger, external borrowing. In addition, changes in the domestic demand for money, reflected in increases in commercial bank loans or demand deposits, may result in changes in the money supply unforeseen by the central bank. To the extent that capital controls are absent or ineffectual, independent monetary policy for the individual countries of UMOA is not possible.

Macroeconomic Crisis and Policy Reforms

Problems of macroeconomic adjustment in the countries of the CFA zone are best understood by focusing on the government budget deficit rather than the balance of payments. Membership in UMOA ensures that Niger's foreign payments obligations will be met, if necessary through credit from the operations account with the French treasury. Instead of a shortage of foreign exchange, the need for adjustment shows up first as rising debt service in the consolidated government accounts. In broad terms, adjustment in the CFA countries is therefore analogous to that in a French province that has no control of money supply and need not be concerned about supply of foreign exchange but is instead concerned with the fiscal resources needed to pay its bills (Dittus 1987).[7] Looked at in this light, Niger's macroeconomic crisis was largely a fiscal problem caused by increasing debt obligations and decreasing revenues.

The Uranium Boom: 1975–80

The large increase in the world price of uranium after 1973 together with increases in the quantity of uranium mined and exported beginning in

---

[7] Although useful, the analogy oversimplifies Niger's situation somewhat. Through trade policy, Niger has some, albeit limited, influence on the real exchange rate (discussed below). Balance-of-payment considerations are also relevant to the credit targets set by the board of the BCEAO.

1978 provided Niger with much-needed revenues for development. The increase in development expenditures was not limited to the additional revenues from current exports, however. Expected earnings from future uranium exports provided the collateral that enabled Niger to borrow money at commercial interest rates for even more investments.

Between 1977 and 1982, annual government revenues averaged CFAF 64.2 billion (1980) (37 percent higher than in 1976), with revenues from the uranium sector directly accounting for 9.2 percent of revenues (compared with 2.4 percent in 1976) (Table 6.5). But Niger's average annual development expenditures for the period were more than four times greater in real terms than in 1976, averaging CFAF 71.8 billion (1980) per year. The increase in development spending was thus more than 300 percent greater than the increase in government revenues. This was possible only because the government of Niger, using its store of yet unmined uranium as collateral, borrowed CFAF 330.3 billion (1980) between 1979 and 1982, equivalent to 71.5 percent of the value of uranium exports between 1977 and 1982. The large increase in export revenues and the inflow of foreign capital spurred income growth as the funds were spent in Niger or imported capital goods were used for productive investments.[8] GDP grew by an average of 6.7 percent per year from 1978 to 1981, in contrast to the negative growth in GDP between 1971 and 1977 due in part to the effects of the 1972–73 drought (Table 6.6). Foreign capital inflows financed a large increase in imports (Table 6.7), and the current account deficit averaged 40.8 percent of GDP between 1978 and 1981, compared with only 2.2 percent of GDP between 1970 and 1977.

Ineffective controls on government borrowing were in part responsible for the crisis. Before 1984/85, development expenditures funding by external loans were not included with other government expenditures in a consolidated budget. Some projects were even begun without actual funding, in anticipation of future foreign resources (USAID 1986b). The government of Niger thus accumulated mounting debt obligations without a specific link of the future repayments to expected revenues. Outstanding long-term foreign debt quadrupled in CFA terms from CFAF 63.5 billion (US$235 million) at the end of 1977 to CFAF 262.4 billion ($287 million) at the end of 1981 (World Bank database).

Despite Niger's membership in the UMOA, inflation increased sharply in the late 1970s. Controls on the size of the government deficit financed by domestic borrowing and limits on central bank loans to commercial banks (rediscounting) could not prevent the inflow of foreign capital from increasing aggregate demand and inflation as these funds were spent domes-

---

[8] Increased foreign capital flows spent on imports of consumer goods would have no direct effect on national production.

Table 6.5. Real government revenue and expenditure, Niger (billion 1980 CFAF)

	1976	1977	1978	1979	1980	1981	1982	1983	1984	1985	1986	1987	1988	1989
Total revenue	44.9	49.0	58.2	70.9	73.3	69.2	61.3	53.5	48.0	48.7	55.2	54.5	48.3	51.0
Budgetary	43.6	47.9	55.7	70.8	73.3	69.2	61.3	53.5	48.0	47.4	53.7	53.2	46.3	49.2
Tax	38.5	39.9	47.5	59.4	65.2	59.0	54.7	48.6	41.9	41.5	46.3	41.7	38.1	40.1
Nontax	5.1	8.0	8.2	11.4	8.1	10.2	6.6	4.8	6.1	5.9	7.4	11.5	8.1	9.1
Annexed budgets, etc.	1.3	1.1	2.5	0.1	0.0	0.0	0.0	0.0	0.0	1.3	1.5	1.3	2.0	1.7
Grants	1.9	0.9	1.9	0.4	0.0	0.0	0.0	8.9	14.7	16.2	24.4	25.0	22.4	23.8
Total rev. grants	46.8	49.9	60.1	71.3	73.3	69.2	61.3	62.3	62.7	65.0	79.5	79.5	70.7	74.8
Memorandum item														
Uranium revenues	1.0	2.0	1.8	4.9	11.2	9.9	5.8	4.7	4.6	—	—	—	—	—
Current expenditures	38.1	35.9	41.7	44.6	45.8	46.7	47.6	43.0	44.6	48.8	55.9	54.8	55.3	58.4
Wages/salaries	12.0	11.9	11.9	14.8	16.8	17.4	16.9	17.2	15.8	17.3	19.2	20.4	21.4	24.5
Interest	2.3	3.6	3.2	6.5	4.5	6.0	5.9	7.1	9.3	11.6	14.5	13.6	13.5	12.9
Net lending	0.9	0.5	5.3	4.4	0.7	0.3	1.2	2.6	2.5	-0.3	-2.0	-3.8	-2.5	-2.2
Capital expenditures	16.2	20.4	28.2	88.6	100.5	113.1	79.9	54.5	36.9	35.7	46.9	43.7	37.8	44.1
Budgetary	4.1	8.7	15.1	18.0	23.9	24.8	9.7	7.4	4.0	2.9	3.9	3.9	3.8	3.9
Loan-financed[a]	12.1	11.7	13.0	70.5	76.6	88.3	70.2	38.2	18.1	16.6	18.6	16.3	14.4	18.7
Grant-financed	—	—	—	—	—	—	—	8.9	14.7	16.2	24.4	23.5	19.6	21.6
Annexed budgets	—	—	—	—	—	—	—	—	—	1.8	2.0	2.4	2.6	3.1
Total expenditure (commitment basis)	55.2	56.9	75.1	137.6	147.0	160.1	128.6	100.1	84.0	86.1	102.8	97.0	93.3	103.4
Overall deficit (commitment basis)	-8.4	-7.0	-15.0	-66.3	-73.7	-90.0	-67.3	-37.8	-21.3	-21.1	-23.3	-17.6	-22.6	-28.6

Sources: IMF 1987d, 1990f; World Bank 1991e.
[a] For the period 1976–82, "Loan-financed" includes all extrabudgetary accounts.

Table 6.6. Macroeconomic summary, Niger

	1970–77	1978–81	1982–84	1985–89
Real GDP (bn 1987 CFAF)	575.4	718.1	694.3	656.5
Real GDP per capita (1987 CFAF)	126.7	132.2	113.6	93.4
Average GDP growth (%)	−0.3ᵃ	6.7	−6.7	1.7
Inflation (% change in GDP deflator)	8.7ᵃ	9.6	10.4	−1.1
Current account balance, inc. transfers/GDP (%)	−2.2	−40.8	−15.0	−5.5
Uranium export revenues (bn CFAF)	11.1	84.0	94.6	85.6
Budget deficit/GDP (%)	−1.9ᵇ	−11.0	−9.1	−9.5
Food production per capita (1987 = 100)	110.2	130.7	113.4	113.9
Cattle population (million)	3.1	3.3	3.0	1.6
Foreign debt outstanding and disbursed, end of period (million US$)	81.0	366.0	639.0	1,259.0ᶜ
Foreign debt outstanding, end of period (million US$)	270.0	913.0	972.0	1,725.0

Sources: World Bank 1991d; World Bank database; IMF 1991b; Ministère du Plan 1991.
Note: All figures are period averages unless otherwise specified.
ᵃ 1971–77.
ᵇ 1976–77.
ᶜ End 1987.

tically. (Increase in demand deposits could also increase commercial banks' loanable funds and enable further money creation, but this appears to have been a small contributing factor to money growth.) Niger's money supply (M2) increased by 37.5 percent per annum on average between 1977 and 1981, and inflation as measured by the GDP deflator averaged 9.6 percent per year in the 1978–81 period.[9]

Fiscal Crisis and the End of the Boom

Uranium export earnings peaked both in quantity and in real value terms in 1981. Export volume then fell by 30 percent between 1981 and 1983, but the implicit price subsidy paid by French producers kept the reduction in real export earnings to only 22 percent. More important,

[9] The increase in money supply is not necessarily the cause of the inflation, however. If financial capital flows between the UMOA countries were perfectly free (as suggested by the rules of the union and the presence of a common currency), the overall money supply of the monetary union would determine the overall price level. In this case, monetary policy in individual countries would affect national price level only insofar as it significantly changed the money supply of the entire union. Since Niger's economy is small relative to that of other union members (Niger's money supply [M2] in 1990 was only 7 percent of total UMOA money supply), changes in Niger's money supply would not be a significant causal factor underlying inflation. Rather, holdings of money in Niger would increase passively, accommodating increases in aggregate demand. As discussed in the next section, inflation in the other UMOA countries was also high in the late 1970s.

Table 6.7. Balance of payments, Niger (billion current CFAF)

	1970	1971	1972	1973	1974	1975	1976	1'
Exports of goods/services	16.7	20.4	23.0	28.2	25.9	36.9	48.5	
Merchandise exports (f.o.b.)	13.0	16.2	17.9	22.2	19.6	29.7	41.0	
Imports of goods/services	24.7	23.8	29.1	38.0	56.8	54.5	75.0	
Merchandise imports (f.o.b.)	15.3	14.7	17.1	25.0	34.8	31.7	47.4	
Net private current transfer	−1.1	−1.6	−2.2	−3.6	−4.0	−4.1	−4.7	
Current account bef. off. transfers	−9.0	−3.8	−8.3	−13.3	−29.5	−15.2	−31.2	
Official transfers	9.0	8.3	11.4	18.8	22.9	17.7	26.5	
Current account after off. transfers	0.0	4.6	3.0	5.5	−6.7	2.6	−4.8	
Net long-term capital	5.8	2.0	1.7	2.3	4.0	7.8	12.5	
Other capital net	−2.9	−3.5	−3.2	−6.6	1.2	−9.1	1.3	
Change in reserves	−2.9	−3.0	−1.6	−1.1	1.5	−1.2	−9.0	

Source: Data from World Bank 1991e.

reductions in the world uranium price (expressed in dollars) reduced the expected future stream of uranium export earnings, and a worldwide cutback in international credit reduced financing of new development expenditures as well as increased the cost of credit needed for repayment of past loans. In 1982 the government of Niger cut development expenditures by 21.8 percent in an attempt to reduce its fiscal deficit. These measures were insufficient to head off a growing crisis, though, and loan agreements were signed with the IMF in 1983, 1984 1985, and 1986. Subsequently, the government of Niger adopted structural adjustment programs supported by World Bank loans in 1986, 1987, and 1989. Two other adverse shocks to its economy in the mid-1980s made Niger's adjustment to adverse changes in the terms of trade much more difficult: The drought of 1984, which devastated the rural economy, and Nigeria's structural adjustment policies beginning in 1984.

The essence of the stabilization program was a reduction in government expenditures and a rescheduling of debt. Public capital expenditures fell by 43.1 percent in real terms between 1982 and 1986, from 14.9 to 9.8 percent of GDP. Government current expenditures also fell. Despite a small drop in tax revenues, the government budget deficit fell sharply. Reducing the government budget deficit automatically reduced the balance-of-

1978	1979	1980	1981	1982	1983	1984	1985	1986	1987	1988	1989
74.1	113.4	136.0	147.5	144.8	147.7	152.1	142.1	140.2	149.0	124.8	117.6
64.9	103.2	121.7	131.7	125.3	127.7	132.5	116.6	114.8	123.8	109.9	99.2
138.3	178.5	214.7	237.4	268.7	205.0	201.9	239.8	142.5	196.8	172.2	182.2
92.7	112.1	143.1	160.8	169.3	126.4	117.9	155.3	107.3	123.1	116.9	117.6
−8.0	−10.6	−12.0	−14.1	−17.5	−18.8	−18.9	−26.4	−15.0	−15.0	−10.4	−12.8
−72.3	−59.3	−90.7	−93.0	−125.8	−75.3	−63.2	−124.0	−17.3	−62.8	−57.8	−77.4
26.9	30.0	32.5	43.6	49.3	51.8	63.7	95.3	51.4	42.7	45.8	42.1
45.3	−29.3	−58.3	−49.3	−76.5	−23.6	0.5	−28.7	34.1	−20.1	−12.0	−35.3
35.3	40.3	47.3	61.0	12.3	17.4	14.0	22.0	27.3	30.8	28.5	30.5
13.1	−11.8	12.2	−10.9	43.4	5.1	−0.9	8.1	59.1	−6.4	−4.9	2.0
−3.1	0.7	−1.2	−0.8	20.8	1.1	−13.7	−1.4	−2.2	−4.2	−11.6	2.7

payments deficit as government imports decreased, and diminished government domestic spending reduced private incomes and private import demand.

The Real Exchange Rate and Trade Policy

In theory, internal and external balance could have been achieved through an exchange rate devaluation and a smaller cut in government expenditures, resulting in a smaller decline in incomes. A devaluation raises the ratio of the price of tradable to nontradable goods, improving incentives for production of tradables and consumption of nontradables, thereby reducing the trade deficit (assuming normal price elasticities for export supply and import demand—the Marshall-Lerner condition). (At the same time, the government budget deficit lessens as revenues from trade taxes and the value of foreign transfers increase.) Some reduction in expenditures is necessary, however, to ensure that the increased demand for nontradables does not increase their price and erode away the effects of the depreciation in the nominal exchange rate.

Niger's membership in UMOA precludes exchange rate devaluation as an adjustment policy. Trade policy (increased import tariffs and export subsidies) could also be used as expenditure-switching instruments, but the structure of Niger's trade limits their effectiveness. Many of the goods imported through official channels are noncompetitive imports; that is, the elasticity of substitution between these imports and domestic goods is nearly zero.[10] Import taxes on these goods do not lead to price-induced changes in expenditure patterns but instead have the effect of a general expenditure tax on consumers of these imports, that is, an expenditure-reducing tax policy. Niger did in fact raise import tariffs and tried to reduce tax avoidance by increasing penalties as part of a tax reform in 1983/84. The policy was reversed in May 1987, however, as tariff rates were lowered on a large number of imported goods in an effort to reduce tax fraud (Jabara 1991). Many of Niger's imports which are closer substitutes to domestically produced goods (e.g., textiles and small manufactures) originate in Nigeria and other bordering countries, making evasion of import taxes relatively easy.

Despite a fixed nominal exchange rate, Niger's real exchange rate varied significantly in the 1980s (Table 6.8 and Figure 6.2). Demand pressures during the uranium boom had boosted Niger's inflation above that of its major European trading partners. As the economy slowed in the early 1980s, domestic inflation fell and the real exchange rate for goods traded through official channels (RER1) depreciated by 15.2 percent between 1981 and 1984. Much of this trade was uranium exports and noncompetitive imports; higher prices of noncompetitive imports did little to increase production of domestic goods. The depreciation of the real exchange rate also tended to increase the cost of intermediate goods imported from France and other developed countries.

Meanwhile, the real exchange rate vis-à-vis the Nigerian naira appreciated 31 percent, reducing the real price of imports from Nigeria. Recession in Nigeria lowered incomes there and, along with the real depreciation of the naira, led to a reduction in the CFA price of goods exported to Nigeria such as livestock and cowpeas. Based on the estimated quantity of exports of livestock and cowpeas in 1987 (Dorosh and Nssah 1991), the 37.7 percent appreciation of the CFA franc relative to the naira on the parallel market between 1985 and 1988 resulted in a direct loss of CFAF 13 billion per year in export revenues, equal to 3.5 percent of rural income in 1987.[11]

---

[10] Devarajan and de Melo (1987) show that under these conditions, and assuming the government saves the additional tax revenues, the demand for nontraded goods falls, lowering their price and inducing a depreciation of the real exchange rate. In the absence of income effects, an import tariff leads to increased demand for nontraded goods and an appreciation of the real exchange rate (Dornbusch 1974).

[11] This example assumes that prices are determined in Nigeria and that the level of Niger's exports does not significantly affect Nigeria's price. Since imported livestock account for only about 10 percent of Nigeria's supply (Cook 1988), these assumptions are a fair approximation of the actual situation.

le 6.8. Nominal and real exchange rates, Niger

	Nominal			Real	
Trade-weighted index[a] (1980=100)	Naira/CFAF official (naira/thousand CFAF)	Naira/CFAF parallel (naira/thousand CFAF)	Trade-weighted index (1980=100)	Naira/CFAF parallel (naira/thousand CFAF)
5	102.6	3.05	4.12	84.9	104.5
'6	102.2	2.62	3.58	95.7	92.7
7	100.3	2.62	4.15	104.1	109.6
'8	102.2	2.82	5.00	104.9	118.8
'9	97.7	2.83	4.93	102.5	118.6
0	100.0	2.58	4.14	100.0	100.0
1	95.7	2.26	3.13	104.8	76.9
2	92.9	2.07	4.02	103.2	102.2
3	90.9	1.96	6.78	89.9	136.6
4	87.8	1.85	8.38	88.9	131.0
5	88.3	2.23	9.25	83.9	135.9
'6	88.4	9.62	13.69	79.0	184.8
7	88.5	13.70	16.37	71.8	187.1
'8	86.9	17.86	25.19	67.7	205.2
9	86.3	22.08	31.15	62.8	183.8
0	89.6	29.07	34.84	59.5	190.5

ources: Berg and Associates 1983; Ministère du Plan 1989c, 1990a; IMF 1985, 1988d, 1990e, 1991b;
ₐDI 1990.
Foreign currency per CFAF index.

Appreciation of the CFA franc also hurt Niger's parallel market exports of textiles to Nigeria, which fell by about 60 percent between the early 1980s and 1989 (Ministère du Plan 1990a).

The absence of effective policy instruments to achieve expenditure switching thus exacerbated the decline in income experienced under the macroeconomic reforms. Niger's difficulties in adjusting to adverse external conditions (i.e., changes in terms of trade) are similar to those experienced by other CFA countries.

In principle, a stable fixed exchanged rate and guaranteed currency convertibility would encourage foreign investment and growth, whereas restrictions on government borrowing in the CFA zone would lead to low inflation and fiscal discipline. Comparing GDP growth rates of CFA countries with other sub-Saharan countries, Devarajan and de Melo (1987) found that the CFA countries had significantly higher growth over the period 1960–82. During this period, the monetary and fiscal discipline imposed by the rules of the CFA zone apparently helped these countries. In a later analysis, however, Devarajan and de Melo (1990) found that the performance of the CFA countries had deteriorated in the 1980s. Their regression estimate for the average GDP growth rate in CFA countries from

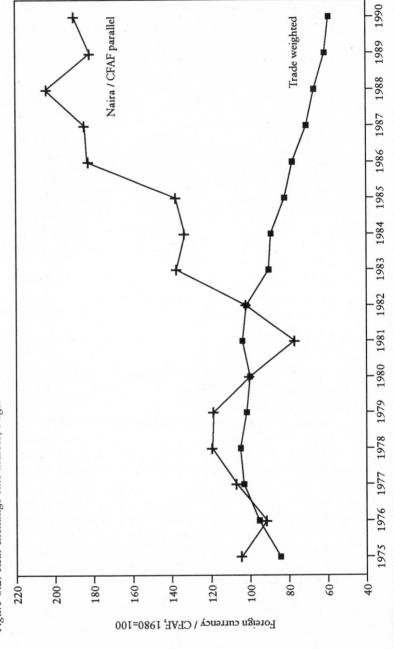

Figure 6.2. Real exchange rate indices, Niger

1982 to 1989 was only 2.1 percent per year, compared with 2.3 percent per year in twenty other sub-Saharan countries. Lack of success in adjusting the real exchange rate to compensate for adverse external shocks and correct trade imbalances was one key factor explaining the lack of growth. On average, the real exchange rate in the non-CFA countries depreciated by about 30 percent in the 1980s and average annual export growth equaled 2.6 percent year, compared with almost no change in the real exchange rate in the CFA countries and annual export growth of only 1.5 percent per year.

Other Policy Reforms

Several other economic reforms, besides changes in trade policy, were carried out in an effort to improve economic efficiency and reduce fiscal costs of government interventions.[12] Tax policy reforms in 1983/84 standardized taxes on income and profits, raised trade and excise taxes, and increased penalties for tax avoidance. A value added tax replaced the existing system of turnover taxes and other producer taxes in January 1986. The largely ineffective export taxes on agricultural and livestock products, which were subject to widespread evasion by traders, were eliminated in 1988.

Reform of public sector enterprises began in 1983 as part of the initial stabilization effort. Much of the Niger's public borrowing had been channeled through parastatals: their total outstanding debt (foreign and domestic) had been about CFAF 120 billion at the end of 1983, equal to 18 percent of GDP (IMF 1987d). Seven of the largest public sector enterprises underwent initial reforms in 1983 with cutbacks in operations, reductions in staff (half the employees of the marketing parastatal COPRO-NIGER were dismissed), and price restructuring. After a study of fifty-four parastatals by the government and the World Bank in 1985, further measures were agreed on and, by 1987, sixteen public enterprises had been privatized, integrated into government services, or liquidated. Decisions to liquidate another three public enterprises were reached in 1988/89.

Initial agricultural policy reforms included reduction in subsidies on agricultural inputs and, in late 1984, a legalization of private trade in key agricultural commodities, thus ending the monopolies on crop marketing held by the parastatals—Office des Produits Vivriers du Niger (OPVN) for millet and sorghum, and Société Nationale de Commercialisation de l'Arachide for cowpeas. In 1985 the role of the OPVN was reduced to management of a food security stock of no more than 80,000 tons of cereals, with purchases and sales of cereals to be conducted on a tender and offer basis.

---

[12] This section summarizes a more detailed account of policy reforms in Jabara 1991.

Restrictions on pricing and marketing of nonagricultural products were also gradually reduced. Import distribution monopolies were abolished in 1985. In the same year, the number of products subject to price ceilings was reduced from twenty-seven to seven; by 1987 only five products had fixed price ceilings. Profit margins were deregulated for twenty-five imported goods in 1987, leaving thirty-nine imported products still with preset profit margins. Finally, 1990 price controls were removed on all products except certain strategic goods and services.

Sectoral Reforms and Microlevel Outcomes of Adjustment

Separating out the effects of policy reforms enacted beginning in 1981 from the adverse effects of the drought and recession in Nigeria is not possible without a formal analytical framework. Efforts at controlling government expenditure successfully reduced the budget deficit so that by 1984, the first year of the drought, the deficit was only 8.2 percent of GDP. Niger's real GDP growth rate was only 1.0 percent per year in the four years before the 1984 drought and GDP dropped by 1.8 percent.

Real GDP fell by 16.9 percent in 1984 and then increased unsteadily by a total of only 8.5 percent by 1989 (averaging 1.7 percent growth per year). Between 1984 and 1989, per capita GDP fell by 7.9 percent. Initial estimates suggest that real GDP fell by another 1.5 percent in 1990 as well.

Microlevel reforms affecting key sectors of the economy (agriculture, the formal sector, and the composition of government expenditures and revenues) were a major part of the adjustment process in Niger. In some cases, the major direct effects of these reforms on household welfare can be ascertained. Many of the reforms had only small impacts; their influence on microlevel outcomes is indistinguishable from the effects of the whole complex of structural adjustment reforms, external shocks, and long-term structural factors in Niger.

Agricultural Reforms

Before the agricultural policy reforms beginning in 1984, government parastatals had been involved in a system of official fixed producer and input prices and restrictions on private trade. Apart from areas connected with irrigation projects (growing mainly rice and cotton), government attempts at market intervention were not generally successful in influencing producer prices. Liberalization of these markets thus had little effect on the vast majority of farmers.

Government (OPVN) purchases of millet and sorghum in years of good harvests between 1971 and 1983 accounted for only 3–6 percent of production. Total marketings on average are variously estimated at between 10 and 30 percent, with OPVN's market share in good harvest years approaching 25 percent.[13] In bad harvest years, OPVN's purchase prices were so low relative to open market prices that their market share dropped to 5 percent or less (Berg and Associates 1983; Borsdorf 1979). Moreover, OPVN's purchases were largely from traders' stocks, a few months after the harvest. To the extent that the spread between the traders' purchase price and their sales price to the OPVN exceeded real storage and handling costs, these sales served mainly as an income transfer to traders and had little effect on producer prices.[14]

Government agricultural policy had a significant impact on incomes of rice farmers, though. Although rice accounts for only 3 percent of total cereal production (5 percent of cereal consumption), much of government agricultural policy revolved around attempts to increase production of irrigated rice, regarded as the most viable option for national food self-sufficiency. All Niger's paddy is grown along the Niger River and, since 1985, more than 75 percent of total production is on large irrigated perimeters along the river.[15] Farmers cultivating land in the large perimeters are required to sell a portion of their paddy output at the official price to the producers' cooperative in order to pay a user fee (*redevance*).

Approximately one-third of irrigated rice is sold to the cooperatives at the official price, another third is consumed by the farmers, and the remaining third is sold in private rural markets. Official producer prices remained in place for rice even after the grain marketing reforms of 1984–85.[16] As world market prices fell, reducing the price of imported rice in Niger, the Société du Riz du Niger (RINI) could not compete with lower-cost imported rice and faced large budget deficits. Several policy changes ensued, including reductions in official producer prices from CFAF 100 per kilogram in 1985 to CFAF 70 per kilogram in 1987, sales of stocks of rice at a discount of 8 to 20 percent, and further restrictions on rice imports.[17]

---

[13] Whereas OPVN's purchases are known with a fair degree of accuracy, estimates of total quantities marketed are uncertain.

[14] If traders replenished their stocks with additional purchases from the stocks of large farmers, total marketed grain may have increased. The size of this increase was probably very small, however (University of Michigan 1988c).

[15] The remaining paddy production (considered "traditional" production) derives from small lowlands and floodplains.

[16] As part of the reforms, the OPVN's monopoly on RINI's milled rice sales was eliminated.

[17] Rice importers were required to purchase an amount equal to 20 percent of their total rice imports from RINI. Later, the purchase requirement was replaced by an import tariff on rice, which grew from 5 percent in 1987 to 22.57 percent in 1989 (Rassas and Loutte 1989.) Import taxes also included a price parity tax (*taxe de péréquation*), which grew from CFAF 5,000 per ton in 1984 to CFAF 25,000 per ton in April 1990 (CERDI 1990).

As world market prices rose from 1987 to 1989 and official producer prices for rice remained unchanged, the nominal rate of protection for producers fell from 68 percent in 1987 to only 4 percent in 1989.[18] Between 1987 and 1989, real producer prices for paddy actually rose slightly as nominal prices changed little and the overall price level fell. Official real producer prices for paddy between 1987 and 1989 were on average 11.8 percent lower than in the 1982–86 period. Import tariffs and other taxes on imports remained high (equal to 33.4 percent of the c.i.f. value of imports in 1989), so consumers continued to be taxed on their consumption of imported rice.

The porous border with Nigeria was a major factor limiting effectiveness of government pricing policy. Reductions in the size of the fertilizer subsidy beginning in 1984 were judged to have little impact on farmers, since open market fertilizer prices at the time were actually determined by the essentially unlimited supply of lower-cost fertilizer from Nigeria (Peterson 1989). Similarly, streamlined administrative procedures will in theory reduce real marketing costs for traders exporting livestock through official channels and could increase farmgate prices. Whether or not official livestock exports were taxed appears to have had little impact on trade volumes in either the official or parallel markets to date (Cook 1988).

Given the limited effectiveness of price policy before structural adjustment, the largest impact of reduced government interventions, in fact, has been a reduction in the size of government subsidies and operational losses. The OPVN's operational losses averaged CFAF 1.3 billion per year between 1975 and 1983 (Berg and Associates 1983), an amount equal to 16.7 percent of capital expenditures in 1983.[19] Although staff and outposts have been reduced, little savings were initially attained as continuing purchases under a tender and bid system resulted in the government purchasing grain at prices higher than market levels in 1985/86.[20] The first true tender for the purchase of grains was finally organized by the OPVN in January 1989 with technical assistance from the World Bank (Louis Berger International 1989b).

[18] Figures for the nominal rate of protection are calculated from data in Rassas and Loutte 1989. These figures are not correct if the c.i.f. Niamey price shown does not already include import duties.

[19] The OPVN's losses were especially high after the good harvest in 1982/83, when large quantities of domestic grain were purchased directly from merchants at recently raised producer prices. OPVN's sale prices to consumers in 1983 were significantly above open market prices, reducing OPVN sales to almost nothing and leaving the OPVN with deteriorating stocks. Subsequently, the IMF included conditions in its stabilization accords with Niger aimed at limiting the financial costs of OPVN's operations (Jabara 1991).

[20] Initially, the new system ran into problems because the lots were too large and there were few bidders. Later, there were problems associated with the timing of the bidding process relative to the harvest (University of Michigan 1998c).

Consumer Prices

Consumer price data from Niamey show steadily rising prices from 1970 to 1984[21] and falling prices beginning with the post-drought year 1985. The rising price trend reflects the boom years for Niger (through 1981) and inflation throughout the UMOA. Niger's later decline in overall prices differs from the patterns in other UMOA countries (Figure 6.3). In part, it was caused by a decline in aggregate demand and less foreign capital inflow, which led to a reduction in the price of nontraded goods and services. In addition, the appreciation of the real exchange rate vis-à-vis the Nigerian naira tended to reduce the price of traded goods competing with Nigerian goods.

Between 1982 and 1989, the food and beverage price index declined by 26 percent in nominal terms and by 42 percent relative to nonfood prices (Table 6.9). Millet prices rose sharply after poor harvests in 1981 and 1984 but have fallen sharply since, and open market prices for millet in Niamey were more than 40 percent below the real price of 1980 in three of four years between 1986 and 1989. Real rice prices from 1982 to 1988 were on average 18 percent below their 1980 level as increased import tariffs failed to outweigh declining world prices. The 1989 real rice price was nearly equal to the level of the late 1970s and early 1980s, however. Although rice is consumed by all household groups in Niamey, consumption of millet per person by lower-income groups is nearly twice that by higher-income groups (CILSS 1989), indicating that low-income urban households have enjoyed relatively greater benefits from these price trends.

Reforms in the Formal Sector

Cutbacks in credit to the formal sector, reduced domestic demand, and the closing of parastatals resulted in a sharp drop in the output of the modern sector under structural adjustment.[22] Real value added fell from CFAF 67.9 billion (1980) in 1981 (12.5 percent of GDP) to only CFAF 33.5 billion (1980) in 1988 (6.8 percent of GDP). Formal sector non-government employment likewise fell from 37,020 in 1981 to only 24,000 in 1988. The real minimum wage fell by 18 percent between 1981 and 1984 but increased in the late 1980s to 96 percent of its 1981 level (12 percent higher than in 1977) (Jabara 1991). Since formal sector employment was equal to only 64,000 people (about 3.5 percent of the labor force) in 1981, these trends in real wages and employment did not directly affect the large majority of the Nigerien population.

---

[21] Consumer prices fell slightly only in 1983.
[22] Excluding mining, water, and electricity.

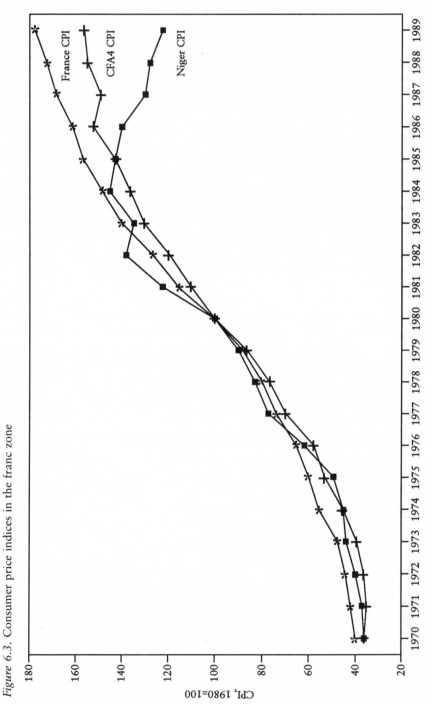

Figure 6.3. Consumer price indices in the franc zone

Source: Data from IMF 1991b.

Table 6.9. Food and nonfood prices, Niger

	Food and beverage CPI index	General CPI African	Food/ nonfood index	Millet (CFAF/kg)	Real millet price (1980=100)	Rice (CFAF/kg)	Real rice price (1980=100)
1970	—	35.1	—	25	73.5	66	113.4
1971	—	36.4	—	28	79.2	70	115.78
1972	—	40.0	—	32	82.4	73	109.8
1973	—	44.7	—	51	117.6	95	128.0
1974	—	46.4	—	40	88.9	92	119.5
1975	—	50.5	—	38	77.5	91	108.5
1976	—	62.4	—	51	84.3	103	99.4
1977	80.1	76.8	114.2	64	85.9	138	108.3
1978	85.7	84.5	104.5	90	109.8	146	104.1
1979	91.7	91.6	100.4	92	103.5	160	105.2
1980	100.0	100.0	100.0	97	100.0	166	100.0
1981	128.9	124.3	112.0	178	147.6	208	100.8
1982	142.3	137.2	111.9	167	125.5	199	87.4
1983	130.1	133.8	92.0	98	75.5	194	87.3
1984	142.6	145.2	94.8	156	110.7	199	82.5
1985	138.9	143.7	90.7	182	130.6	201	84.3
1986	131.0	139.1	84.4	73	54.1	171	74.1
1987	116.0	129.8	73.8	65	51.6	162	75.2
1988	111.2	127.9	68.9	95	76.6	183	86.2
1989	105.1	124.3	64.6	72	59.7	200	96.9
1990	105.0	123.3	65.6	86	71.9	200	97.7

Sources: CERDI 1990; Berg and Associates 1983; Ministère du Plan 1990a.

Government Budgetary Policies

Government trade and tax reforms have had relatively modest effects on total revenues. There were no increases in the rates for personal income taxes (*l'impôt sur les traitements et salaires* and *l'impôt général sur les revenues*), but as salaries have increased in nominal terms more workers are subject to pay taxes under the progressive tax rate structure. The share of total revenues from taxes on salaries and employee benefits increased from 8.8 percent of total revenue in 1982 to 13.9 percent in 1988 (Louis Berger International 1989a). These increases in revenues from personal income taxes have helped offset the loss in uranium tax revenues in the 1980s. More important, foreign grants increased from an average of only CFAF 1.0 billion per year in 1976–79 to CFAF 33.0 billion per year (45.8 percent of total government revenues) in 1986–89 (Table 6.5 and Figure 6.4).

Major changes in government budgetary policies have taken place on the expenditure side, in particular the major reduction in development expenditures noted above. Annual development expenditures fell by 68.2 percent in real terms between 1980 and 1984, yet they were still 128 percent higher

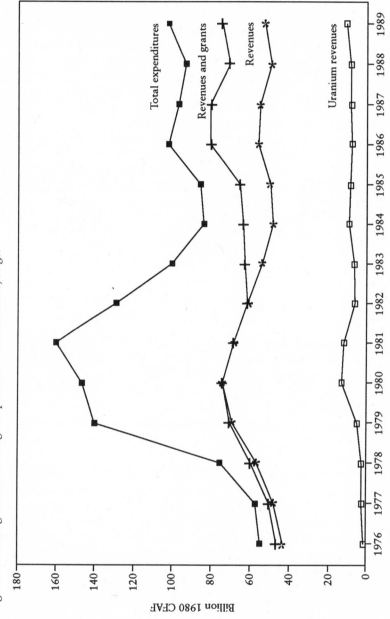

Figure 6.4. Real government budget expenditures and revenues, Niger

Sources: IMF 1985, 1987b, 1990f; World Bank 1991e.

than in 1976, mainly because of a large influx of foreign grants. Although much of this spending was for imported goods (cutbacks here directly reduced the balance-of-trade deficit), part of the expenditure was on local costs of construction, including wages. In addition to its negative implications for long-term growth, reduced development expenditures thus directly reduced current incomes for workers involved in investment activities.

Government current expenditures increased in real terms every year from 1977 to 1989, except for slight drops in 1983 and 1984. Much of this increase was debt service, which more than tripled in real terms in the decade and in 1989 accounted for an average of 24.1 percent of current expenditures from 1985 to 1989. Apart from debt service, the level of real current expenditures stagnated between 1981 and 1985 and then increased by 22.6 percent from 1985 to 1989 (Figure 6.5).

The pattern of personnel expenditures suggests that the Niger government placed significant emphasis on direct employment creation, which partially offset the decline in employment in the private formal sector.[23] Government expenditures on wages and salaries rose by 76.0 percent in real terms between 1981 and 1989, and the number of people employed increased by 36.7 percent. Combined with decline in formal private sector employment, the decline in total (government and private formal sector) employment was 6 percent between 1981 and 1988 (Jabara 1991). Average real wage rates increased by only 28.8 percent in the same period. With overall nondebt service expenditures rising only slowly, expenditures on materials and supporting services fell in real terms by 34.0 percent from 1981 to 1989, from CFAF 861,000 to CFAF 568,000 per government worker.

Health and Education Expenditures

Current expenditures on health rose by 50 percent from 1981 to 1987, but capital expenditures declined such that total real expenditures on health were roughly the same in 1989 as in 1981. The overall trend masks a severe cutback in capital expenditures from 1983 to 1985, however, when capital expenditures averaged only CFAF 42.4 billion and total health expenditures averaged 23 percent below the 1989 level.

Cutbacks in government health expenditures may have had little impact on the majority of Niger's underserved population. In the rural areas, the average population per rural dispensary increased from 21,311 in 1982 to about 28,000 in 1989, a figure higher than in 1978 (26,297 people per

[23] One benefit of such a strategy would be to limit political dissent by university students and recent graduates who otherwise might be unable to find jobs during the structural adjustment period.

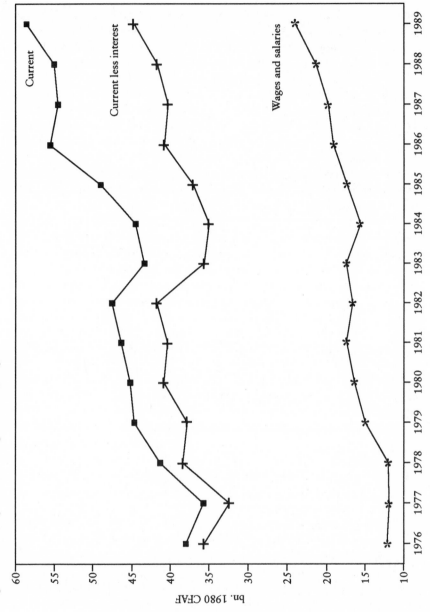

Figure 6.5. Real government expenditures, Niger

Sources: IMF 1985, 1987b, 1990f; World Bank 1991e.

rural dispensary).[24] Much of the available funds are spent in urban areas for hospital and specialized care rather than for basic health services. In 1989, 46 percent of the Ministry of Health's material expenses were allocated to the three national hospitals and the departmental hospital centers, compared with less than 30 percent for rural health supplies (Tinguiri 1990). As a condition for a USAID health sector loan in 1986, greater efforts at cost recovery were instituted. Before the reforms, the rate of cost recovery for the entire health system in 1984 was only 4.5 percent, whereas the cost recovery rate for hospitals (the only part of the health system in which fees were collected) was 10 percent (USAID 1986a). Beginning in October 1985, hospitals began to require outpatients to pay for services before being examined. As a result, the recovery rate for outpatient treatment rose from 7 percent in 1986 to 40 percent in 1988. These reforms have been criticized, however, for misapplication of the new fee system to lower-income patients who in theory are not required to pay. In addition, the fee requirement may be encouraging a bias by medical staff against referrals of lower-income, nonpaying clients to the hospitals (Tinguiri 1990).

For most of the 1980s, recurrent government expenditures on education remained unchanged in real terms, accounting for 20–25 percent of total current expenditures (Tables 6.10 and 6.11). Development expenditures were much more affected: government investment on education fell by more than 80 percent in real terms after 1982.

The share of recurrent budget education expenditures devoted to primary education has gradually increased from 33 percent in 1981 to 44 percent in 1989. Under a series of education reforms agreed to by the government of Niger and the World Bank in 1989, the trend toward increasing emphasis on primary education will continue. Budgetary resources allocated to primary schools are to increase by 7 percent per year from 1988 to 1995 (Jabara 1991).

Conclusions

Niger's economy suffered a series of massive adverse shocks in the 1980s: a drop in world uranium prices, drought, and trade and exchange rate problems related to adjustment policies in Nigeria. These negative factors only compounded the country's structural problems: a poor natural resource base, a fragile environment for crops and livestock, and a rapidly

[24] Each rural dispensary is manned by a nurse and is designed to provide services for all inhabitants within a 10-mile radius.

Table 6.10. Real government expenditures, by sector, Niger (million 1980 CFAF)

	1980	1981	1982	1983	1984	1985	1986	1987	1988	1989
Recurrent expenditure										
Agriculture	2,656	2,458	2,465	2,458	2,440	1,484	1,660	1,553	1,751	2,412
Mining	657	648	525	378	324	116	127	113	128	142
Transport./commun.	2,317	2,489	2,176	1,997	2,056	1,765	1,138	1,076	1,175	619
Education	11,409	10,717	11,888	11,715	10,983	10,708	11,486	11,141	1,222	13,134
Health	3,143	3,114	3,010	3,122	2,938	3,078	3,668	3,615	3,694	3,955
Social services	551	525	478	507	458	492	589	492	526	611
Debt service	4,545	5,978	5,893	7,179	9,997	11,628	14,533	13,592	13,514	12,865
Defense	3,219	3,340	3,149	3,266	3,032	3,235	3,642	3,593	3,751	4,118
Other	17,299	17,408	17,977	12,455	11,350	16,322	19,042	19,318	18,566	20,510
TOTAL	45,796	46,678	47,561	43,075	43,578	48,827	55,884	54,767	55,331	58,365
Development expenditure										
Agriculture	13,400	15,452	12,871	8,702	9,312	10,361	15,226	14,807	14,845	14,050
Mining/industry/energy	19,700	15,728	9,550	11,732	3,903	2,764	1,804	1,166	842	2,369
Commerce	2,500	3,127	3,322	0	137	2,312	3,294	547	488	945
Transport	26,900	24,558	17,854	8,313	7,053	8,564	6,857	8,897	6,948	9,689
Telecommun.	4,900	15,912	3,820	932	1,096	1,351	1,565	1,245	248	472
Water supply	3,400	4,047	4,650	1,865	2,876	5,361	5,783	5,867	4,960	7,377
Education	14,300	13,429	13,037	1,787	1,575	2,068	2,482	2,152	1,083	2,762
Health	1,700	3,403	2,657	0	753	738	1,975	1,886	1,875	1,897
Housing	5,300	3,679	996	155	0	1,574	1,081	1,073	531	1,207
Other	8,400	13,797	11,127	15,850	5,204	1,497	5,418	5,543	6,021	3,365
TOTAL	100,500	113,131	79,885	49,337	31,090	36,589	45,485	43,183	37,840	44,134

Souces: IMF 1985, 1990f; World Book 1991e.
Note: Niger's fiscal year ends in September; e.g., 1980 denotes 1979/80. Deflated by the implicit GDP deflator, 1980–100.

Table 6.11. Share of real government expenditures, by sector, Niger (percentage)

	1980	1981	1982	1983	1984	1985	1986	1987	1988	1989
Recurrent expenditure										
Agriculture	5.8	5.3	5.2	5.7	5.6	3.0	3.0	2.8	3.2	4.1
Mining	1.4	1.4	1.1	0.9	0.7	0.2	0.2	0.2	0.2	0.2
Transport./commun.	5.1	5.3	4.6	4.6	4.7	3.6	2.0	2.0	2.1	1.1
Education	24.9	23.0	25.0	27.2	25.2	21.9	20.6	20.8	22.1	22.5
Health	6.9	6.7	6.3	7.2	6.7	6.3	6.6	6.6	6.7	6.8
Social services	1.2	1.1	1.0	1.2	1.1	1.0	1.1	0.9	1.0	1.0
Debt service	9.9	12.8	12.4	16.7	22.9	23.8	26.0	24.8	24.4	22.0
Defense	7.0	7.2	6.6	7.6	7.0	6.6	6.5	6.6	6.8	7.1
Other	37.8	37.3	37.8	28.9	26.0	33.4	34.1	35.3	33.6	35.1
Development expenditure										
Agriculture	13.3	13.7	16.1	17.6	29.2	28.3	33.5	34.3	39.2	31.8
Mining/industry/energy	19.6	13.9	12.0	23.8	12.2	7.6	4.0	2.7	2.2	5.4
Commerce	2.5	2.8	4.2	0.0	0.4	6.3	7.2	1.3	1.3	2.1
Transport	26.8	21.7	22.3	16.9	22.1	23.4	15.1	20.6	18.4	22.0
Telecommun.	4.9	14.1	4.8	1.9	3.4	3.7	3.4	2.9	0.7	1.1
Water supply	3.4	3.6	5.8	3.8	9.0	14.7	12.7	13.6	13.1	16.7
Education	14.2	11.9	16.3	3.6	4.9	5.7	5.5	5.0	2.9	6.3
Health	1.7	3.0	3.3	0.0	2.4	2.0	4.3	4.4	5.0	4.3
Housing	5.3	3.3	1.2	0.3	0.0	4.3	2.4	2.5	1.4	2.7
Other	8.4	12.2	13.9	32.1	16.3	4.1	11.9	12.8	15.9	7.6

Sources: IMF 1985, 1990f; World Bank 1991e.
Note: Niger's fiscal year ends in September; e.g., 1980 denotes 1979/80. Deflated by the implicit GDP deflator, 1980=100.

growing, mostly uneducated population. In spite of six years of stabilization and structural adjustment beginning in 1983, GDP per capita in 1989 was 10.4 percent below its level of 1980. Initial estimates for 1990 likewise were not encouraging, showing a further 1.6 percent fall in real GDP.

The initial need for stabilization and structural adjustment policies arose in 1982 because the government of Niger was unable to pay off international debts arising from its surge of development expenditures in the previous four years. Niger's uranium export revenues fell by only 18.6 percent in real CFA franc terms in 1982, but with tighter world capital markets additional foreign financing was unavailable.

Fiscal policy reforms have been at the heart of Niger's macroeconomic adjustment. Most of the reduction in government budget deficits was achieved by cutting back public investments financed by foreign borrowing. Current expenditures were much less affected and, apart from a few years, real expenditures on personnel were protected from cuts. Efforts at increasing revenues have met with little success.

The reforms undertaken appear to have improved efficiency in marketing and reduced some wasteful government expenditures. Government efforts at intervention in agricultural markets before the reforms had been largely ineffective in maintaining low consumer prices or guaranteed producer prices. Market liberalization thus resulted in substantial government savings without a major effect on farmgate or consumer prices. Removal of price controls on manufactured and consumer goods also coincided with reforms of parastatals and closing or state divestiture of several firms, again reducing government expenditures.

Part of Niger's disappointing economic performance in the 1980s can be traced to the appreciation of the CFA franc relative to the Nigerian naira after 1981. Membership in the CFA zone prevented a nominal exchange rate devaluation relative to the naira, which would have boosted incomes of producers of exportable goods (livestock and cowpeas) and helped Nigerien industries compete with imported goods from Nigeria. Trade policies, which could in theory substitute for nominal exchange rate policies to influence the real exchange rate, were of very limited usefulness since most trade with Nigeria took place outside legal channels. With no means of inducing expenditure switching, deeper expenditure cuts were necessary.

It is doubtful, however, that exit from the CFA zone would be beneficial for Niger. Nominal exchange rate devaluation in itself is insufficient to bring about a change in the real exchange rate. Without restrictive fiscal policies or trade policy changes, a nominal exchange rate devaluation would lead to inflation, preventing a depreciation of the real exchange rate. Niger has benefited from relatively low levels of inflation thanks to the monetary and financial discipline imposed as rules of membership in the CFA zone. More important, a withdrawal from the CFA zone would likely hurt investor confidence and lead to greater capital flight from Niger because of greater inflation and exchange rate uncertainty. On balance, though membership in the CFA zone makes adjustment to external shocks more difficult, the long-term growth prospects are likely to be greater with continued membership.

For long-term growth, both the quantity and efficiency of investment in Niger must increase. A major policy concern remains funding for government development expenditures. Uranium export revenues fell slightly in 1989, and Niger is already heavily dependent on foreign aid, which equaled 31.9 percent of total government revenues in 1989 and 5.0 percent of GDP. The challenge is to institute an efficient way of taxing the informal sector without damaging producer incentives. Equally important is the debate on expenditure priorities. Nearly a quarter of total government spending is devoted to current account expenditures on personnel; investment expenditures accounted for only 42.7 percent of total government expenditures in 1989. The choice between investments largely targeted toward an urban elite (expenditures on universities or on curative health care) or the rural population (primary education and preventive health care) will help determine the distribution of benefits from development. Finally, though government investments currently play the largest role in development efforts in Niger, maintaining an environment favorable to private investment will ultimately be crucial for long-term sustained growth.

Structural adjustment in Niger cannot be said to have failed, but significant adjustments have taken place almost exclusively within the small formal sector of the economy. Much of the economy remains outside the scope of effective government influence and is still extremely vulnerable to weather and external shocks (in particular, economic conditions in Nigeria). The reforms undertaken to date are only a modest start at addressing the serious constraints on long-term development in Niger; a most difficult task remains ahead.

7

Adjustment without Structural Change: The Case of Malawi

David E. Sahn and Jehan Arulpragasam

In 1970 Malawi was among the poorest nations in sub-Saharan Africa. Its GNP was only US$60 per capita, in contrast with the region's average value of $150.[1] Its economy based on agriculture, Malawi's export crops of tobacco and tea came primarily from estates largely owned or operated by expatriates. This estate subsector existed, however, side-by-side with a largely subsistence smallholder subsector, the source of income and sustenance for the vast majority of Malawians.

Malawi's strong economic performance in the years after independence initially led to considerable optimism about the country's future. During the 1960s and 1970s, GDP growth was 4.9 and 6.3 percent, respectively. Exports valued at current prices grew at a rate of 15 percent during the 1960s and 22.5 percent between 1973 and 1977. In contrast, the average growth rate for sub-Saharan Africa was −1.9 percent. Malawi's impressive performance was largely fueled by growth in agriculture. Malawi, unlike many of its neighbors, did not eschew its most abundant natural resource, land, in lieu of "modernization" through an industrial-led, import-substitution development strategy. Rather, Malawi's post-independence period was characterized by exploitation of its comparative advantage in agriculture. This strategy in turn helped to mobilize both domestic and foreign resources (Acharya and Johnston 1978) and contributed to the perception that Malawi was a country to emulate in the region.

[1] A more poignant contrast would be with countries served relatively well by the colonial period, such as Côte d'Ivoire, Ghana, or Senegal, whose GDPs per capita were US$270, $250, and $220, respectively, at independence.

Malawi's policymakers can be credited for following a relatively liberalized trade and exchange rate regime. Import duties were kept low, quantitative restrictions were limited, licensing for imports was generally not practiced, and the plethora of controls that existed in many countries in the region were kept to a minimum. This allowed Malawi to take advantage of an external environment that witnessed favorable prices for its agricultural exports. The demand for tobacco was exceptionally buoyant in the 1970s because of the economic embargo imposed on the former Rhodesia, a major tobacco producer and competitor. Consequently, estate owners fared well.[2] A combination of policy and exogenous events contributed to the GDP growth rate exceeding consumption in the 1970s. By 1979 private and public consumption declined to approximately 80 percent of GDP. Savings and investment were the beneficiaries. Fiscal policy restraint was displayed in post-independence Malawi, at least relative to other countries in sub-Saharan Africa.

Two periods can be distinguished with respect to fiscal policy from independence to the onset of Malawi's economic crisis of the 1980s. In the 1960s the annual rate of government expenditure increased 10–15 percent, mainly because of the expansion of civil service. In that period, 70–80 percent of the government budget was for recurrent expenditures. By the early 1970s, however, a push toward public sector–driven investment was under way. Infrastructure development drove the share of the capital budget to nearly 40 percent by the end of the 1970s.

At the same time, public enterprises grew during the 1970s. Most important of these were the agricultural marketing parastatal (ADMARC), the Malawi Development Corporation, and Press Holdings, Ltd., a private enterprise with close state affiliation and subsidiaries in virtually all sectors of the economy. In their early days, the parastatals and Press were in fact considered growth poles, extracting some of the surplus from what was a thriving economy and becoming key shareholders in what was supposedly Malawi's unfettered private sector. In addition, the parastatals and Press assumed the prominent role as the conduit through which public investment was channeled. They were straining the economy through their heavy borrowing and almost unrestrained access to foreign and domestic financial resources. The inefficiencies and inappropriate activities of these institutions were, however, conveniently overlooked. In fact, many parastatals—in particular ADMARC—were being sanctioned by the same international organizations that were to work so hard for their elimination in the years to follow.

---

[2] Estate owners who had remained from the colonial period or had moved to Malawi from war-torn Rhodesia were favored. Their gains were, however, at the expense of the smallholder sector (see below).

Impending Crisis and Its Economic Manifestations

The fragility of the foundations of Malawi's early growth, including its strong export performance and outward-looking strategy, was revealed in the late 1970s as exogenous shocks began to buffet the economy. Of equal importance, the thin veneer of growth, which in fact was accompanied by a marginalization of the peasantry that made up nearly 75 percent of the population, was also eroding as the economy began to turn downward.

Malawi's export performance became increasingly subject to the vagaries of tobacco production and demand as tobacco's share of exports grew from 32 percent in the years immediately after independence to 55 percent in the period 1977–79. Likewise, by 1977, 50 percent of Malawi's imports consisted of auxiliary and intermediate goods, imports thus becoming key to the economy's vitality. At the same time, the burgeoning investment budget was reliant on foreign financing, with interest rates on the rise from the commercial lenders to whom Malawi increasingly turned. It was also becoming clear that a large share of Malawi's investments being financed from abroad were for projects with low rates of return, such as an international airport, military aircraft, palaces, and colleges (Roe and Johnston 1988). Finally, little attention was paid to the development of human capital, further contributing to Malawi's economic vulnerability. This was manifested by low levels of investment in social service infrastructure. In particular, less than 13 percent of the development budget was allocated to all social services in 1977/78. The share of the adult population that was illiterate was 75 percent, and the infant mortality rate was 169. In contrast, illiteracy and infant mortality rates among other countries in sub-Saharan Africa during the late 1970s were 72 and 122, respectively (World Bank 1981b, 1989g).

These factors, which were in part policy induced, combined to prevent Malawi from responding to the shocks it was to confront. Beginning in 1977, the terms of trade, which had been improving at a rate of 0.7 percent annually thus far in the decade, fell at 15.5 percent per year for the remainder of the decade. The result was that by 1980 the terms of trade was less than 0.56 percent of the 1970 level (Table 7.1). This fall was partially due to the fall in the world market price of tobacco, but it also resulted from the second oil shock, which increased petroleum's value share of imports to a record 10 percent. Exacerbating the problem was the influx of refugees from war-torn Mozambique and the cutting off of Malawi's traditional transport corridor. The incredible increase in the transport margin, which rose by 35 percent between 1980 and 1983, contributed to the growing deficit on the nonfactor services account, which was over 10 percent of GDP by the end of the 1970s, twice the level observed a decade earlier.

Table 7.1. Terms of trade and quantum indices of exports and imports, Malawi

	Terms of trade					Quantum indices	
	Tobacco	Tea	Sugar	Groundnuts	Aggregate	Exports	Imports
1967	1.60	2.18	—	0.73	1.53	—	—
1968	1.51	2.13	—	0.91	1.59	—	—
1969	1.83	2.03	0.95	0.92	1.64	—	—
1970	2.20	2.43	1.01	1.14	1.77	—	—
1971	1.87	0.93	1.14	1.00	2.29	36.30	51.10
1972	2.12	2.16	0.80	1.08	1.65	53.00	87.80
1973	1.97	1.85	1.39	1.01	1.59	56.30	83.10
1974	1.86	1.64	3.11	0.87	1.49	55.80	85.10
1975	2.00	1.63	1.81	0.72	1.44	57.50	96.70
1976	1.97	1.48	2.24	1.08	1.40	60.90	70.70
1977	2.11	2.04	0.95	1.30	1.66	68.50	82.40
1978	1.93	1.38	0.93	1.52	1.49	65.40	108.30
1979	1.36	1.26	0.91	1.26	1.18	83.70	109.30
1980	1.00	1.00	1.00	1.00	1.00	100.00	100.00
1981	1.33	0.91	1.07	1.34	1.22	79.30	84.20
1982	1.63	1.03	0.66	0.83	1.23	82.40	80.10
1983	1.42	1.18	0.58	0.81	1.13	109.20	80.40
1984	1.24	1.94	0.60	0.85	1.18	79.40	64.10
1985	1.04	1.47	0.75	0.61	1.01	94.80	81.30
1986	1.11	0.80	0.50	0.60	0.88	98.93	78.99
1987	1.10	0.63	0.63	0.38	0.85	104.21	66.41
1988	—	—	—	—	0.82	103.61	77.66
1989	—	—	—	—	0.77	93.75	84.21
1990	—	—	—	—	0.71	107.01	84.78

Sources: Reserve Bank of Malawi 1987, 1988; Malawi Government, various years, 1990.

Adverse movements in the factor services account during this period also exacerbated the balance-of-payment situation, primarily because of increasing interest burdens on the accumulated foreign debt coupled with the repatriation of wages and profits earned by foreign capital and labor employed in Malawi (Reserve Bank of Malawi 1987, 1988). Finally, the drought of 1980/81 severely reduced agricultural production, affecting the sector that comprised close to 40 percent of the country's GDP.[3]

The performance of the economy worsened dramatically in 1980, led by a 6.5 percent drop in agricultural GDP (Figure 7.1). Conditions deteriorated, with a 8.16 percent decline in agricultural GDP in 1981, which contributed to an overall −5.25 percent GDP growth rate. After a growth rate in GDP per capita that was virtually unparalleled in Africa from

---

[3] At independence agriculture comprised around half the country's GDP, and industry only 14 percent. By 1978 agriculture's share fell to less than 40 percent, industry's increased to 18 percent, and services' rose from around 37 to 45 percent. These numbers suggest that Malawi is not underindustrialized, and that the services' share is actually quite typical for a relatively industrialized country.

Figure 7.1. GDP by services, industry, and agriculture sectors, Malawi

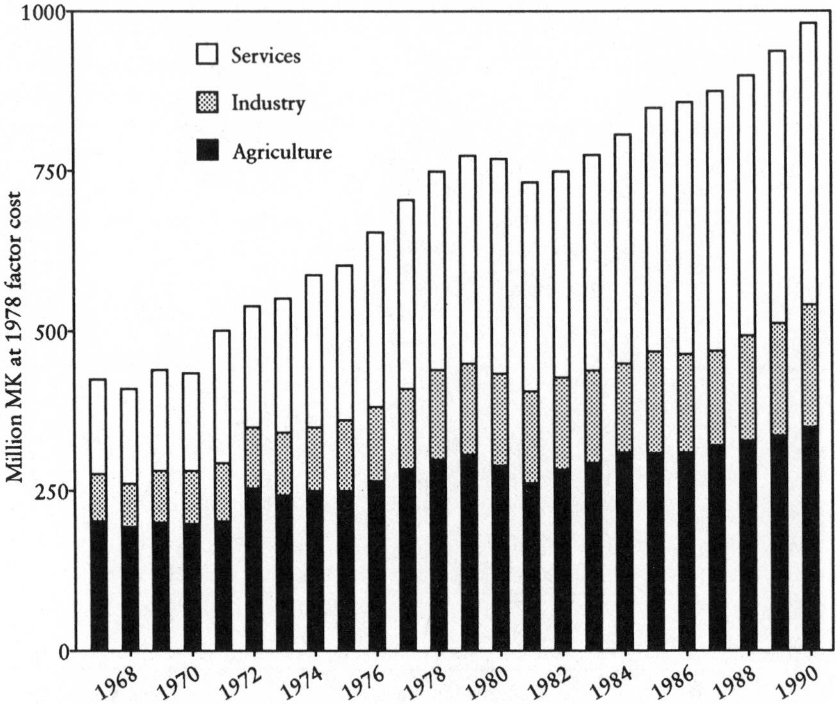

Sources: Reserve Bank of Malawi 1987, 1988; Malawi Government Economic Reports.

independence until 1979, the fall of around 10 percent in two years represented lost ground that had yet to be recovered a decade later.

One obvious manifestation of the inherent weakness of the government's
management of the economy was that Press Holdings, ADMARC, and
other public enterprises, which enjoyed the fruits of the surpluses generated
in a growing economy, were unable to respond to the developing adver-
sities. This is partially explained by the severity of the shocks. The large
deficits experienced, especially by ADMARC, were also due to institutional rigidities, coupled with poor management.

Adverse external factors and domestic policy contributed not only to a
decline in GDP but also to a budget deficit that in 1981 reached over 10
percent of GDP, the largest ever recorded. This deficit was due not only to
recurrent expenditures on personnel, maintenance, and subsidies to parastatals but also to the increasing debt burden. Debt had grown to nearly

two-thirds the size of GNP. Furthermore, problems in the external accounts were worsening. Export revenue was falling despite the fact that import's share of nominal GDP had increased from 32 percent in 1967 to 40.8 percent in 1979. Thus the current account deficit worsened by 100 percent between 1977 and 1978 and by another 45 percent in the following year. By 1980, as GDP registered a negative growth rate, the current account deficit was one-fifth of GDP.

The need for action was obvious. How to cope with a declining rate of growth brought on by policy and external conditions thus became the major issue the government and donors alike would face during the 1980s.

Policy Reform Measures

Given its deteriorating economic health, Malawi began an adjustment program late in 1979, commencing with a standby facility loan from the IMF. The components of the program were typical: diversification of revenue sources, reduction in government spending, reduction in new credit, diversification of credit to the private sector, and rationalization of interest rates. This initial attempt at stabilizing the economy was not a great success, partially because of the transport shock resulting from the closure of routes through Mozambique which hit shortly after the accord with the IMF. Credit ceilings were exceeded and other targets were not achieved, a situation aggravated by the need to borrow to import food during the drought emergency.

A second standby loan was thus negotiated, this time to last through mid-1982. Once again, however, efforts at stabilizing the economy did not meet with great success. The continued import of food, the lack of success in generating new revenues, the high level of interest payments, and the insolvency of parastatals which had commercial banks teetering on the verge of failure all contributed to a continued crisis in the balance-of-payments and fiscal deficits.

Despite the failure of Malawi's early stabilization policies, the government was still successful in reaching accord with the World Bank, which provided the first tranche of its first, US$45 million, structural adjustment loan (SAL I). Interestingly, many of the objectives of the loan were similar to the suspended IMF standbys, including restoring equilibrium in the internal and external account balances. Nonetheless, the structural adjustment loan further emphasized restoring growth through resource management, institution building, and price policy and marketing arrangements.

To improve balance of payments, increased exports of smallholder crops were targeted. This was to be achieved primarily through price-oriented adjustment, as prices on export crops were to be raised. Measures were also to be taken to make ADMARC, the parastatal charged with agricultural marketing, more efficient. No mention was made of the estate sector, presumably because it had generated most of the economic growth during the previous decade. The enforced duality between estate leaseholds and customary land under cultivation was not viewed as a problem.

Just as the pricing of smallholder crops was a priority for Malawi's adjustment program, the prices of numerous commodities and of wages were also to be reviewed with a view to improving flexibility. Public utility rates were to be raised to address the growing dependence on imported fuels. In conjunction with IMF standby requirements, the exchange rate and interest rates were to be frequently reviewed. With respect to resource management, external borrowing by domestic banks and the government was to be monitored. The government investment program called for revised recurrent and development expenditure targets that increased commitment to the agricultural and other key economic sectors such as education, health, and housing.

Several other IMF facilities supplementing SAL I were arranged in 1982 and 1983. Nevertheless, initial implementation of SAL I was slow in most areas. In the face of the problems encountered, the signing of the SAL II of US$55 million was delayed until January 1984. Revising the targeted growth rate to a more realistic 3.4 percent over the next five years, the second phase of the adjustment program was intended to continue the reforms initiated in the first phase.

SAL II addressed two important issued not addressed in SAL I. The first was fertilizers. SAL II committed the government to a phased removal of its entire fertilizer subsidy by 1985/86, with a 50 percent reduction in 1983. Moreover, the government committed to procure and distribute fertilizer to smallholders and to contribute to the establishment of a fertilizer revolving fund. A second significant aspect of phase two of the adjustment program was the government's commitment to implement important measures to improve the operation of ADMARC. These included reducing ADMARC's marketing costs by cutting the number of markets in which it operated. limiting ADMARC's investments to those related to marketing and processing, and increasing the role of the private sector in this regard with a view to improving crop marketing and distribution.

The government's adherence to SAL II guidelines, coupled with GDP growth rates of 3.6 percent in 1983 and 4.5 percent in 1984, helped expedite the approval of SAL III in November 1985. New commitments were made to complete the agricultural price liberalization program so as to serve the goals of food self-sufficiency, export promotion, and crop diversification. The fertilizer subsidy, which was maintained to compensate for the longer transportation routes, was to be eliminated by 1989/90. For the first time, the role of the estate sector was part of the conditionality: it was to be supported with a pilot scheme to provide medium- and long-term credit, with an extension and management training program. While the active exchange rate policy was continued, measures were to be taken to complete the development of an export promotion policy and to set up an export financing facility. The third adjustment load also committed the government to adopt strategies for restructuring the tax system and improving public sector management.

In 1986, however, the economy once again showed signs of weakness, and progress faltered. Under these circumstances, after the termination of the extended fund facility in September 1986, the Malawi government and IMF could not reach any agreement on the extension of further facilities to Malawi for 1986 or 1987. The World Bank, however, extended a SAL III supplemental credit to Malawi in January 1987. Several rectifying measures were subsequently taken by the government, in consultation with the Bank and the IMF. A new fiscal program restraining expenditure and generating revenues was also designed.

By mid-1988 the government, once again working with the World Bank and IMF, adopted a shadow stabilization program designed to reduce the fiscal deficit. The World Bank, meanwhile, had expressed an intention to "move away from broad-based SALs to a series of policy-based sectoral operations, designed to address remaining structural constraints in the key productive sectors" (World Bank 1988e). In making this move, the Bank approved an industrial and trade policy adjustment credit for US$70 million in 1988. This program focuses on the liberalization of the foreign exchange allocation system, the promotion of appropriate exchange rate policy, and the establishment of a duty-drawback system and an export revolving fund to benefit exporters. The removal of legal provisions that inhibit the entry of new firms in industry was also committed to under this agreement.

Finally, an agriculture sector adjustment credit approved in 1990 includes agreements to legalize the production of burley tobacco on a limited basis among smallholders and to discourage the transfer of land from the smallholders to the estate sector. The proposals also include measures that would raise rents on leasehold land, partially privatize the distribution of fertilizer, and partially liberalize official maize prices to reflect transportation prices to and from ADMARC's twenty-two main depots. The agricultural credit also stresses research leading to the development of a maize variety with acceptable storage and processing qualities.

In sum, the reform program undertaken by Malawi has dictated national economic policy for an entire decade. Combining World Bank structural

adjustment measures with IMF stabilization prescriptions, the program
was structured to achieve multiple ends. First, it sought to attain both
internal and external balance. Second, by manipulating demand and restructuring supply, policy reform also targeted growth.

Reform and Macroeconomic Performance

Malawi's first decade under adjustment was characterized by a fluctuation in most indicators of performance. GDP growth, inflation, the current
account and budgetary deficits were generally unstable and lacked any
secular trends that suggest dramatic improvements. GDP grew at only 2.4
percent between 1980 and 1987. For the period 1988–90, however, there
is some tentative evidence that GDP growth improved, averaging 4.2 percent (Figure 7.1). This growth is largely attributable to estate agriculture,
although manufacturing also showed some strength. The smallholder sector performance continued to decline due to a combination of structural
problems and poor weather. Thus, the majority of the rural population
failed to benefit from the recently improved GDP performance.

Efforts to get inflation under control have been unsuccessful. More specifically, in contrast to the average rate of inflation of 9.9 percent between
1976 and 1988, between 1983 and 1988 inflation averaged 17.4 percent,
with the latter year seeing inflation surge to over 30 percent. This reflected
a combination of the devaluations of the kwacha, relaxation of some government price controls, monetary expansion, and increased demand for
food due to the influx of refugees. Since then, inflation has moderated,
falling to 15.7 percent in 1989 and to 11.7 percent in 1990. The figures
over 1990 show inflation rising from 9.2 percent in the first quarter to 14.7
percent in the last quarter.

In the early years of adjustment, the deficit on the current account
reached a level in excess of 10 percent of GDP in 1982 and 1983 (Table
7.2), reflecting a large negative trade balance resulting primarily from the
growing costs of transportation.[4] Also contributing to the deteriorating
current account balance was the net outflow of factor and nonfactor services. Interest payments on Malawi's debt accumulated in large amounts at
commercial terms during the period 1978–81 and approached one-fifth of
total export revenue in 1983.

The external account experienced a significant improvement in 1984.
Several factors accounted for the turnaround. First, Malawi saw a marginal

---

[4] Landlocked Malawi's gradual loss of access to the coast, with the closure of rail lines,
meant that the average distance to seaports had risen from 800 to 3,500 kilometers since the
1970s. This meant that the c.i.f. margin had escalated to 35 percent of total import costs.

improvement in its terms of trade despite the long-term secular deterioration in this measure which has been observed both before and after adjustment (Table 7.1).[5] Second, even though the c.i.f. margin continued to rise, reaching 40 percent of the total cost of imports, total imports c.i.f. actually fell. Third, private transfers including remittances from Malawian workers overseas continued to augment the current account. Fourth, with the signing of SAL II, Malawi enjoyed a one-shot capital inflow of SDR 52 million. Fifth, for the third year in a row Malawi rescheduled debt, garnering relief amounting to SDR 23 million. As a result of these factors, in 1984 Malawi had a surplus on its overall balance-of-payments account, enabling the country to restock its gross official reserves.

The gains did not persist, however. In 1985 external accounts began another slide. Even the 15 percent devaluation of the kwacha did not prevent a sharp downturn in the terms of trade.[6] The balance-of-payments situation worsened in 1986. Import compression continued to characterize the merchandise trade balance and contributed to the general economic stall. Non-maize imports dropped 22 percent from the previous year. Interest payments continued to dominate the services account and, at a value of SDR 44.4 million, swamped the continued inflow of private transfers, which amounted to SDR 21.6 million. The salient factor in 1986, though, was the negative balance on the capital account. The deterioration came despite an SDR 63.9 million inflow of SAL-related funds and has been attributed to large debt-servicing payments and to short-term, unidentified capital outflows.[7]

As a result, in 1986 the overall balance reached SDR −67 million, or −6.4 percent of GDP. Again, reserves were almost completely depleted to finance this deficit. With foreign exchange reserves amounting to less than one month's worth of imports, Malawian authorities imposed quantitative restrictions on the allocation of foreign exchange. Still unable to bridge the gap, the bank was forced to accumulate import-related arrears.

The balance-of-payment situation improved marginally in 1987. The continued improvement in the overall balance of payments in 1988 was due largely to new financial arrangements extended to Malawi by the nation's primary creditors. Rescheduling agreements with both Paris and London

---

[5] Led by a 64 percent increase in the export price index of tea and a 3 percent increase in the export price index of sugar, the aggregate terms of trade appreciated by 4.4 percent between 1983 and 1984. The 34 percent increase in export value was also the result of the movement of tobacco and sugar stocks out of the country.

[6] International tea prices plummeted. The export price index for tea fell by close to 60 percent in the next two years, that of tobacco dropped by 10 percent, and the aggregate terms of trade declined by 25 percent.

[7] This has been said to include a large expenditure on security-related imports (World Bank 1988b).

Table 7.2. Balance of payments, Malawi (million SDR)

	1982	1983	1984	1985	1986	1987	1988	1989	1990
Current balance	-115.4	-124.9	-16.2	-95.3	-63.8	-46.0	-86.7	-133.8	-119.3
Trade balance	-59.8	-58.7	44.7	-36.3	-7.3	-13.5	-85.3	-159.4	-126.0
Exports, f.o.b.	217.1	230.3	308.0	246.2	211.6	215.4	218.4	221.0	287.8
Imports, c.i.f.	-276.9	-289.9	-263.3	-282.5	-218.9	-228.9	-303.7	-380.4	-413.8
Non-maize imports, f.o.b.	-179.9	-188.0	-158.0	-169.4	-131.4	-134.0	-167.3	-191.6	-215.0
c.i.f. margin	-97.1	-100.9	-105.3	-113.1	-87.6	-89.3	-102.5	-117.4	-127.4
Maize imports, c.i.f.	—	—	—	—	—	-5.6	—	—	—
Services and private transfers	-55.6	-66.2	-60.9	-59.0	-56.6	-36.2	-35.3	-45.8	-64.6
Nonfactor services	-10.2	-14.0	-21.7	-17.2	-18.4	-7.4	-9.3	-10.0	-13.3
Receipts	21.2	22.0	26.4	26.0	19.0	17.5	—	—	—
Payments	-31.4	-35.9	-48.1	-43.2	-37.2	-24.9	—	—	—
Factor services	-63.6	-60.0	-54.4	-52.1	-50.4	-44.0	-40.1	-41.3	-39.2
Receipts	1.4	1.3	3.6	5.4	2.8	2.5	—	—	—
Payments	-65.0	-61.2	-58.0	-57.5	-53.2	-46.5	—	—	—
Interest	-38.5	-41.4	-43.3	-44.1	-44.4	-39.0	-38.7	-37.1	-35.6
Other	-10.9	-19.8	-14.6	-13.4	-8.8	-7.5	—	—	—
Private transfers	18.2	7.7	15.3	10.3	12.2	15.2	14.1	5.5	-12.1
Receipts	28.7	22.5	26.5	21.2	21.6	26.6	—	—	—
Payments	-10.6	-14.7	-11.3	-10.9	-9.3	-11.3	—	—	—
Capital account	67.1	42.5	58.4	49.6	-3.2	90.0	149.1	100.7	135.5
Long-term net	37.8	34.6	59.7	42.0	60.1	76.6	117.6	100.0	161.9
Government transfers	32.9	27.6	33.8	24.1	24.9	67.4	68.2	42.9	66.1
Credit	33.8	28.6	25.7	26.0	26.5	25.5	44.3	13.0	32.9
SAL-related grants	—	—	—	—	9.3	6.4	—	—	—
Grants for maize	—	—	—	—	—	3.7	—	—	—
Debit	-0.9	-1.0	-1.9	-1.8	-1.6	-2.2	—	—	—
Government loans[a]	20.3	22.0	51.1	27.4	40.5	49.0	47.2	45.0	93.7
Credit	61.5	60.8	90.5	62.0	89.6	86.2	—	—	—
SAL-related loans	18.1	—	52.0	7.7	63.9	40.8	30.4	38.7	52.1
Debit	-41.2	-38.8	-39.4	-34.6	-49.1	-37.2	—	—	—

Public enterprises	−18.4	−28.4	−12.9	−15.2	−9.3	3.6	—	—	—
Credit	6.9	0.7	0.7	0.7	0.2	9.0	—	—	—
Private sector	3.0	13.5	−2.3	5.6	3.9	0.7	2.1	12.2	2.0
Credit	11.0	19.5	6.7	9.9	11.6	4.8	—	—	—
Debit	−8.0	−6.1	−9.0	−4.2	−7.6	−4.1	—	—	—
Short-term unidentified	29.3	7.9	−1.3	7.6	−63.2	13.4	31.6	0.7	−26.4
Overall balance	−48.3	−82.3	42.2	−45.7	−67.0	44.0	62.4	−33.1	16.2
Financing	48.3	82.3	−42.2	45.7	−67.0	44.0	62.4	−33.1	16.2
Official net foreign assets (increase −)	31.4	27.0	−65.2	39.0	20.8	−34.8	−74.9	19.2	−16.2
Gross official reserves (increase −)	25.4	7.7	−45.2	18.2	24.1	−18.7	—	—	—
IMF purchases	14.7	34.2	37.8	23.0	—	—	—	—	—
IMF repurchases	−12.6	−10.3	−20.4	−16.0	−20.6	−23.6	—	—	—
Change to other liabilities, net	3.9	−4.6	−37.4	13.8	17.3	7.5	—	—	—
Change in arrears	—	—	—	—	43.8	−9.7	−34.1	—	—
Import related	—	—	—	—	43.8	−26.4	−17.4	—	—
Debt service-related	—	—	—	—	—	16.7[b]	−16.7	—	—
Debt relief	16.9	55.3	23.0	6.7	2.4	0.4	46.7	13.9	—
Memorandum items									
Current account (% of GDP)									
Exc. official transfers and emergency maize imports	−10.6	−10.9	−1.4	−8.2	−6.1	−4.8	−8.4	−10.8	−8.8
Including official transfers	−7.7	−8.5	0.6	−6.1	−4.6	−2.4	−1.8	−7.4	−3.9
Overall balance (% of GDP)	−4.5	−7.2	3.6	−3.9	−6.4	4.6	—	—	—
Gross official reserves	20.5	12.8	58.0	39.8	15.7	34.4	107.4	76.0	96.6
End-period stock									
In weeks of c.i.f. non-maize imports	3.9	2.3	11.5	7.3	3.7	7.2	19.2	12.0	13.6
C.i.f. margin (in % of c.i.f. imports)	35.0	35.0	40.0	40.0	40.0	40.0	38.0	38.0	37.2

Source: Data from IMF 1991b.

a Includes net loans to public enterprises for 1988–90.

b Estimated debt service payments to Paris Club creditors and principal payments to London Club creditors suspended since end of August 1987 pending Malawi's request for debt rescheduling.

club creditors explain the large element of debt relief that also helped improve the balance-of-payments situation. These factors, however, hide the continued decline in the trade balance, estimated to have deteriorated by over 500 percent. The 1988 devaluation, together with the liberalization of foreign exchange allocation that year, led to a 34 percent increase in imports that was unmatched by exports. In 1989 the current account deficit worsened further, reflecting the import liberalization program. The situation in 1990 improved slightly, as the current account deficit as a share of GDP fell to 8.8 percent from the previous year's peak of 10.8 percent. This improvement in 1990 was primarily due to the unexpected, unusually high exports of estate crops.

Just as external developments saw no improvement over the decade, despite considerable fluctuations, the budget deficit likewise showed considerable instability. During the four years before adjustment (1976/77–80/81), the budget deficit excluding grants rose from 7.3 to 15.5 percent of GDP. This was attributable to an increase in both development and recurrent spending, in excess of increased revenues. The overall deficit then commenced to decline to a low of 5.8 percent projected for 1990/91. This fall in the deficit during the adjustment period is attributable to a return to more sustainable levels of development expenditures. Recurrent expenditures, however, have not shown any sign of falling in the 1980s after their increase in the years before adjustment.

As important as the overall deficit figures is the sectoral allocation of expenditures. The share of recurrent expenditures, which averaged 23.2 percent of the total between 1977 and 1979, declined to below 20 percent between 1982 and 1986. In the years 1987–91, the average once again climbed to 22.1 percent, despite considerable year-to-year fluctuation. The share of development expenditures allocated to the social sector increased throughout the adjustment period, once again despite considerable inter-year instability. This applies to all categories of social spending, although the greatest increase was in education, where spending in 1989–91 comprised 13.2 percent of the total, as opposed to 1977–79, when education's share of development expenditures was only 4.7 percent. Recurrent expenditures on social services increased markedly between 1977 and 1981 in real terms. Thereafter, no secular trend is observed during the years under adjustment, despite considerable year-to-year instability.

Reform and Agricultural Sector Performance

The emphasis of policy reform in Malawi during the early 1980s has been on restoring growth in agriculture. This was to be achieved through

Table 7.3. Estate and smallholder production as share of real agricultural GDP, Malawi

	Real agricultural GDP (million MK)	Estate (% of total)	Smallholder (% of total)
1978	294.90	17	83
1979	304.10	17	83
1980	284.20	19	81
1981	261.00	19	81
1982	277.60	22	78
1983	289.90	23	77
1984	306.50	21	79
1985	307.40	21	79
1986	308.00	21	79
1987	312.50	22	78
1988	318.70	24	76
1989	329.70	25	75
1990	346.50	23	77

Sources: Reserve Bank of Malawi 1987, 1988; Malawi Government 1990.

altering the structure of incentives and improving the efficiency of markets. Changing relative prices and removing obstacles to the proper functioning of markets through reducing state interference were the pillars of Malawi's reform program.

In Malawi, agriculture is dualistic in structure. On the one hand is the subsistence-oriented smallholder sector, operating in land administered under customary law, which accounted for close to 80 percent of Malawi's agricultural production in 1980 (Table 7.3).[8] Smallholder production is concentrated on maize, cassava, and other subsistence crops, as well as cash crops such as cotton, groundnuts, and oriental, sun-, and air-cured tobacco. Crop choice, however, is not based solely on the farmer's calculus. Indeed, the law has forbidden the smallholder sector from growing certain export-oriented cash crops: burley and flue-cured tobacco, tea, and sugar.

A large portion of smallholder production is consumed domestically, but surplus production is subject to marketing and pricing regulations. Smallholders have traditionally had little option but to sell their produce to the state-owned marketing agency, ADMARC, which is also their primary source of inputs such as fertilizer. Thus the setting of producer and input prices is a powerful policy tool for the government and an important determinate of cropping patterns and production, as well as of marketed levels and of household food security.

The estate sector differs from the smallholder sector on every count.

[8] The share of estate sector production in agricultural GDP increased from about 17 percent in 1978 to close to 23 percent in 1983. After a slight retreat in subsequent years, it again rose to over 23 percent in 1988 (Table 7.3).

Having historically targeted export crop production as the vehicle for growth and the estate sector as the pole on which to hinge such growth, the government has attempted to institute favorable production, pricing, marketing, and land tenure policies accordingly. Given these initial rules of the game, how the two subsectors have interfaced is particularly relevant. They compete for the same resources and are linked through the input, land, and labor markets.

Structural adjustment policies have thus been applied to an agricultural sector with strongly defined institutional cleavages. In particular, as can be gleaned from the overview of reform programs discussed in the section above, adjustment has concentrated on pricing policy in the smallholder sector, virtually ignoring the enforced dualism in general, and the estate sector specifically.

Three major agricultural sector reforms have been prominent in the smallholder sector. First, producer prices of agricultural commodities were to have been increased. Second, the adjustment program has aimed at removing the subsidy on fertilizers for smallholders. Third, adjustment has meant the privatization of the important grain marketing function traditionally carried out by ADMARC.

Product Price Reform

Price reform in Malawi has had two objectives: increasing agricultural producer prices in general, and raising the relative prices of export commodities in particular. Both aim at raising incomes in the agricultural sector, specifically among smallholders. Both objectives are also characteristically based on the elimination of the high levels of taxation on smallholder production which were prevalent in preadjustment pricing (Christiansen and Southworth 1988). Movements of real prices and the level of taxation on agriculture reveal that, although the sharp nominal price increases for maize, tobacco, and groundnuts resulted in rising real prices in the early 1980s, they have not kept pace with inflation since 1982 (see Table 7.4).[9] As a result, through 1987 all three crops experienced declining real producer prices. Although between 1988 and 1990 prices for maize and tobacco showed signs of some improvement, they still remained at levels lower than those of the 1970s and the early 1980s. Meanwhile, groundnuts, rice, and haricot beans continued to experience falling prices.

The decline in real prices for the main agricultural commodities raises some questions concerning the success of price-oriented reform. Declining CPI-deflated prices of export commodities are contrary to the expectation

[9] The real price of rice, to the contrary, fell between 1980 and 1983 and then rose slightly between 1983 and 1985, although never regaining its 1980 value.

Table 7.4. Nominal and real producer prices, Malawi

	Nominal prices (current tambala/kg)					Real price (1980 tambala/kg)				
	Maize	Rice	Tobacco	Groundnuts	Haricot beans	Maize	Rice	Tobacco	Groundnuts	Haricot beans
1975	3.90	10.00	23.76	18.70	—	6.09	15.63	37.13	29.22	—
1976	5.00	10.00	27.53	19.80	—	7.46	14.93	41.09	29.55	—
1977	5.00	10.00	30.96	20.00	—	7.14	14.29	41.09	29.55	—
1978	5.00	10.00	40.74	22.00	—	6.58	13.16	53.61	28.95	—
1979	5.00	10.00	40.23	33.00	—	5.88	11.76	47.33	38.82	—
1980	6.60	10.00	42.12	33.00	—	6.60	10.00	42.12	33.00	—
1981	6.60	10.00	42.61	33.84	13.94	5.89	8.93	38.04	30.21	12.45
1982	11.00	10.00	45.07	51.85	14.50	8.94	8.13	36.64	42.15	11.79
1983	11.00	11.50	75.87	59.46	30.00	7.91	8.27	54.58	42.78	21.58
1984	12.20	15.00	74.62	69.28	40.00	7.31	8.98	44.68	41.49	23.95
1985	12.20	17.00	89.38	73.76	42.00	6.59	9.19	48.31	39.87	22.70
1986	12.20	19.00	84.56	73.76	44.00	5.78	9.00	40.08	34.96	20.85
1987	12.20	22.00	90.62	73.76	44.00	4.62	8.33	34.33	27.94	16.67
1988	16.70	27.00	103.56	75.00	48.00	5.22	8.44	32.36	23.44	15.00
1989	24.00	30.00	142.00	85.00	48.00	6.03	7.53	35.68	21.34	12.05
1990	26.00	35.00	180.00	95.00	60.00	5.91	7.95	40.91	21.59	13.64
1991	27.00	37.00	—	100.0	70.00	—	—	—	—	—

Sources: Christiansen and Southworth 1988; Malawi Government 1987a and Economic Reports; Harrigan 1988; World Bank 1986e.

of rising prices of tradables relative to nontradables. Moreover, if falling real producer prices signify falling real incomes from agricultural production, then the smallholder producers have likely not gained from planned price reform.

The discussion of producer prices leads to the more complex issue of the extent and impact of policy reforms on taxation of agricultural products in Malawi. A major motivation for price reform was to reduce the high levels of implicit taxation on export crop production incurred by the Malawian smallholder so as to allow the prices of Malawian agricultural commodities to reflect their true opportunity costs (Harrigan 1988; Christiansen and Southworth 1988). SAL II, in fact, explicitly called for the increase of export crop producer prices on the basis of export parity pricing criteria. Several observations can be made with regard to the ensuing experience. First, the increase in cash crop producer prices commencing in 1982 was initially associated with falling implicit taxes on smallholders (rising nominal protection coefficients, NPCs) for groundnuts, tobacco, and rice (Table 7.5).[10] In the case of tobacco, for example, the NPC, calculated as the ratio of producer price to auction price on sun- and air-cured tobacco, increased from 0.17 percent in 1982 to 0.89 in 1985 before falling again. The NPC for groundnuts increased from 0.37 in 1981 to 0.92 in 1983 and 1984. The NPC for rice rose from 0.29 in 1981 to 0.73 in 1985 in one estimate and from 0.28 to 0.39 in another.[11] Thus, between 1982 and 1985, implicit taxes on the main smallholder cash crops declined.

The second observation is that, although NPCs increased in 1985, the falling levels of taxation were due not to an increase in real producer prices but rather to a fall in real world prices. Indeed, the nominal producer price increases in the early 1980s at best served to stabilize real producer prices in the face of falling world prices.

Thus the initial optimism regarding the success of pricing policy in reducing the rate of implicit taxation on smallholders (see, e.g., Christiansen and Southworth 1988) may have been premature and misguided. Evidence from recent years indicates that real producer prices continue to be upwardly sticky. A sharp increase in the export parity price of tobacco since 1985, for example, was not matched with a corresponding increase in producer prices. In fact, the NPC fell from 0.89 to 0.24. Similarly for rice, the estimate based on Malawian trade data (2) shows the real export parity

[10] The NPCs cited in this paragraph are calculated on the basis of export parity prices denominated in Malawi kwacha and converted at the official exchange rate.

[11] The first estimate is based on rice prices f.o.b. Bangkok net of international and domestic transportation, handling, and marketing costs. The second estimate is based on the f.o.b. unit value of Malawian rice exports from Malawian trade statistics. For a further discussion on estimation methods, see Sahn, Arulpragasam, and Merid 1990.

le 7.5. Nominal protection coefficient (NPC), selected crops, Malawi

	Rice				Groundnuts		Tobacco	
	NPC (1) at official rate	NPC (2) at official rate	NPC (1) at shadow rate	NPC (2) at shadow rate	NPC at official rate	NPC at shadow rate	NPC at official rate	NPC at shadow rate
'5	0.38	—	0.28	—	1.96	1.41	0.17	0.13
'6	0.62	0.32	0.40	0.21	0.48	0.33	0.17	0.12
'7	0.63	0.57	0.39	0.36	0.36	0.23	0.17	0.11
'8	0.42	0.41	0.31	0.30	0.33	0.25	0.52	0.38
'9	0.53	0.34	0.38	0.25	0.54	0.40	0.52	0.38
0	0.40	0.36	0.27	0.25	0.56	0.39	0.46	0.32
1	0.29	0.28	0.19	0.18	0.37	0.25	0.31	0.20
2	0.41	0.20	0.26	0.13	0.88	0.58	0.17	0.12
3	0.56	0.30	0.35	0.20	0.92	0.60	0.34	0.22
4	0.66	0.29	0.39	0.18	0.92	0.57	0.39	0.24
5	0.73	0.39	0.44	0.24	1.42	0.88	0.89	0.55
6	0.73	0.23	0.44	0.15	0.84	0.53	0.49	0.31
7	0.91	0.21	0.58	0.14	1.05	0.70	0.35	0.24
8	—	0.20	—	—	1.02	—	0.26	—
9	—	—	—	—	—	—	0.24	—

ources: Christiansen and Southworth 1988 and Malawi Government *Economic Reports* (1986–90). Calcu-
ons of NPCs based on producer prices and estimated border prices as follows: rice border price estimate (1)
m data in *International Financial Statistics* (IMF, 1986–90); estimate (2) from data in *Annual Statement of
ernal Trade* (Malawi Government, various years); groundnuts border price estimates from data in *Annual
ement of External Trade* (Malawi Government, various years).

price rising by MK 195 per metric ton between 1985 and 1988 and the real
producer price falling by approximately MK 6 per metric ton in those
years.[12] This represents an NPC decrease from 0.39 to 0.20 (Table 7.5).
Thus pricing policy rules per se appear not to have undergone a lasting
change during the past decade.

A third observation is that, although these implicit tax levels were declining, estimates of such taxes at the official exchange rate consistently underestimate the magnitude of actual taxation as measured at the shadow
exchange rate. This is because the former estimates capture only the direct
component of taxation. Kwacha-denominated estimates of the world price
converted at the official exchange rate, however, disregard the important
indirect taxation of smallholders through an overvalued exchange rate.
Hence the export parity prices measured at the shadow exchange rate lie

[12] Estimate (2) in Table 7.5, in being based on data from Malawian trade statistics, is
presumably a more accurate indicator of actual revenue generated by Malawi given its export
markets. The discrepancy between (1) and (2) (i.e., the higher value and volatility of export
parity prices based on Malawian trade data over that based on international price data) is
probably a function of the isolated markets to which Malawi exports (in Zambia and war-
torn Mozambique) and the selected times at which it exports (on the occurrence of unforeseen
shortages).

above those converted at the official exchange rate.[13] In 1985, for example, whereas the NPC for tobacco was 0.89 using the official exchange rate to convert world prices, use of a shadow rate (which more appropriately captures the true value of the kwacha) indicated an NPC of only 0.55. Similarly for groundnuts, estimates based on the official exchange rate resulted in an NPC of 0.92 in 1984, but estimated based on the shadow exchange rate reveal an NPC of 0.57.[14]

Fertilizer Subsidy Removal

Another integral component of the planned agricultural sector reform in Malawi has been the fertilizer subsidy removal program (FSRP). The marketing and pricing of fertilizer in the smallholder sector is, as always, under the direct control of the Ministry of Agriculture and ADMARC. Unlike the estate sector, which purchases its fertilizers primarily through Optichem at market-determined prices, the government has set up the smallholder farmers' fertilizer revolving fund, working in conjunction with ADMARC and the Agricultural Development Districts not only to supply fertilizer to the smallholder but to achieve a variety of related social and economic objectives. The key element of achieving those social objectives is subsidizing the price of smallholder fertilizer, ostensibly to encourage use of fertilizer with the objective of raising agricultural output of maize, the dominant smallholder crop.

In 1983 the government committed itself to the FSRP, a program for the gradual but complete removal of this subsidy. The program stipulated a reduction of the subsidy rate to 22.6 percent in 1985/86. In 1986/87 the subsidy was targeted to decline to 17 percent and was to be down to 12 percent in 1987/88. Complete removal of the subsidy was scheduled for 1988/89. The FSRP was instituted primarily because the subsidy had been a major cause of ADMARC's financial troubles. It required large portions of a limited government budget. Thus the FSRP was an important component of the policy package aimed at rectifying Malawi's fiscal imbalance.

[13] The terms "direct" and "indirect" used here are in keeping with the terminology discussed in Krueger, Schiff, and Valdés 1988. The "indirect effect" of Krueger et al., however, includes both distortions in the exchange rate and deviations in the ratio of prices of agricultural goods to those of nonagricultural goods that would exist in the absence of intervention. Here we are referring solely to the first component of the Krueger et al. indirect effect. At least one study (Dorosh and Valdés 1989) has shown that the second component, disregarded here, is in fact empirically small in magnitude. The shadow exchange rate series for Malawi was computed using the methodology outline by Krueger, Schiff, and Valdés (1988) and discussed by Sahn, Arulpragasam, and Merid (1990).

[14] It is interesting to note, moreover, that the margin of distortion between the official and shadow exchange rates has not been altered significantly between 1975 and 1988. Adjustment apparently has not brought official and shadow exchange rates any closer into alignment. The ratio of shadow to official exchange rate rose from 1.48 in 1981 to 1.57 in 1984 before falling back to 1.45 in 1987.

ble 7.6. Fertilizer subsidy, Malawi

	1983/84	1984/85	1985/86	1986/87	1987/88	1988/89
TAL PURCHASED/ DONATED (MT)	55,231	66,994	70,925	63,341	82,361	82,882
TAL SOLD (MT)	57,009	69,222	63,977	67,303	72,500	—
livery to mkt. cost (MK)	23,085,603	30,048,115	32,893,214	34,809,936	47,993,593	70,674,566
es value (MK)	16,051,370	22,313,580	24,949,060	27,484,430	38,508,540	49,719,740
volving fund trading deficit (MK)	7,034,233	7,734,535	7,944,154	7,325,506	9,485,053	20,954,826
t interest received (MK)	94,961	471,768	639,892	1,347,284	2,937,079	—
t subsidy required (MK)	6,939,272	7,262,767	7,304,262	5,978,222	6,547,974	—
t subsidy as % of budget deficit	5.06	6.24	7.91	2.75	−9.59	—
t subsidy as % of GDP	0.41	0.37	0.33	0.22	0.18	—
gregate fertilizer subsidy rate (%)	30.47	25.74	24.15	21.04	19.76	29.65
volving fund target subsidy rate (%)	—	—	22.60	17.00	12.00	0.00

Source: Sahn, Arulpragasam, and Merid 1990, Courtesy of Cornell Food and Nutrition Policy Program.
Notes: "Net subsidy required" is derived by subtracting sales revenue for a given year (determined by ultiplying the quantity of fertilizer purchased by the sales price to farmers) from the cost of delivering to the urket the quantity of fertilizer sold that year. "Delivery to market cost" is the cost of all fertilizer (both rchased and donated) sold to smallholders nationally in a given year at the going c.i.f. cost that year, plus stoms levy, depot and storage costs, internal transport charges, and rebagging costs. We refer to the difference tween the sales revenue and cost of delivery to the market as the trading deficit. From this trading deficit is btracted the net interest earned by the fertilizer revolving fund on accumulated funds on deposit with the serve Bank of Malawi to arrive at the net subsidy.

Table 7.6 presents figures on the net subsidy needed annually to cover the fertilizer revolving fund's financial deficit. The figures show that the FSRP stemmed the drain on the treasury only marginally. After rising from MK 7.03 million in 1983/84 to MK 7.94 million in 1985/86, the fund's trading deficit fell slightly to MK 7.32 million in 1986/87. In 1987/88, though, it increased to MK 9.48 million. Then, with the formal termination of the FSRP, the trading deficit more than doubled to MK 20.95 million in 1988/89. This trend was somewhat tempered by rapidly growing interest earned by the revolving funds on deposit in the reserve bank. As a result, the actual net subsidy required from the treasury rose to MK 7.30 million, or 0.33 percent of GDP, in 1985/86 before falling to MK 5.98 million, or 0.22 percent of GDP, in 1986/87. Despite the more than doubling of interest payments received from the funds on deposit in 1987/88, the net required subsidy began to increase in that year, rising to MK 6.55 million, 0.18 percent of GDP. The dramatic escalation of the trading deficit in 1988/89, furthermore, indicates that the treasury subvention would have been much higher in that year. The numbers thus reveal that the FSRP made little headway in meeting one of its primary objectives.

An examination of the subsidy rate on fertilizer further reinforces that the FSRP failed to bring about planned change. The aggregate subsidy rate, defined as the weighted average of the difference between the sale price and the delivered-to-market costs of fertilizer where the weights are the shares of the different types of fertilizer, indicates that in the year before the FSRP went into effect the subsidy rate dropped from 30.47 to 25.74 percent. Then, for the first three years of the FSRP, the subsidy rate continued to decline: to 24.15 percent in 1985/86, 21.04 percent in 1986/87, and 19.76 percent in 1987/88 (Table 7.6). Reluctant to pass on exogenously escalating costs to smallholders, however, the government felt compelled to abandon the FSRP in 1987. This policy slippage is evident in the numbers. During 1988/89 the subsidy rate shot back up to 29.65 percent, its original pre-FSRP level of 1983/84.[15]

The FSRP failed to achieve a sustained decline in the aggregate subsidy, despite price increases to the smallholder, primarily because of escalating freight costs. Freight charges have not, however, been the only source of the increased cost of fertilizer to Malawi. By 1988 the kwacha had been devalued by close to 82 percent of its 1984 value. Compounded with rising world prices of fertilizer, this translated into higher kwacha costs f.o.r. Other costs also contributed to the increasing cost of fertilizer. For example, the decision to provide fertilizer in smaller bags in order to improve its access to smallholders meant additional rebagging costs domestically— MK 437,000 in 1987/88.

The original justifications for instituting a fertilizer subsidy and the subsequent reasons for not complying with the terms of the FSRP were that higher prices would reduce smallholder uptake and production. This being the case, the crucial questions are these: what is the role of pricing on smallholders' fertilizer use? who are the beneficiaries of the subsidy? and were there in fact sound food-security and distributional reasons for failing to meet the terms of the FSRP?

Contrary to expectations, fertilizer uptake in the smallholder sector *increased* with price increases after the initiation of the FSRP (Figure 7.2). Fertilizer uptake rose every year after 1985, averaging an annual growth rate of 7.11 percent. Given the larger quantities of high-analysis fertilizer in the uptake, the average annual growth rate of nutrient uptake over the same period was 12.89 percent.[16] This apparent paradox suggests examining the price of fertilizer relative to output prices. The real price of

---

[15] Moreover, in 1988/89 the customs levy was no longer included as a cost component to fertilizer, the government having evidently withdrawn the levy on fertilizer. This would imply an even greater transfer of resources away from the government as a result.

[16] To the extent that these amounts are really supplemented by cheaper fertilizer smuggled in from Zambia—a pervasive phenomenon (IFDC 1989)—actual uptake in the smallholder sector may be even larger.

Figure 7.2. Fertilizer utilization, Malawi

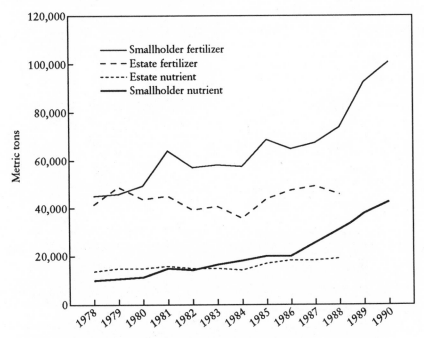

Sources: International Fertilizer Development Center 1989; Atukorala et al. 1990.
Note: Nutrients included are nitrogen, phosphate, and potash.

fertilizer for our purposes may be deflated by the producer price of the prominent, fertilizer-using, smallholder crops. The most important indicator is the ratio of fertilizer to maize prices. After falling dramatically from 1978/79 to 1981/82, this index rose steadily until 1986/87. By then it was 17 percent higher than before the initiation of the FSRP two years earlier. It is noteworthy, however, that given the increase in maize producer prices in 1987/88 and 1988/89, as well as the reestablishment of a high subsidy rate on fertilizer, the ratio dropped in these last two years, only to increase slightly again in 1989/90 (Table 7.7). As for other crops, between 1984/85 and 1987/88 the real price of fertilizer had risen as measured by the fertilizer/tobacco price ratios (Table 7.7). In 1987/88 the smallholder fertilizer/tobacco price ratio was higher than it had ever been since the mid-1970s and 40 percent higher than in 1984/85, before the initiation of the FSRP. It fell somewhat in 1988/89 and 1989/90, however, commensurate with higher tobacco prices. The fertilizer/rice price ratio was 5 percent

Table 7.7. Price indices of rice, tobacco, and maize relative to fertilizer, Malawi

	Fertilizer/rice	Fertilizer/tobacco	Fertilizer/maize
1976/77	0.83	1.41	1.20
1977/78	0.83	1.07	1.20
1978/79	0.83	1.08	1.20
1979/80	0.83	1.04	0.91
1980/81	1.00	1.23	1.09
1981/82	1.02	1.19	0.67
1982/83	1.17	0.93	0.88
1983/84	0.99	1.04	0.87
1984/85	1.00	1.00	1.00
1985/86	1.08	1.27	1.21
1986/87	0.98	1.25	1.27
1987/88	1.03	1.40	1.19
1988/89	1.05	1.17	0.95
1989/90	1.10	1.12	1.06

Sources: Christiansen and Souththworth 1988; International Fertilizer Development Center 1989; Malawi Government *Economic Reports*.

higher in 1988/89 than it was in 1984/85, although in general it remained quite steady throughout the 1980s.

The results indicate that the price of fertilizer relative to farmgate prices generally increased between 1983/84 and 1986/87, not explaining the increasing application of fertilizer. But of greater interest is the result of the value-to-cost ratio, a more appropriate metric by which to gauge the incentive structure faced by smallholders. This ratio is calculated as the incremental output per kilogram of nutrient divided by the value of incremental cost per kilogram of nutrient. Maize, being the dominant crop, is used as the reference output. Ratios for local maize using low analysis fertilizer reveal that increasing maize prices have generally not compensated for increasing nutrient prices. The ratio has in fact steadily declined from its value of 3.28 in 1981/82 to a low of 1.71 in 1986/87, before rising marginally in 1987/88 and 1988/89 as a result of maize price increases in those years. Since the initiation of the FSRP after 1984/85, the ratio has averaged 1.93. This level is below the ratio of 2.00–3.00 generally thought to be required to encourage farmers to invest in fertilizer (International Fertilizer Development Center [IFDC] 1989). Not only have value-to-cost ratios been falling for local maize during the time in which fertilizer utilization was increasing, but they apparently even fell below this threshold minimum.

The increased fertilizer uptake is likely not a function of prices as much as a reflection of the increase in the quantity of fertilizer imported and rationed by the government. Since demand exceeds the supply of fertilizer to the smallholder sector at the subsidized price, aggregate uptake of fertilizer by smallholders is dictated by government decisions regarding the allocation of foreign exchange for the purchase of smallholder fertilizer as well as the quantity of fertilizer received in the form of aid. The claim that maintaining the subsidy was necessary to ensure continued uptake, therefore, does not seem borne out by the evidence. In fact, if a low subsidy rate would have allowed more imports, then use may have actually grown faster.[17]

Furthermore, the observed increase in uptake of fertilizer and nutrients in the face of declining value-to-cost ratios relative to maize also points to the possibility of increased leakage to the estate sector despite the moderate and temporary reduction in subsidy rates between 1983/84 and 1987/88. The leakage phenomenon can be readily understood in noting that AD-MARC's (subsidized) fertilizer selling price is between 25 and 51 percent less than Optichem's selling price for comparable products (IFDC 1989). Estimates of the extent of leakage vary. The Ministry of Agriculture estimates the amount of leakage at 17–19 percent of the total annual fertilizer sales to the smallholder sector. Other sources have estimated the magnitude of the illicit trade at 25–35 percent of the fertilizer intended for the smallholder sector (Robert R. Nathan Associates 1987). More recently, one Agricultural Development District official claimed that up to 50 percent of ADMARC fertilizer sales in his district ended up in the estate sector (IFDC 1989).[18] One survey found that 59.1 percent of sampled estates obtained some fertilizer from ADMARC (Mkandawire, Jaffee, and Bertoli 1990). Although it is unclear whether leakage increased or decreased with a reduction in the subsidy rate and an increase in the smallholder price of fertilizer, it is evident that the phenomenon continues to be prominent, implying that one of the major justifications for the program—its positive distributional implications for the smallholders—is not all it appears at first glance.

---

[17] A contributing factor to fertilizer use not being price-responsive could be that many farmers cultivating customary lands were demand- rather than supply-constrained. In particular, if smallholders' access to credit has improved, the increase in fertilizer utilization by this group would still be consistent with increased prices, especially given the high prevailing value-to-cost ratios. There is no evidence, however, that these demand constraints, as mediated through access to credit, were significantly relaxed between 1984 and 1989.

[18] Indeed, the gap between the amount of ADMARC sales to the smallholder sector and Optichem's fertilizer sales to the estate sector generally grew throughout the 1980s. Optichem sales to estates slowed considerably after 1985/86 in terms of weight and nutrient uptake and commenced falling in 1987/88. This is despite a general increase in hectarage under the estate sector and an increase in estate sector output over this period.

Table 7.8. Fertilizer use, by holding size, Malawi, 1984/85

Holding size (ha) (1)	% of smallholder cultivated area (2)	Mean holding size (ha) (3)	Fertilizer use (kg/ha) (4)	% of smallholder fertilizer use (5)	Fertilizer use by mean holding (kg (6)
<0.50	6.20	0.31	6.18	0.63	1.92
0.50–0.99	20.90	0.74	12.44	4.26	9.21
1.00–1.49	21.31	1.23	33.12	11.56	40.74
1.50–1.99	16.30	1.71	59.16	15.80	101.16
2.00–2.49	12.20	2.22	84.74	16.94	188.12
2.50–2.99	8.30	2.73	113.59	15.45	310.10
≥3.00	14.80	4.00	145.82	35.36	583.28

Source: Data from Kandoole 1990.
Note: Fertilizer uptake by holding size data collected from extension participants. Column (5) is computed a the share by holding size of [column (2) × column (4)]. Column (6) is computed as [column (3) × column (4)

In fact, compounding the issue of leakage and counterbalancing productivity arguments in favor of the subsidy are distributional considerations. The fact that small farms use limited amounts of fertilizer implies that removing the subsidy will therefore have only a marginal effect on their incomes and production. The 55 percent of all Malawian smallholder landholdings that are less than 1 hectare in size (and that constitute over 27 percent of land cultivated by the smallholder sector) use less than 5 percent of all fertilizer applied within the sector (Table 7.8). Similarly, 86 percent of all Malawian smallholder landholdings are less than 2 hectares in size and account for 65 percent of all land cultivated within the smallholder sector. Yet these holdings apply only 32 percent of all fertilizer used by the smallholder sector. This is less than the amount of fertilizer used by the 4 percent of all smallholdings that are 3 hectares and greater in size.

Estate Sector Dominance

Since Malawi began adjusting its domestic policy to an unfavorable external environment, it has continued to reassert a strategy in which the estate sector represents the engine of growth. Whereas many countries in sub-Saharan Africa viewed promoting manufacturing and industry as the path to development during the post-independence period, Malawi has stuck with an agriculture-based strategy that features support of the enclave export-oriented estate sector. The legal and institutional framework has favored estates by facilitating their access to leaseholds and protected them from competition from smallholders by legislating sole domain to lucrative crops; it has offered them world market prices for products; and it has ensured a low-cost supply of labor by taxing smallholders and therefore reducing the reservation wage. These features of Malawi's economic past went largely unaddressed through its first decade of macroeconomic and sectoral adjustment.

Although efforts to reform Malawi's agriculture were focused on smallholders, the 1980s witnessed the continuation of some important trends. The concentration ratio of exports in the major estate crops grew. Whereas tobacco, tea, and sugar comprised 60 percent of all exports in 1968, their share advanced to 85 percent in 1988. At the same time, most of the growth in agricultural GDP was attributable to estate, not smallholder, production. In 1978 the share of estate production in agricultural GDP was 13 percent; by 1988 it was 23 percent. Furthermore, the estate sector was the source of large increases in employment during the years under adjustment, with most of the new jobs created for tenants on burley estates.

The estate sector has outperformed the smallholder sector during the period of economic reforms, but the figures on growth in outputs, exports, employment, and so forth are simply a continuation of a trend, left unaffected by reforms. These trends, however, are a result not of increases in production or intensity of production but rather of the expansion of land under the estate sector in general. The appropriation of smallholder (i.e., customary) land has continued during adjustment, with a recent estimate suggesting that during the 1980s the number of estates increased from 1,200, covering 300,000 hectares, to 14,700, covering 843,000 hectares (Mkandawire, Jaffee, and Bertoli 1990). This phenomenal rate of increase is no doubt partially a consequence of "progressive" smallholders converting their land to leaseholds in an effort to benefit from the high world market prices received, especially from the most important leasehold crop, burley tobacco.

Several concerns remain, however, despite the fact that estates have been an important source of growth and employment generation and that a share of smallholder farmers have been able to convert their land to legally produce high-value export crops. These concerns center, first, on the fact that productivity on estates stagnated during the 1980s. Second, there is evidence that a significant portion of estate land is underutilized. Even on land under cultivation, the levels of investments and capital improvements are, in general, extremely low. In fact, one survey suggests that over one-third of the large estates cultivate less than 15 percent of their total land (Mkandawire, Jaffee, and Bertoli 1990). This underutilization is partially due to the fact that rents are far below the economic value of land. The fixed rent on land is MK 10 per hectare, regardless of land quality and potential returns. Moreover, these meager rents are often not even collected, thereby encouraging the view that land is a virtually costless asset for those who have the connections to gain access to a leasehold. The low level of investment is also partially a reflection of the lack of incentives for estate owners to invests. More important, however, estate tenants have no incentives to pay the high implicit interest rates associated with the purchase of inputs from owners and managers, especially since the structure of the relationship has the tenant shouldering all the risk of a production shortfall or crop failure.

An additional factor detrimental to productivity is that, despite the increased employment opportunities on estates, wages and remuneration for laborers and tenants are extremely low, especially relative to the profits of estate owners. Furthermore, living and working conditions are abysmal. This applies to everything from household, to schooling, to health, as well as to extension support and the provision of implements.

These points suggest some underlying problems that have recently found a place on the policy reform agenda as Malawi enters the 1990s. In essence, there is an increasing realization that, during the first decade of adjustment, product and factor market failures, which promoted the duality in agriculture, have gone unaddressed. Overlooked too was the extensification of estate holdings, with no corresponding productivity growth.

A recent recognition that dualistic policies have perpetuated rent-seeking behavior and impeded the growth of the smallholder and estate sectors alike has led to strong calls for reforms that go beyond price adjustments and address the very structures that perpetuate the cleavages in agriculture. Specifically, the newest structural adjustment credit sponsored by the World Bank in coordination with other bilateral donors such as USAID places considerable emphasis on allowing smallholders to produce burley tobacco and halting the transfer of land from smallholders to estates. Related market reforms in land, for example, as well as input and product prices are key aspects of the new efforts to address the fundamental constraints to agricultural growth in general and rural welfare specifically. In addition, efforts to address the underutilization of estate lands due to factors such as the lack of capital and weak management skills of newly formed estates are also crucial to the new reforms being planned. So too are attempts to put a greater emphasis on the efficiency of agricultural expenditures, such as correcting for the extremely low ratio of nonwage to personnel recurrent costs in extension.

In combination, the conditions associated with the newest adjustment loan finally begin to go to the heart of the issues in reforming agricultural policy. To date, the neglect of these issues has perpetuated a narrow export base, a stagnant smallholder sector, a parastatal structure that is a fiscal drain, a pattern of government expenditures that has focused on subsidies and failed to promote technological change, and a generally inefficient utilization of Malawi's resources. The question of how well the government will adhere to the stated objectives and conditions of this new path to agricultural adjustment remains, as do questions on the ultimate efficacy of the actions. It is, however, quite clear that, after a decade of marginal structural reforms, the policy agenda in the 1990s appears to be at least on a sounder analytical footing.

Reform and Industrial and Service Sector Performance

The adjustment program in Malawi has given relatively little attention to the industrial and service sectors. But a brief examination of some available aggregate data does not indicate that the response has been as one might predict, or desire.

Industrial Sector

Malawi's industrial sector is characterized by high concentration ratios that suggest both inefficiencies and monopoly rents. For example, eight of twenty-one subsectors at the three-digit level International Standard Industrial Code classification have less than three firms (World Bank 1988f). This high ratio is partly explained by concentration in ownership. Public conglomerates, such as Press Holdings, the Malawi Development Corporation, and ADMARC, together with branch operations of multinational enterprises (owned either privately or jointly with the public enterprises), form the hub of the manufacturing sector. Press, ADMARC, and Malawi Development not only hold a large part of total equity but also have close ties with the financial sector. The latter two own 80 percent of the National Bank of Malawi and 70 percent of the Commercial Bank of Malawi (World Bank 1988f). High concentration of formal industry is true not only of ownership but also of geographic location. Anecdotal evidence indicates that the manufacturing sector is biased toward a few large, urban-based operations. Yet it should be noted that many of these statistics probably underestimate the role and magnitude of rural manufacturing and small-scale enterprise, due to the lack of a comprehensive national registry of such ventures.

The industrial sector's annual rate of growth averaged 7 percent in the 1970s but dropped to slightly greater than 1 percent between 1980 and 1987 (Table 7.9). There are, however, some signs for cautious optimism with growth's return to the 7 percent level between 1988 and 1990. In seeking to explain the poor performance of Malawi's industrial sector in the early 1980s, we find that exchange rate and trade policies are perhaps most important. Nominal exchange rate devaluations and the imposition

Table 7.9. GDP and rate of growth, by three major sectors, Malawi

	Sector growth (million MK at 1978 factor cost)				Growth rate (%)			
	GDP	Agricultural DP	Industrial DP	Services DP	GDP	ADP	IDP	SI
1978	742.50	294.90	143.50	304.10	—	—	—	
1979	767.30	304.10	144.40	318.80	3.34	3.12	0.63	4.
1980	764.40	284.20	146.50	333.70	−0.38	−6.54	1.45	4.
1981	724.30	261.00	142.20	321.10	−5.25	−8.16	−2.94	−3.
1982	744.90	277.60	142.40	324.90	2.84	6.36	0.14	1.
1983	771.70	289.90	147.00	334.30	3.53	4.43	3.23	2.
1984	841.40	308.00	157.40	376.00	4.46	0.49	7.59	6.
1985	841.40	308.00	157.40	376,00	4.46	0.49	7.59	6.
1986	850.60	309.90	154.60	386.10	1.09	0.62	−1.78	2.
1987	868.20	312.50	156.90	398.80	2.07	0.84	1.49	3.
1988	896.80	318.70	168.40	409.70	3.29	1.98	8.33	2.
1989	935.40	329.70	181.90	423.80	4.30	3.45	8.02	3.
1990	979.20	346.50	192.20	440.50	4.68	5.10	5.66	3.

Source: Reserve Bank of Malawi 1987, 1988; Malawi Government 1990.
Note: Service has been calculated as a residual and debited "unallocable finance charges."

of quotas for the allocation of foreign exchange hurt an industrial sector that relies on foreign inputs but markets its products domestically. Indeed, the production process in Malawi is heavily dependent on the external sector. Two-thirds of raw material inputs are imported, and there are few interindustry domestic linkages (World Bank 1988f). Hence the increased costs of production due to progressive devaluations of the exchange rate and foreign exchange restrictions have slowed industrial production. Moreover, the potential gain from devaluation through an expansion in exports has been inapplicable to Malawi's industrial sector. With a domestic resource cost ratio of about 1.2, the sector exports only 3 percent of its total sales (World Bank 1988f). With the production of import-substituting products and the processing of agricultural products representing a large fraction of output, the industrial sector caters primarily to the domestic market. With food, beverages, and textiles accounting for two-thirds of total product (World Bank 1988f), Malawian consumers have faced a contraction in output from the domestic industrial sector and the pass-through of increased costs of production.

Industrial value added was not doing well through much of the 1980s, so it is somewhat surprising to find that the service sector, dominated by government services and by retail and wholesale distribution, was in fact growing during the years since Malawi began adjusting. In particular, between 1982 and 1988 the sector grew, both in absolute terms and relative to other economic sectors (Figure 7.1), at an average annual rate of 3.6 percent. As a result, the sector's share of GDP increased progressively, climbing from 31 percent in 1960, to 35 percent in 1967, to over 45 percent in 1988. This is somewhat puzzling. The traditional structural adjustment package of macro policies pursued by the government should theoretically have raised the relative share of the traded goods sector (viz., agriculture and industry) by liberalizing the prices of traded goods. Instead, the shares of both agriculture and industry lost ground to the services sector, raising questions about the effectiveness of recent policy, or at least about its implementation in Malawi.[19]

Employment and Earnings

The industrial and service sectors can represent an important source of employment and earnings for Malawi's low-income populations. For urban low-income households, as well as for marginalized agricultural households, wage earnings from manufacturing, small-scale enterprise, and formal industry (in either urban or rural areas) may be a sole or supplementary source of income. This is particularly true in Malawi, where the pressure on agricultural land has been increasing.

Statistics reveal that employment fell in both the industrial and services sectors during the recession of 1980–82 but then rose in both sectors until 1988 (Table 7.10). Between 1982 and 1986, the number of people formally employed in the industrial sector increased from 60,400 to 101,480 specifically through higher employment in manufacturing. Within the service sector, employment rose from 103,800 to 140,890. Approximately 60 percent of this increase was a rise in employment in the wholesale and retail trade subsector.

Employment statistics paint a positive picture of changes in wage-earning opportunities outside agriculture, but available information on changes in wage rates is particularly worthwhile examining in Malawi given the relatively low level of distortions in the labor market. An examination of statutory minimum wage rates in the face of inflation, which by 1987 had risen to an annual rate of over 25 percent, points to the extreme vulnerability of low-income wage earners. As seen in Figure 7.3, real minimum wages declined throughout the 1970s, were pushed back to the levels of the early 1970s as a result of the nominal wage increases of 1980–83, but then continued to drop through 1988 (except in 1987). This decline in real wages is also evident in the average real wage per subsector found in formal employment and earning data from the Reserve Bank (1987).[20]

---

[19] There are two possible explanations. First, government sector growth, surfacing here in the service sector account, has not been bridled by policy to the extent desired. Second, due to exogenous factors, domestic transportation services have increased.

[20] The picture of dramatically rising employment and falling real wages is noteworthy. Although this brings into question the reliability of the data, it is nonetheless not implausible; nominal wages did increase quite rapidly, but not enough to keep pace with inflation.

Table 7.10 Number employed, by sector, Malawi (thousand persons)

	Agriculture			Industrial			
	Agriculture	Mining	Total	Manufacturing	Utilities	Construction	Total
1968	42.20	0.50	42.70	21.20	1.50	15.30	38.00
1969	48.30	0.80	49.10	17.70	1.50	17.40	36.60
1970	53.70	0.50	54.20	19.50	1.70	18.50	39.70
1971	57.40	0.60	58.00	21.70	2.20	17.70	41.60
1972	63.70	0.80	64.50	23.10	2.40	18.20	43.70
1973	76.30	0.70	77.00	25.70	2.90	21.10	49.70
1974	80.40	0.80	81.20	26.80	2.50	22.80	52.10
1975	93.00	0.90	93.90	31.40	2.70	21.10	55.20
1976	103.90	1.10	105.00	36.00	3.00	21.10	60.10
1977	154.70	0.60	155.30	33.50	2.80	23.30	59.60
1978	168.90	0.60	169.50	35.80	2.90	31.60	70.30
1979	182.30	0.60	182.90	37.10	3.50	33.40	74.00
1980	181.10	0.60	181.70	39.70	4.00	32.70	76.40
1981	157.20	0.60	157.80	35.40	4.10	24.70	64.20
1982	158.10	0.60	158.70	31.40	4.30	24.70	60.40
1983	197.20	0.50	197.70	47.60	5.40	23.40	76.40
1984	177.70	0.30	178.00	49.20	4.90	25.90	80.00
1985	189.30	0.30	189.60	59.90	4.50	23.10	87.50
1986	185.14	0.31	185.45	68.00	4.68	28.80	101.48
1987	179.85	0.30	180.14	49.63	5.52	30.71	85.85
1988	197.84	0.29	198.13	53.68	5.20	31.47	90.35

Sources: Pre-1978, World Bank 1982a; post-1978, Reserve Bank of Malawi 1987, 1988; 1987 and 1988, Malawi Government 1990.

Between 1982 and 1985, a decline in average real wages was experienced in both the industrial and service sectors (Figure 7.4) as well as in both the private and public sectors. This decline corresponds to the fall in the administered minimum real wage over the same period. Within the industrial sector, this fall in real wages appears to have been most drastic in the manufacturing subsector. Between the public and the private sector, real wages dropped the most in the latter, reflecting the lack of gains in productivity.

Civil servants have also seen a deterioration in average real salaries since the beginning of austerity measures connected with adjustment, although the effects have been tempered by nominal salary increases in this subsector. On aggregate, the index of civil service salaries increased by only 80 percent between 1980 and 1987, compared to a 170 percent increase in the cost of living index during that period (Roe and Johnston 1988). Moreover, indications are that, due to nominal salary increases being regressive, the fall in real income has especially hurt the lower-paid civil servants.

Recognizing the large-scale erosion of real earnings among minimum

	Service				All		
Wholesale/ retail trade	Transport/ commun.	Finance/ Business	Commun./ pers. services	Total	Private sector	Government sector	Total
9.40	8.20	0.90	37.40	55.90	889.60	44.90	134.50
11.00	8.40	1.10	40.30	60.80	99.90	46.60	146.50
12.30	8.50	1.20	43.40	65.40	110.10	49.20	159.30
13.80	9.20	1.40	48.30	72.70	119.40	52.80	172.20
15.80	9.80	1.40	54.10	81.10	130.50	59.00	189.50
18.40	10.40	1.90	57.90	88.60	150.10	65.20	215.30
20.90	11.40	2.30	59.00	93.60	160.50	66.40	226.90
19.90	11.90	2.80	61.20	95.80	176.20	68.50	244.70
20.70	12.90	3.40	61.90	98.90	194.00	70.10	264.10
25.20	16.60	6.60	45.50	93.90	240.20	68.60	308.80
27.50	17.80	6.80	47.40	99.50	271.30	68.00	339.30
28.30	18.40	8.40	48.10	103.20	290.40	69.60	360.00
26.30	17.20	12.10	53.60	109.20	290.90	76.40	367.30
23.60	17.00	10.60	54.60	105.80	251.50	76.10	327.60
21.80	16.70	10.00	55.30	103.80	249.30	77.20	326.50
24.80	21.80	11.20	55.30	113.10	307.50	80.00	387.50
31.70	22.00	11.50	57.60	122.80	301.70	79.10	380.80
38.60	23.90	12.70	57.00	132.20	328.60	80.70	409.30
38.47	26.31	13.03	63.08	140.89	343.55	84.29	427.83
34.35	24.61	12.79	69.64	141.40	316.36	91.03	407.39
35.02	25.18	12.78	66.67	139.64	342.22	86.90	429.12

wage earners, the government increased administered wage rates substantially in 1989. The real wage rate increased by 57 percent in Blantyre, Lilongwe, and Mzuzu and by 81.6 percent in rural areas. To the extent that minimum wages effectively act as a floor, this move would be expected to have a dramatic positive impact on earnings. Given Malawi's increasing inflation rate, however, the risk of wage erosion, especially among Malawi's urban poor, will persist unless adjustments are made to nominal rates on a more frequent annual basis.

Unfortunately, this discussion cannot tell the whole story, especially given that the relationship between statutory wages and those actually received by workers in private enterprises has been shown to be unpredictable in other countries. Moreover, poorer households may be able to protect themselves from declining real wages in the formal sector by participating in the informal sector, where wages have little if any relationship to administered wage levels.

Not enough is know about Malawi's informal sector and, in particular, how policy reform measures bring about new economic opportunities for

Figure 7.3. Real minimum wages, Malawi

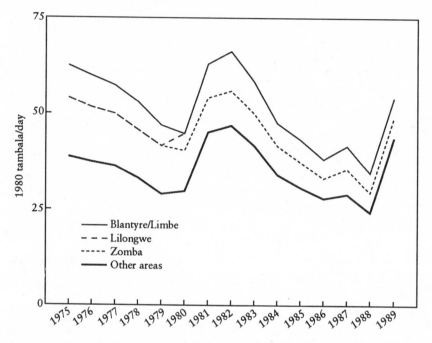

Sources: World Bank 1986e; Malawi Government 1988.

small-scale entrepreneurs. It appears, however, that the informal sector remains smaller in Malawi than elsewhere in Africa. The limited data also raise doubts as to how important informal sector activity is to the urban poor and whether liberalization of markets and moves toward privatization offer any real potential for generating employment opportunities for the poor. The 1977 population census, for example, shows that most participants in the informal sector are well educated and that 85 percent of them are based in rural areas. Yet the census may be expected to underestimate the extent of participation, particularly among the urban poor. To the extent that people participate in the parallel market or want to hide their income for taxation purposes, they do not reveal their participation in this sector. Moreover, the census counted the number of self-employed and may have undercounted informal sector participation as a result. In fact, indica-

Figure 7.4. Average monthly wages in the private sector, Malawi

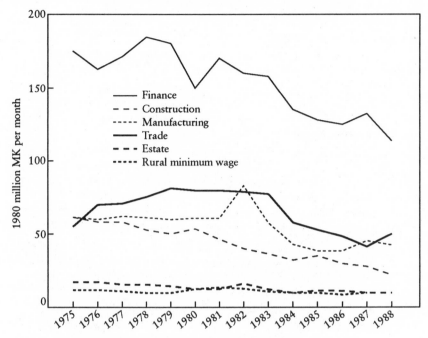

Sources: Malawi Government 1990; World Bank 1982a, 1985b.

tions are that, even if the urban poor did work in the informal sector, that job would not provide the primary source of income and is probably an activity most would engage in only part time or irregularly. The majority of enterprises not recorded in official employment statistics are in fact one-person operations, according to one report (World Bank 1988d).

Policies aimed at market liberalization, by reducing controls and licenses, for example, offer the prospect of addressing impediments to the activities of traders and entrepreneurs. Similarly, the removal of administered prices has the potential for raising small-scale entrepreneurial activity. Questions remain, however, as to whether policy reform will create a climate of confidence and stability and whether reform, including measures such as credit schemes, will foster growth of small-scale enterprises and off-farm employment generation.

Conclusions

A decade of donor-sponsored efforts at economic reform in Malawi has met with only limited success in terms of the intended scope of the adjustment process, its degree of implementation, and its consequences. In particular, a look at Malawi's economic performance during the 1980s leads one to the conclusion that adjustment has not addressed the country's structural weaknesses and has by and large failed to position the country to better cope with the changes, both internal and external, it will confront during the years ahead.

This conclusion comes as somewhat of a surprise, given the fact that Malawi has often been characterized as a relatively strong adjuster and a country where, at least in comparison with its neighbors, distortions were limited. Indeed, Malawi did not suffer some of the post-independence excesses of other African countries, in particular having never embraced an industrial-led, import-substitution development strategy such as led to egregious distortions elsewhere. In fact, the years of the first decade after independence were good years in Malawi. The estate sector-led development strategy paid off in the face of strong world prices and demand, industry and manufacturing displayed a respectable rate of growth, and tight party control kept the rate of growth of government under control. Still, the high level of export concentration in estate crops, the neglect of smallholder agriculture, the failure to make high-return investments, the low priority accorded to human resource development, and the imposition of economic controls that impeded the development of markets and entrepreneurial activities left Malawi unprepared to respond to the exogenous shocks that were to hit in the late 1970s.

The negative terms-of-trade shock of the late 1970s rippled through the economy, setting off a series of events that quickly drew the attention of sympathetic donors. The 1980s saw Malawi's precarious position being bolstered by a stream of IMF standbys, World Bank adjustment loans, and bilateral, policy-based loans. The experience of Malawi in meeting the conditions outlined in the numerous financing agreements was highly variable. The less than stellar performance in terms of meeting certain components of the conditionality, however, did not seem in any way to affect the financial flows. And this is to say nothing of the issue of the adequacy of the conditions themselves as the basis for reforming Malawi's economic structure.

Some conditions that accompanied concessional financing brought about positive changes, such as reducing the rate of parastatals such as the Malawi Development Corporation as well as the pervasive influence of Press Holdings. Likewise, gains were noted in terms of rationalizing the public investment program and improving the commercial banking sector. Despite these accomplishments, many weaknesses in the Malawian economy remained unaddressed. Heavy reliance on fuelwood continues to be a threat to Malawi's future economic prospects and to the environment. Weak interindustry linkages and heavy import dependence remain. Capital markets remain inefficient and underutilized. The civil service remains technically weak. Parastatals, oligopolies, and monopolies continue to predominate and are generally not commercially viable without continued indirect subsidies from the government.

Two further issues, however, are of overriding importance. First, reform has witnessed little improvement in terms of addressing the inadequacy of human capital investment. There has been little improvement in terms of low education achievements, high levels of malnutrition and disease, and the general lack of entrepreneurial talent to lead Malawi's fledgling commercial, manufacturing, and industrial sectors. Second, the fundamental features of Malawi's agricultural sector, the backbone of the economy and the source of income and employment for the vast majority of workers, have hardly been altered. Although the donor community is to be credited with providing the financing that prevented Malawi's economy from collapsing, it has failed to work effectively to design and implement a realistic strategy to alter the structural impediments that contribute to Malawi's ailing agricultural economy.

Adjustment has worked only on the margins in affecting change in agriculture for several reasons. For one, even the fundamentals of price-oriented structural adjustment seem to have been given lip service, without any true conviction to promoting policy change. This is the case, for example, in the smallholder sector, where the requisite reforms of input and product prices have not occurred. Subsidies to fertilizer continue to provide rents to the better-off farmers. Subsidies to consumers continue to be out of the reach of many of the poor. Taxation of smallholder crops continues to hurt the prospects for a buoyant agricultural sector while depressing returns to labor of the poor.

Reforms also largely neglected the control of land and assets and the use of natural resources. Most poignant in this regard is the banning of smallholders' production of the most lucrative crops, which is limited to the estate sector producers, who are also favored in terms of prices and marketing arrangements. The prohibition against the production of remunerative crops needs to be addressed, as does the entire set of issues regarding land policy. Policy must rectify the continued favoring of the estate sector in the appropriation of smallholder land. So too must policy revise the absurdly low lease payment made by the estates to the government, the owner of all leasehold land. Policy must also eliminate the other institutional obstacles that prevent land from finding its most productive use. Once again, such policy will require addressing the cleavages between smallholder and estate agriculture, an artifact of the state intervening on behalf of the power elite.

The remaining and significant scope of price-related adjustment initiatives in raising output and productivity—and consequently incomes—also needs to be accompanied by other reform measures to raise and improve public inputs into the production process. Any adjustment program that intends to increase output and promote a supply response must not ignore the reality that public inputs, whether for education, physical infrastructure, agricultural research, or credit institutions, are essential for ensuring the success of incentive policies to raise the level and productivity of private inputs.

Arguably, then, reform in Malawi must include efforts to raise labor productivity, especially among the smallholder sector, through more than the price mechanism. Low yields, stemming from the depletion of nutrients in the soils and cultivation of marginal lands as population pressures grow, coupled with the failure to innovate, are both cause and manifestation of the stagnation of Malawi's economy and the poverty of its people. This is true both on estates and on customary lands. Indeed, the green revolution has to date failed in Malawi. Although the battle should continue, perhaps the promise of the future lies in biotechnology, which offers considerable hope for circumventing the shortcomings of traditional efforts at plant breeding and reducing the costs of related investments, such as irrigation, which are often required complements to the use of improved seeds. Policy reform to address the mismanagement of Malawi's system of agricultural research (Kydd 1989) is thus an important element of any long-term development strategy.

Once again, however, initiatives such as improved technology or more credit should not be construed as magic bullets that will transform on-farm agricultural activities and thereby enable them to be the major source of income growth for rural households in the future. Despite the need for improved access to credit in order to enhance fertilizer uptake and the adoption of improved agricultural practices among smallholders, the limitations of this strategy should be recognized. For the growing number of smallholders, those with less than half a hectare of land, for example, there is little prospect of generating a sustained surplus above and beyond their own food requirements.

Increased attention, therefore, has to be focused on raising rural wages, both in the traditional arena of hired labor on large smallholder plots and estates and through the identification and development of nontraditional and alternative employment opportunities. The large and growing role of wage employment as a source of income for the poor, partially precipitated by the increasing land pressures, sets the stage for Malawi to follow the pattern of other countries whereby the increased commercialization of agriculture contributes to the economic transformation that raises the importance of wage, labor, and nonfarm incomes. As population pressures grow and the competition between producing for home consumption and for exports mounts, the answer to the poverty problem will increasingly be found in investment that raises productivity of the land and encourages the development of nonfarm enterprises that generate employment and rural incomes. This is especially the case for the small landholders and the landless, for whom equity-enhancing, off-farm employment is especially important. In addition, nonfarm income is often countercyclical to agricultural incomes, another important dimension to raising food security given the pronounced seasonalities of agricultural employment, earnings, and prices. Whether it be in areas such as food preparation and processing or in marketing of traditional crafts, there is little question that enhanced employment and higher wages through rural enterprise are based on the increased commercialization and growth of agriculture.

Efforts are furthermore needed to better service the needs of the informal sector. At present, the informal sector in Malawi appears to be much smaller than in other African countries. But if lessons elsewhere be our guide, in order to foster growth and productivity in the informal sector and smallholder enterprises there is a need for economic reform measures to end the implicit discrimination that favors the formal modern sector. The steps to facilitate small-scale business development are similar to the efforts required to encourage adoption of improved agricultural technology. Priorities must include improving infrastructure and access to capital, developing a system of advice and services to the small-scale entrepreneur, ending excessive regulatory constraints, and stimulating private sector initiatives in areas previously controlled by parastatals or dominant private entrepreneurs.

In the final analysis, there is little question that major reforms are still required to get Malawi's economy on a long-term growth path. Likewise, favorable external conditions, such as a lowering of transport costs that would result from a resolution of the conflict in Mozambique, would go a long way to improving Malawi's economic performance. Nevertheless, continued concessional financing from multilateral and bilateral agencies clearly remains prerequisite to a sustained recovery in Malawi. The appropriate mix of policy-based versus project-specific lending is certainly open to debate. Moreover, although there remains considerable scope for fundamental policy change such as getting prices right, this is not sufficient. Price-oriented adjustment needs to be accompanied by structural reforms in the roles and functioning of institutions and in the economic rules and regulations that will lead Malawi into the next century.

8

Structural Adjustment in a Country
at War: The Case of Mozambique

Steven Kyle

Mozambique, twice the size of California and with a population of approximately 15 million, was listed in the 1990 World Development Report as the poorest country in the world, with per capita income of $100 in 1988 (World Bank 1990g). Never a rich country, the economic declines of the late 1970s and 1980s resulted in income levels about half of those achieved before independence in 1975.

This decline is the result of both internal and external factors, among which are war, natural disasters, and economic policies pursued by the central government. A large part of the problem can be traced to structural imbalances inherited from the colonial period and exacerbated by the disruptions surrounding independence and the departure of the Portuguese colonial regime. An understanding of the colonial legacy and the relationships between Portugal and its African colonies is essential to an understanding of the subsequent performance of the economy.

The Colonial Legacy

Though Portuguese explorers first rounded the southern tip of Africa in the late 1400s and started moving into the Indian Ocean by 1500, it was many years before Portugal could make its presence felt in the interior of the country. After initially establishing settlements on the coast of Sofala, in Angoche, and on Mozambique Island, the Portuguese established outposts on the Zambezi in Sena and Tete and in the latter half of the sixteenth century moved to control the large indigenous states that dominated the interior.[1]

There were two major indigenous powers before the arrival of the Portuguese: the kingdom of Muenemutapa, which extended from the banks of the Zambezi south to the Save River and west to the highlands of Zimbabwe; and the Malawi Confederation immediately to the north, which controlled the area from Lake Niassa (Lake Malawi) to the coast. Both of these states had been brought under Portuguese domination by the end of the 1630s, allowing Lisbon to distribute land to white settlers in the form of *prazos*, or land grants, which were intended to serve as the basis of a permanent European presence.

In spite of these early successes in pursuing an expansionist policy, the second half of the seventeenth century saw the rekindling of resistance to the European invaders and the expulsion of the Portuguese from the Zambezi valley and most other areas except a few coastal enclaves. Apart from the reestablishment of a token presence along the Zambezi, this situation was maintained for the next two centuries. In spite of repeated armed expeditions, the Portuguese controlled no more than a small radius of a few kilometers around each of their coastal settlements.

Nevertheless, some of the economic characteristics evident in modern Mozambique were clearly present throughout most of the precolonial and colonial periods. First was the importance of primary commodity exports to the outside world. Gold from the kingdom of Muenemutapa was dominant during the sixteenth and seventeenth centuries, and ivory became more important thereafter. Second was the role of Mozambique as a supplier of labor. The slave trade grew to become the most important economic activity in terms of the international economy by the beginning of the nineteenth century and remained important even after the official abolition of the trade in 1836. Exports of laborers continued into the twentieth century in the form of mine workers for South Africa, for which the colonial government (but not the workers) received payment in gold. Mozambicans were also exploited as corvee laborers to work on European-owned plantations.

After the Portuguese achieved the subjugation of indigenous states in the twentieth century, several policies were designed to force Africans to work as contract laborers off-farm. Taxes, which could be paid only in cash, were an effective means of forcing the population into wage labor. In addition, production requirements for cotton and rice forced those remaining on the

[1] See Isaacman and Isaacman 1983 for a more complete account of the precolonial and colonial periods.

farm into the cash economy. Closely controlled at all stages by the colonial authorities, these crops were intended to provide food and raw materials for Portugal, which monopolized production of manufactured goods.

An important aspect of the Portuguese regime, and one that has had profound effects in the period after independence, was the extreme reluctance of the authorities to countenance any education or training of Mozambicans beyond a very basic level together with a policy of restricting all jobs requiring even minimal skills to Europeans. The policy of encouraging immigration of Portuguese peasants meant that it was extremely difficult for Africans to gain any but the most menial employment.

The reliance of Mozambique on labor remittances and service payments in the balance of payments was a structural characteristic of the colonial period that foreshadowed the need for structural adjustment policies in later years. Merchandise exports never amounted to more than half of merchandise imports, with receipts from laborers (mainly in South Africa), transportation, and tourism playing a very important role. Transportation links to Malawi, Zimbabwe, and the Transvaal provided much needed foreign exchange, and tourism from these countries accounted for a large share of employment in coastal cities of Mozambique, particularly Lourenço Marques (now Maputo) and Beira.

Independence and Aftermath

The fight for independence arose from various groups, the most important of which joined to form the Frente de Libertação Moçambicano (FRELIMO) in the 1960s. Though approximately one-third of the country was under FRELIMO control by the time of independence in 1975, the final separation was a result of political events in Portugal itself rather than the armed conflict in Mozambique. The departure of the colonial regime was accompanied by the departure of about 90 percent of all European settlers and resulted in widespread economic disruption and dislocation as the lack of trained workers brought many activities to a halt. The abandonment of large plantations and the sabotage of facilities that could not be removed left a great deal of productive capacity either destroyed or unused as the state apparatus took over management and sought to regain former output levels. Furthermore, the tendency of the new government to organize production in the form of large and highly mechanized state farms proved ill suited to Mozambican conditions.

Sharp declines were registered as a result not only of abandonment of productive units but also of the collapse of the rural trading and marketing network. This network, which once provided a means for peasants to

Table 8.1. Growth of global social product, Mozambique (billion 1985 meticais)

	Agriculture	Industry/ commerce	Construction	Transport./ commun.	Commerce/ other	Total
1981	89.6	43.6	6.8	23.0	25.2	188.2
1982	88.8	37.6	7.1	21.1	25.2	179.8
1983	69.1	30.1	7.2	16.8	23.8	147.0
1984	70.2	23.8	6.5	13.0	25.5	139.0
1985	70.8	19.4	6.2	11.5	23.8	131.6
1986	70.4	18.6	9.0	12.2	23.8	133.7
1987	75.2	19.7	7.5	11.7	24.4	138.4
1988	78.7	20.7	7.5	12.1	25.1	144.1
1989	80.3	22.1	7.5	12.6	26.0	149.0

Source: Comissão Nacional do Plano 1988; Comissão Nacional do Plano data bank.
Note: Global social product is the income aggregate used by Mozambique authorities. Though based on physical outputs and not on conventional income accounting methods, the rate and composition of growth of this measure is close to that estimated for GDP.

market agricultural surplus, was rendered virtually nonexistent by the abrupt departure of the Portuguese who previously monopolized it. The uncertainties and dangers of the armed conflict further promoted the disintegration of economic links within the country.

It is difficult to exaggerate the damage and dislocation caused by the armed conflict between the FRELIMO government and the opposing Resistencia Nacional Moçambicana (RENAMO). Attacks by RENAMO have been targeted on infrastructure, particularly roads, hospitals, and schools, as well as on the rural population more generally in an effort to disrupt food production as much as possible. Although active since the mid-1970s, rebel attacks increased markedly after 1981.

The costs have been heavy in economic terms (Table 8.1) but are perhaps most damaging in human terms. More than 4 million people in rural areas have been forced to leave their land or are so adversely affected by the war that they can provide less than half their food needs. More than a million more have fled entirely and are now refugees in surrounding countries (Table 8.2).

All these problems were accompanied by several natural disasters that would by themselves have resulted in major economic setbacks. Southern Mozambique, along with much of the region, suffered a prolonged drought in the late 1970s and early 1980s which affected different parts of the area more or less severely in any given year. Floods and typhoons also caused much damage.

These factors all contributed to a drastic worsening in the structural imbalance in Mozambique's international accounts inherited from the colonial era. Table 8.3 shows that exports, which had consisted primarily of

Table 8.2. Population of Mozambique, 1988

Urban	2,600,000
Rural	12,600,000
Dislocated[a]	1,700,000
Affected[b]	2,800,000
Other	8,000,000
Refugees abroad	1,200,000
Employed abroad	100,000
TOTAL	16,500,000

Source: Ministry of Commerce data bank.
Note: Total may not add due to rounding.
[a]Internal refugees.
[b]Approximately 40 percent self-sufficient
in food production.

agricultural products, suffered as production in rural areas collapsed due to rebel activities, natural disasters and low prices. Receipts from invisibles also collapsed as the closing of rail links to interior countries eliminated the previous through shipments to Zimbabwe, Malawi, and South Africa. Tourism shrank to zero as the war spread throughout the country, making safe travel impossible. Import requirements grew as the ability to produce sufficient food for the population decreased. By 1988 domestic production represented only 13 percent of marketed supplies (Table 8.4).

Mozambique's present economic problems cannot, however, be attributed to external causes only; past economic policies were also an important factor in the decline and an obstacle to improved performance. With a highly centralized and planned vision of economic organization, the new government administered prices and allocation of most basic goods, in-

Table 8.3. Volume of major exports, Mozambique (thousand metric tons)

	Cashew nut	Cashew oil	Sugar	Shrimp	Tea	Cotton	Copra
1973	29.6	14.8	281.2	2.7	17.5	51.0	57.8
1976	21.1	8.3	148.0	3.9	12.6	16.4	49.7
1977	17.0	10.0	778.2	2.1	12.3	6.3	41.8
1978	18.3	7.9	76.8	2.1	12.3	6.3	41.8
1979	17.1	6.3	183.6	4.0	23.3	16.1	32.2
1980	15.6	2.0	63.8	5.0	30.0	5.7	19.4
1981	12.2	4.8	63.1	7.6	16.0	15.0	12.2
1982	16.7	7.1	28.5	5.9	25.1	13.7	12.2
1983	5.8	3.0	25.0	4.8	13.3	13.2	6.0
1984	4.1	0.7	16.4	4.4	7.7	5.9	4.2
1985	3.1	1.0	16.8	5.4	1.8	4.7	12.7
1986	3.1	1.0	19.5	5.4	1.5	0.8	11.7
1987	6.0	3.5	10.4	4.9	0.7	3.8	14.8

Source: Comissão Nacional do Plano 1988.

Table 8.4. Grain supply, Mozambique, 1988

	Supply (thousand tons)	% of total
Marketing of domestic maize	60	13
Marketing of domestic rice	41	9
Imports of food grains	360	78
Total market supply	461	100
Cost of food imports	US$98.9 million	

Source: Comissão Nacional do Plano data bank.

cluding marketing of agricultural products. But the imbalances evident at independence proved intractable. Table 8.5 shows the extent to which the balance of payments had deteriorated by the mid-1980s, with current account deficits of nearly US$500 million. Revenues covered less than half of government expenditures at this time (Kyle 1991).

Agriculture, the occupation of more than three-quarters of the labor force, is the focus of reform efforts aimed at resolving the crisis, and the success achieved in this sector will determine the overall success of any policy package. Agricultural production is the key to both the food deficit and the deficit in the balance of payments. Since much of the government budget is dedicated to subsidies and other expenditures necessitated by inadequate production, a reactivation of agriculture would go far toward resolving this problem as well.

Agriculture and Food

Structure and Policies

At independence in 1975, approximately 80 percent of the population was engaged in agriculture, most of them subsistence farmers growing various staple crops. The most important of these is maize, which is grown at higher elevations in the north and west and in dry land areas in the south. Cassava is also grown widely on lower elevations, particularly in the north, and millet and sorghum are cultivated in various areas. Rice is grown in coastal areas near the Zambezi delta.

The two main export tree crops, coconut and cashew, are grown in the eastern regions of the country, and cotton is grown mostly at medium elevations in the north. The departing Portuguese abandoned approximately two thousand estates growing mainly tea, sisal, and sugar cane (Isaacman and Isaacman 1983). These crops, together with cotton and cashews grown by smallholders, constituted the main exports from the agricultural sector.

Table 8.5. Balance of payments, Mozambique (million US$)

	1981	1982	1983	1984	1985	1986	1987	1988	1989
Trade balance	-520.3	-606.8	-504.8	-444.0	-347.1	-463.6	-528.0	-612.0	-674.0
Exports (f.o.b.)	280.8	229.2	131.6	95.7	76.6	79.1	96.9	103.0	101.1
Imports (c.i.f.)	801.1	835.9	636.4	539.7	423.7	542.7	625.0	715.0	775.1
Service balance	55.8	30.9	0.1	-32.2	-93.0	-15.87	-147.8	-124.4	-77.9
Current account	-464.5	-575.9	-504.9	-476.2	-440.1	-622.0	-675.8	-736.4	-751.9
Debt amort. (scheduled)	309.2	329.3	296.5	337.8	278.5	335.5	384.1	378.1	291.9
Export/import ratio	0.35	0.27	0.20	0.17	0.18	0.14	0.15	0.14	0.13
Exchange rate (Mt/US$)	35.35	37.77	40.18	42.44	43.18	40.42	400.00[a]	620.00[a]	840.00[a]

Source: Banco de Moçambique 1989.
[a]Year end; rate = 1,500 as of mid-1991.

Table 8.6. Land use in 1983, Mozambique

	Area (thousand ha)	% of marketed production	% of agricultural exports
Family	1,929.7	51.2	19.4
State	118.6	43.4	65.6
Cooperatives	7.9	1.5	0.1
Private	41.2	3.8	14.9
TOTAL	2,097.5	100.0	100.0

Sources: National Directorate of Agricultural Economics data bank, Azam and Faucher 1988.

Table 8.6 shows the relative sizes and contributions to production of the principal types of farms in 1983. Family farms were by far the most important in terms of land area, but state farms contributed disproportionately to marketed surplus and exports. Private farms were also significant producers for the internal and export markets, whereas cooperatives accounted for a very small share of land and production.

The state farm focus of the government's rural strategy in the years immediately after independence emphasized rapid mechanization and expansion of the area under cultivation. In accord with this plan, massive equipment purchases, including 1,200 tractors and more than 500 combines, were made in 1977, followed by additional acquisitions in 1978. Overall, 90 percent of centrally planned agricultural investment went to the state sector before 1983 (see Tarp 1984). Agronomists from socialist countries were brought in to assist in coordinating state farm production. All these efforts succeeded in reversing the production declines of the early 1970s, though almost exclusively through area expansion rather than productivity increases.[2]

The emphasis on state farms encountered various problems that eventually caused the government to rethink its strategy. One serious problem was an extreme lack of managerial and administrative personnel in what were some of the largest plantations in Africa. In addition, much of the machinery purchased was used only seasonally, which together with repair problems left it idle a large percentage of the year. Extreme seasonality of labor requirements on large monocropped units led to difficulties as well. Finally, state farms proved to be vulnerable targets for rebel attacks after 1981.

Marketing. The departure of all but 20,000 of an original 250,000 Portuguese settlers at independence left the government in charge of not

---

[2] See Barker 1985 and Isaacman and Isaacman 1983 for discussions of state farm organization and production.

only the plantation sector but most of the rural trading network as well, since the departure of many traders left large areas without any network for distribution and marketing of goods (Isaacman and Isaacman 1983, 145). This situation reinforced the government's vision of a centrally directed development in which provision of goods to rural populations could be controlled. The need for rural marketing of consumer goods was met by a network of government stores called *lojas do povo*.

Though the government set up its own trading organizations, it was recognized from the outset that private traders would continue to operate at the retail level. In fact, this was necessary since the system of government stores did not prove adequate in many areas. Nevertheless, traders' margins were regulated and the government retained a monopoly on wholesale trade, which was under the direction of the Ministry of Internal Commerce.

Exclusive responsibility for crop marketing was vested in government organizations. The Ministry of Internal Commerce was responsible for all agricultural trading and the supply of goods to rural areas. Cashews and cotton were each controlled by separate parastatal organizations, the Secretaria de Cajú and the Secretaria de Algodão. The Ministry of Agriculture was responsible for marketing state farm output.

In 1981 all marketing of crops (except cashew and cotton) and distribution of goods to rural areas was consolidated in AGRICOM, a new government enterprise under the control of the Ministry of Internal Commerce. AGRICOM is obliged to buy all crops offered to it at the stated government price. Throughout the post-independence period until the reforms of the mid-1980s, the government viewed the terms of trade of agriculture vis-à-vis non-agriculture in barter terms. That is, policy and planning were based on the quantity of consumer goods required in exchange for marketed surpluses; the fact that money changed hands in the process was of secondary importance to planners.[3]

Table 8.7 presents data on crops marketed in the 1980s. It is clear that maize and rice were by far the most important marketed food crops. It is important to note that these figures, as with virtually all agricultural statistics in Mozambique, are of very uncertain quality. Nevertheless, the broad outlines of the data seem to conform to observable trends.

Price Policy. Before the reforms of the 1980s, prices for virtually all crops were set centrally by the National Price Commission and the National Planning Commission. A high priority in early years was the setting of

[3] See Mackintosh 1986 and 1988 for a description of agricultural marketing during this period.

e 8.7. Marketed agricultural production, Mozambique (thousand metric tons)

	1980	1981	1982	1983	1984	1985	1986	1987	1988	1989
d crops										
aize	65.0	87.3	89.2	55.8	82.6	58.6	21.5	27.3	44.6	79.8
ce	43.6	28.9	41.5	17.3	19.1	17.9	19.0	31.5	31.7	21.1
rghum	0.6	1.0	1.6	1.3	2.1	1.8	0.6	0.5	0.9	2.2
ans	9.6	14.9	6.9	4.8	3.5	4.6	4.0	9.2	7.1	13.1
assava	8.8	10.9	9.5	8.5	6.9	6.4	6.0	7.3	12.3	11.8
roundnuts	6.3	5.0	1.5	0.7	2.0	2.0	0.9	2.1	1.8	2.1
getables	6.4	6.8	5.6	7.9	20.0	33.9	23.9	20.7	33.1	NA
ort crops										
ashew	87.6	90.1	57.0	18.1	25.3	30.4	40.1	37.5	43.6	51.3
opra	37.1	54.4	36.6	30.7	24.8	24.0	28.6	25.5	22.8	13.3
ottonseed	64.9	73.7	60.7	24.7	19.7	5.7	10.8	27.3	33.3	NA
sal	298.0	233.8	139.9	122.4	136.6	78.8	22.4	4.4	4.6	NA
a	90.2	99.2	109.7	51.1	59.8	25.0	6.4	1.8	0.0	NA
er crops										
inflower	11.8	12.1	10.8	7.3	5.0	5.7	1.0	1.1	1.0	1.0
same	0.0	0.5	0.9	0.3	0.3	0.3	0.1	0.3	0.4	0.3
afurra	0.0	3.8	6.4	5.7	5.3	2.6	2.2	9.3	1.9	7.9
bacco	1.4	0.8	0.9	0.7	0.8	0.3	0.4	0.1	0.1	NA
mato	6.4	2.1	6.0	3.4	16.6	13.4	23.6	8.0	17.9	NA

urces: 1980–87 data from Comissão Nacional do Plano, *Informação Estatistica;* 1988–89 data from
istry of Agriculture, National Directorate of Agricultural Economics data bank.
ote: NA, not available.

prices on a pan-territorial basis. These prices were left unchanged for long
periods of time. Official prices for major crops in the post-independence
period are shown in Table 8.8, where it is clear that prices were sometimes
left constant for as long as seven years.[4]

To present these prices in real terms is difficult in Mozambican conditions, given the repressed nature of inflation resulting from the highly
controlled economic structure. Nevertheless, it is clear from the CPI in
Table 8.8 that the relative terms of trade of rural producers declined drastically in the 1980s. Table 8.9 shows producer prices deflated by the CPI. It
is clear that by the mid-1980s real terms of trade had deteriorated significantly for most producers. Even so, in the years preceding the sharp rise in
the CPI, the purchasing power of the metical dropped in rural areas more
as a result of extreme scarcity of goods to buy than as a result of price
inflation per se. When goods are available they are often sold outside
official channels at prices well above official levels.

By the end of 1980, production in many sectors had been restored to

[4] The accuracy of Mozambican price data and information on marketed output is uncertain. Various sources conflict, especially for the years immediately after independence. The
data presented here are the most comprehensive and consistent series available.

Table 8.8. Producer prices, Mozambique (meticais per kilogram)

	1976	1977	1978	1979	1980	1981	1982	1983	1984	1985	1986	1987	1988	1989
Food crops														
Maize	2.50	3.20	3.20	3.20	4.00	4.00	6.00	6.00	6.00	13.00	13.00	40.00	65.00	110.00
Rice	5.50	6.20	6.20	6.20	6.20	6.20	10.00	10.00	10.00	16.00	16.00	48.00	75.00	95.00
Wheat	4.40	4.40	4.40	4.40	4.40	4.40	4.40	4.40	4.40	11.50	11.50	50.00	80.00	—
Sorghum	2.30	3.00	3.00	3.00	3.00	4.00	5.00	5.00	5.00	12.00	12.00	35.00	50.00	95.00
Beans[a]	6.50	10.00	10.00	11.00	15.00	15.00	15.00	15.00	15.00	23.50	23.50	100.00	150.00	230.00
Cassava	NA	NA	NA	NA	2.00	3.00	4.50	4.50	4.50	free	free	free	free	free
Groundnuts	8.50	10.00	10.00	10.00	10.00	13.50	15.00	15.00	15.00	20.00	20.00	100.00	150.00	255.00
Sugar[b]	NA	NA	NA	NA	12.72	12.72	12.72	12.72	12.72	12.72	12.72	18.30	148.50	178.00
Export crops														
Cashew	3.50	3.50	3.50	3.50	5.00	5.00	5.00	5.00	10.00	10.00	10.00	60.00	105.00	165.00
Copra	3.20	4.70	4.80	4.85	4.85	5.00	5.00	5.00	5.50	5.50	5.50	30.00	60.00	100.00
Cottonseed	8.00	8.00	8.00	8.00	11.00	11.00	11.00	12.50	12.50	16.00	16.00	65.00	104.00	175.00
Other crops														
Sunflower	7.00	8.50	8.50	8.50	8.50	8.50	10.50	10.50	10.50	15.00	15.00	50.00	75.00	130.00
Mafura	2.20	2.20	3.00	3.00	3.00	3.00	5.00	5.00	5.00	8.00	8.00	30.00	40.00	70.00
Tobacco[c]	57.00	57.00	57.00	57.00	57.00	57.00	63.00	63.00	63.00	63.00	320.00	320.00	512.00	830.00
CPI (1975=100)[d]	101	103	104	106	107	109	120	165	216	278	325	531	796	1,035

Sources: Comissão Nacional do Plano 1988; Ministry of Agriculture, National Directorate of Agricultural Economics data bank.
[a]Type 1.
[b]Yellow.
[c]1st grade.
[d]This index applies to the capital, Maputo.

Table 8.9. Real producer prices, Mozambique (1980=100)

	1976	1977	1978	1979	1980	1981	1982	1983	1984	1985	1986	1987	1988	1989
Food crops														
Maize	66	83	82	81	100	98	124	97	74	125	107	202	218	284
Rice	94	104	103	101	100	98	134	105	80	99	85	156	163	158
Wheat	106	104	103	101	100	98	83	65	50	101	86	229	244	—
Sorghum	81	104	103	101	100	131	138	108	83	154	132	235	224	327
Beans	46	69	69	74	100	98	83	65	50	60	52	134	134	159
Cassava	—	—	—	177	100	147	187	146	111	—	—	—	—	—
Groundnuts	90	104	103	101	100	133	124	97	74	77	66	202	202	264
Sugar	—	—	—	—	100	98	83	65	50	38	33	29	157	145
Export crops														
Cashew	74	73	72	71	100	98	83	65	99	77	66	242	282	341
Copra	70	101	102	101	100	101	86	67	56	44	37	125	166	213
Cottonseed	77	76	75	73	100	93	83	74	56	56	48	119	127	164
Other crops														
Sunflower	87	104	103	101	100	98	102	80	61	68	58	119	119	158
Mafura	78	76	103	101	100	98	138	108	83	103	88	202	179	241
Tobacco	106	104	103	101	100	98	92	72	55	43	185	113	121	151

Source: Table 8.8.

1975 levels but were still below levels achieved in 1970, before the struggle for independence began to disrupt economic activity. In addition, large external deficits of the years since independence resulted in a large external debt, with an associated debt service well in excess of export earnings (see Munslow 1984).

Food Consumption

The collapse in income and inability to produce food for subsistence caused by the war and economic collapse resulted in a drastic drop in consumption and a consequent increase in malnutrition. Measures of malnutrition such as height for age, arm circumference, and birth weight show that Mozambique scores at the low end of the range in sub-Saharan Africa (World Bank 1989i). Though the urban rationing system has helped to some extent, malnutrition is widespread over virtually the entire country. It has been estimated that only 10 percent of the rural population could be counted as "food secure"—that is, assured of an adequate diet through employment or own production—in 1988. Another 35 percent were dependent on emergency relief or marketed distribution to meet requirements (World Bank 1989i). These deliveries can be highly unreliable, with some areas not receiving food for month at a time (see, e.g., Finnegan 1989).

The situation in urban areas is different in that home production can supply only a very small portion of food requirements. Marketed supplies and food relief must meet virtually all needs. Given the low or nonexistent incomes earned by the majority of urban dwellers, availability of marketed supplies does not guarantee an adequate diet, even with the rationing system. The average urban household spends more than 60 percent of income on food, and a large number of families are in a far worse situation than this average would suggest, with expenditures on food reaching 90 percent of income for the poorest groups (World Bank 1989i).

Policy Reforms and Economic Rehabilitation

Initial Reforms

The government recognized in the 1983 FRELIMO party congress that its own economic policies were a contributing factor in the economic collapse and that a significant reorientation was required, even though the problem of rural insecurity could clearly be expected to persist for the foreseeable future. The attempt to promote rural development on an Eastern European model centered on large state farms and centralized allocation and pricing was not successful. Concentration of available capital on low-productivity state farms prevented a more efficient use of resources, and family sector production for the market declined in the face of fixed prices and lack of goods to buy in rural areas.

In addition, efforts to organize the rural population in communal villages had an added disruptive effect. By 1982 nearly 20 percent of the farm population lived in these villages (see Isaacman and Isaacman 1983, 155). Yet most villages produced little if anything on a communal basis; family plots remained the preferred unit of production. The lack of incentive for communal production was exacerbated by government inability to provide needed inputs or services on time. Reversion to subsistence production occurred on a widespread basis as the crop marketing network failed.[5]

Overall, the situation in both peasant and state farm sectors resulted in a sharp decline in marketed surplus. An important contributing factor to this problem was the extreme imbalance in foreign accounts, where imports exceeded exports by a factor of six by 1986, partly because of the need to import food. This prevented imports of consumer goods as well as the raw materials needed to manufacture those produced locally. Of particular importance was the lack of agricultural implements—hoes and plows—which hindered family sector production. Azam and Faucher (1988) present estimates of manufactured goods supplies to rural areas and find that these supplies are a significant determinant of marketed agricultural production, especially in the cotton and cashew growing area.

One of the first steps toward liberalizing agricultural pricing and marketing was the freeing of vegetable and fruit prices from administrative controls in 1985. As shown in Table 8.7, this liberalization resulted in a substantial increase in production, primarily in the relatively secure "green zones" surrounding the principal urban markets of Maputo, Beira, and Inhambane. The success achieved with this initial effort paved the way for more radical subsequent reforms in response to continuing declines in marketed output.

Structural Adjustment and Aid Flows

The government's initial reform effort laid a foundation for subsequent agreement with the World Bank on a package of policy reforms to be supported by loans and official debt relief. The contribution of World Bank loans to financing the overall gap in requirements has, however, never been dominant. Table 8.10 shows total net disbursements of official develop-

---

[5] See Harris 1980 and Roesch 1984 for a discussion of the establishment of communal villages.

Table 8.10. Net disbursement of ODA, Mozambique (million US$)

	1983	1984	1985	1986	1987	1988	1989	Per capita $ (1989)	As % of G 1989
Total	211	259	300	422	651	893	759	49.4	59.2
World Bank (IDA)	0	0	5	30	87	127	176	—	—

Source: Data from World Bank 1991e.

ment assistance together with World Bank lending, all of which has been done on the standard concessional terms for International Development Association (IDA) loans. Nevertheless, this money represents a large part of the cash available to the government, with the majority of the remaining gap being closed by debt forgiveness or rescheduling together with commodity aid. The World Bank has been instrumental in coordinating and facilitating this aid, particularly in its role of promoting structural adjustment.

The economic rehabilitation program (ERP) implemented in 1987 and subsequent years succeeded in arresting the drastic decline in economic activity of the preceding years and in restoring moderate growth in 1987 and 1988. GDP stopped declining in 1986 and GDP per capita posted a modest increase in 1987. The broad objectives enunciated by the government were to reverse the decline in production, to ensure a minimum level of consumption, to strengthen the balance of payments, and to lay the foundation for economic growth. More specifically, the plan aimed to (1) increase agricultural production for domestic consumption, export, and agroindustries, (2) increase industrial production to support agricultural marketing, for import substitution, and to stimulate the development of exports such as minerals and marine products, (3) rehabilitate physical infrastructure and industrial capacity, in particular for the support of directly productive activities, (4) increase international rail and port traffic, and (5) mobilize external resources, both in kind and in the form of grants and debt relief.

In contrast to policy before 1984, which placed emphasis on increasing production in the state farm sector, the new strategy was directed toward increasing incentives to market production from the family farm sector, which accounted for more than three-quarters of the population. In particular, it was recognized that the system of fixed prices for agricultural products had gradually eroded incentives for marketing output from the smallholder sector. Also, the extreme scarcity of consumer goods in rural areas represented a significant constraint to the effectiveness of price incentives.

Overall, the ERP was designed to promote growth based on Mozambique's abundant agricultural resources through a program of liberalization of marketing and prices. Though the government relinquished direct administrative control of much of the agricultural production and marketing systems, it remains committed to centralized distribution of consumer goods and government-guaranteed markets for smallholder production since the security situation prevents private reestablishment of the rural trading system. Both the lack of capital and the high risk relative to return of rural trading hinder private sector expansion into trading and distribution.

The first steps toward realigning prices and restoring incentives were two large devaluations of the metical: by 80 percent in dollar terms in early 1987, and by another 50 percent in June of the same year. These moves had an immediate effect both on the official foreign exchange market and on the parallel market, where rates for the U.S. dollar exhibited a tendency to converge to the newly devalued official rate from about 1,500–1,800 meticais per dollar to about 1,200 by 1988. Though quite variable, the parallel rate remained at a level about 100–120 percent higher than the official rate through 1990. By mid-1991 the official rate had reached 1,500 meticais per dollar, with a parallel rate about 75 percent higher.

Concurrent with the devaluations, steps were taken to restore fiscal balance in order to end financing of government deficits through monetary expansion. The increase in the money supply of 50 percent in 1987 and of 43 percent in 1988 represents a substantial decrease in the real supply of money after the large price increases are taken into account. The periodic devaluations of the metical since 1987 have provided additional real depreciation.

Nevertheless, it was recognized from the outset that expenditures would outstrip revenues for the foreseeable future, necessitating continued external support for the government's budget. Within this overall context, a significant reallocation of expenditures was programmed. In agriculture, spending was shifted from the state farms toward the provision of support services for the family sector, and large new irrigation projects were deemphasized in favor of low-cost small-scale irritation. Transport expenditures were directed toward maintaining and rehabilitating existing roads and railways, particularly since rail transhipment to interior countries has the potential to generate large amounts of foreign exchange. Overall, capital expenditures to complete ongoing projects were given priority over new ones, and new projects were to be judged on the ability to maintain them and sustain recurrent costs rather than on simple availability of financing. To ensure adequate planning and evaluation, the Ministry of Finance and the National Planning Commission initiated annual preparation of investment programs with a three-year horizon.

In addition to raising official prices for agricultural products, the number of commodities with fixed prices was reduced from forty-four at year-end 1986 to twenty-eight by mid-1988 and to fewer than twenty by the middle of 1989. At the same time, the number of goods subject to distribution by central allocations was reduced from forty-three at the end of 1986 to thirty by the middle of 1988 and to eleven by the middle of 1989. Distribution of some goods at fixed prices will continue in order to ensure minimal levels of supply to all areas in the absence of an efficient trading system. In particular, a rationing system was put into effect in the two largest cities of Beira and Maputo to counteract the effects of the drastic price increases on the large numbers of refugees and poor in those areas.

These efforts have been supported by large amounts of food aid from the United States and European countries. Indeed, this aid has been essential in preventing mass starvation, as can be seen from the fact that 87 percent of marketed grain supplies were imported in 1988, much of this in the form of food aid. Overall, grain imports amounted to more than 85–90 percent of total marketed supplies over the period of the ERP.

Urban Food Rationing

The government initiated a rationing system in Maputo in 1981 and in Beira in 1986.[6] The primary function of the rationing system, called the Novo Sistema de Abastecimento (NSA), at its inception was to assure an equitable distribution of the limited food available. Prices were fixed by the Comissão Nacional de Salarios e Preços. These prices failed, however, to respond to inflation and became progressively more distorted measures of opportunity costs.

In 1981 approximately 160,000 households were enrolled in the NSA in Maputo. At that time, the coverage reached 750,000 individuals, virtually the entire population of the city. By 1991 the coverage in Maputo had increased to 1.0 million people, though population had reached at least 1.5 million. The increase was mainly due to the influx of a large number of new migrants who have fled war-torn areas.

Thus a rationing scheme that was set up to be a universal entitlement for rich and poor alike failed to respond to changing circumstances. There are no survey data by which to assess the number of families or individuals without access to goods at the ration shops, but many *deslocados*—who are presumably among the poorest of the poor—appear not to have ration cards.[7] Originally intended to provide equity in distribution of limited

---

[6] This section is based on Alderman, Sahn, and Arulpragasam 1991.

[7] Furthermore, the nature of the purchase requirement may also limit access to the system by the poor. Many families' income streams are so thin that they lack the discrete sum to make the ration purchase.

quantities of food, the system now functions more as an income support program that is not well targeted to those in need.

The level of benefits provided by the NSA has varied through time, with both fixed prices and distorted exchange rates contributing to subsidization (Table 8.11). The markup of food grain from border prices is evident in Table 8.12. The marketing parastatal and the treasury wholesalers and retailers together are responsible for marketing margins of 50 percent and more, thus limiting the subsidy that can be passed on to consumers. Thus, while many poor consumers are not covered by the system, a substantial transfer is made to traders and marketers.

Table 8.12 shows the value of food aid at both world and internal prices. It is clear that the government is valuing the food at far lower prices than those on world markets, allowing an implicit subsidy to be passed along to the marketing system or consumers. Such "hidden" subsidies (in the sense that they do not appear in the budget) are often at the center of the disagreements between donors and recipients over food aid countervalues; this observation is true in the Mozambican case. Though valuation closer to border prices would result in greater revenue for the treasury, the implicit nature of the subsidy allows the government to maintain greater transfers to beneficiaries of the system.

Successes and Failures

The effort to restore price incentives for agricultural production has resulted in substantially higher prices both nominally and in real terms. The change in the trend of producer prices is readily evident in Tables 8.8 and 8.9 where it can be seen that real and nominal producer prices were revised sharply upward starting in 1987. What is striking is not only the sharp upward revisions experienced from 1987 to 1990 but the realignment of incentives away from food crops and toward export crops.

The real prices of principal agricultural export crops suffered both in absolute terms and also relative to important food crops before 1987. As can be seen in Table 8.9, real prices for the three export crops shown had declined by between 34 and 67 percent between 1980 and 1986, whereas the prices for all important grain crops had either risen somewhat or declined by a lesser amount.

These adverse incentives contributed to the drastic fall in marketed production shown in Table 8.7. For cashews, widely grown by smallholders in coastal areas, incentives were so poor—both in terms of prices and because few consumer goods were available for purchase—that harvest was neglected or left for children to harvest while adults worked at growing food crops for subsistence consumption.

Table 8.11. Income transfer through the NSA, Mozambique

	February 1991[a]	August 1990	April 1990	December 1988	April 1988	December 1987	January 1987 (Pre-ERP)
Maize flour[b]							
Official price (Mt/kg)	250	250	250	145	145	38	9.5
Parallel price (Mt/kg)	637	511	422	225	176.5	400	NA
Ration quota (kg)	0	2.5	2.5	2.5	2	2.5	2
Income transfer (Mt)	0	652	430	200	63	906	NA
Rice							
Official price (Mt/kg)	756	756	756	271	271	40	13.5
Parallel price (Mt/kg)	1,305	1,214	1,149	467	389	600	NA
Ration quota (kg)	1.5	1.5	1.5	1.5	2	2.5	2
Income transfer (Mt)	824	687	589	294	237	1,400	NA
Sugar							
Official price (Mt/kg)	569	569	569	311	264	50	18
Parallel price (Mt/kg)	883	939	1,307	450	350	700	NA
Ration quota (kg)	1	0.5	0.5	1	1	1	1
Income transfer (Mt)	314	185	369	139	86	650	NA
Oil[c]							
Official price (Mt/l)	2,751	2,751	2,751	540	540	360	58.5
Parallel price (Mt/l)	2,417	2,875	3,500	916	800	1,300	NA
Ration quota (l)	0.5	0.5	0.17[a]	0.17[a]	0.5	0.5	0.5
Income transfer (Mt)	83	62	127	64	130	470	NA
Beans							
Official price (Mt/kg)	NA	461	461	260	260	260	27
Parallel price (Mt/kg)	1,914	1,073	1,069	400	NA	400	NA
Ration quota (kg)	0	0	0	NA	NA	NA	1
Income transfer (Mt)	0	0	0	NA	NA	NA	NA

Dry fish							
Official price (Mt/kg)	NA	NA	NA	560	335	335	50
Parallel price (Mt/kg)	NA	NA	NA	800	NA	500	NA
Ration quota (kg)	0	0	0	NA	NA	NA	2
Income transfer (Mt)	0	0	0	NA	NA	NA	NA
Salt							
Official price (Mt/kg)	—	—	—	100	NA	200	7
Parallel price (Mt/kg)	241	150	150	100	NA	200	NA
Ration quota (kg)	0	0	0	NA	NA	NA	1
Income transfer (Mt)	0	0	0	NA	NA	NA	NA
Total							
Nominal income transfer on food							
Per cap/month	1,221	1,587	1,516	696	516	3,426	NA
Per cap/year	14,652	19,038	18,190	8,358	6,188	41,115	NA
Per family (of 6.5)/month	7,937	10,312	9,853	4,527	3,352	22,271	NA
Real income transfer on food (1987=100)							
Per cap/month	NA	589	629	449	446	3,426	NA
Per family (of 6.5)/month	NA	3,830	4,086	2,921	2,899	22,271	NA

Source: Alderman, Sahn, and Arulpragasam 1991. This table was first published in Food Policy 16, no. 5 (October 1991): 395–404 and is reproduced here with the permission of Butterworth-Heinemann, Oxford, UK.

aThe planned allocation for maize flour was 2.5 kg. due to milling problems, however, maize flour was not available. Whole grain maize was distributed, however, in the canico, or more peripheral areas of the city. For those households that received whole grain, the income transfer value was Mt 525, corresponding to 2.5 kg., at an official price of Mt 190 compared to the parallel market price of Mt 400.

bSome households, especially in the canico part of the city, receive whole maize grain rather than maize flour. For them, the value of the income transfer is less, since the difference between the NSA and parallel market price is greater for maize flour than whole grain maize.

cOil distribution according to family size: 0.5l:1–3 people; 1.0l:4–6 people; 1.5l:7–9 people; 2.0l:10+ people. We assume a family of 6.5. Ration quotas for December 1987 and 1988 are those applies by the Gabinete de Organização do Abastecimento a Cidade de Maputo in November of those years.

Table 8.12. Food pricing, Mozambique, August 1990

	Maize grain	Rice extra	Rice corrente	Rice trinca	Wheat flour	Oil
World price ($/Mt)	185	362	317	287	278	875
Mt/kg at official rate	172	337	295	267	259	814
Mt/kg at parallel fate	444	868.8	760.8	688.8	667.2	2,100
To treasury (Mt/kg)	94.34	470.34	259.65	121.82	NA	NA
as % of c.i.f. official	55	140	88	46	NA	NA
as % of c.i.f. parallel	21	54	34	18	NA	NA
To IMBEC (Mt/kg)	99.30	495.09	273.32	128.23	NA	NA
as % of c.i.f. official	58	147	93	48	NA	NA
as % of c.i.f. parallel	22	57	36	19	NA	NA
to warehouses (Mt/kg)	132.00	576.00	339.00	169.50	NA	NA
as % of c.i.f. official	77	171	115	64	NA	NA
as % of c.i.f. parallel	30	66	45	25	NA	NA
To retailers (Mt/kg)	151.00	642.50	390.00	204.50	NA	NA
as % of c.i.f. official	88	191	132	77	NA	NA
as % of c.i.f. parallel	34	74	51	30	NA	NA
To public (NSA) (Mt/kg)	190.00	756.00	471.00	253.00	NA	2,751
as % of c.i.f. official	110	225	160	95	NA	338
as % of c.i.f. parallel	43	87	62	37	NA	131
To public (obs. parallel) (Mt/kg)	368.75	1,214.00	1,180.25	NA	1,900.00	2,875
as % of c.i.f. official	214	361	400	NA	735	353
as % of c.i.f. parallel	83	140	155	NA	285	137
as % of official price	194	161	251	NA	NA	105

Source: Alderman, Sahn, and Arulpragasam 1991. This table was first published in *Food Policy* 16, n (October 1991): 395–404 and is reproduced here with the permission of Butterworth-Heinemann, Oxfc UK.

Notes: Countervalue is calculated as 95 percent of the price set by the State Enterprise for Import Consumer Goods (IMBEC) (and does not include turnover tax). The official exchange rate used here is Mt per U.S. dollar and the parallel exchange rate used is Mt 2,400 per dollar. NA, not available.

Clearly the sharp rise in exports of smallholder crops, cashew and cotton, are in large part due to restoration of price and material incentives. The failure of state farm production of tea and sisal on former plantations is a testimony to the effectiveness of guerrilla raids against such concentrated targets as well as the inefficiency of these production units. Current policy promotes conversion of many of these large units into small holdings, allowing what in many areas are the best lands to be farmed by peasants willing to return to devastated areas.

The increases in marketed production of maize in 1988 are quite impressive, though levels achieved are still well below those of earlier years. Further increases will be dependent on coordination with massive amounts of food aid distributed at low prices. Reports on marketing of maize in the 1989/90 season were favorable, but adverse weather conditions contributed to what is reported to be a poor performance in the 1990/91 season. This situation underscores the limits of policy in what are often poor physical conditions for production.

Overall, however, it is clear that the restoration of price and material incentives for agricultural production has been effective in terms of supply response. The elasticity of marketed supply with respect to the real price (deflated by the urban CPI) cannot be determined with any certainty given the quality of the data from Mozambique and the short time period, but based on Table 8.7 it is clear that the response of maize and rice marketings to the policy package implemented in 1987 has been positive. Those for export crops have been less so, a result that is somewhat surprising given the clear tilt in incentives toward these crops in terms of real prices.

Other sectors of the economy have recovered somewhat in that production has stopped falling and in some cases has increased; nevertheless, the value of output overall has increased at a relatively slow pace. Overall, global social product, the aggregate output measure used by the government, as well as estimates of GDP have grown at a rate of about 4–5 percent since the implementation of the ERP at the end of 1986.

Manufacturing output remains dependent on imports of intermediate inputs and raw materials, which are subject to sporadic shortages due to foreign exchange constraints. Manufacturing, like grain production, has been subject to some adverse influences from foreign aid. The Ministry of Commerce has reported that sales of hoes and plows, considered priority items, have suffered due to large imports of these items from foreign donors who have unfortunately tied donations to procurement at home rather than in Mozambique.

In general, the international organizations have reinforced the government's overall strategy of restoring price incentives and dismantling administrative control; in this case, the need for liberalization was clearly seen by the government. Nevertheless, it is clear that the international organizations played a significant role in the formulation of the program. Even so, the government's determination to pursue its strategy has, if anything, meant that it has outpaced its own implementation capacity with the speed of its reforms.

This discussion touches on what is perhaps the most striking aspect of the policy reforms as supported by the international organizations—their heavy emphasis on price incentives to reactivate production. Removal of administrative controls is seen as a necessary condition for resources to flow where newly restored price incentives direct. Physical barriers to production and resource flows have received a lesser emphasis both in government plans and in the loan programs of the World Bank and the IMF.

There are several reasons for the heavy emphasis on price incentives. One is that prices had been fixed at low levels with few major changes since 1975. Neither absolute nor relative prices bore any relation to market conditions. Prices for some crops remained fixed in nominal terms for periods as long as nine years (see Table 8.8). A second major reason for emphasis on prices is the fact that they could be changed relatively quickly. The war conditions affecting large parts of the countryside make investment in physical infrastructure or capital extremely difficult to achieve and, if achieved, difficult to maintain. Progress remains slow in rural areas. Of particular importance are road building and the reconstruction and expansion of warehouses and rural stores to serve agricultural producers. Many of the rural stores constructed by the government in the late 1970s, as well as those stores built in the early 1980s, were destroyed by rebel forces along with schools and rural health posts. This highlights the fact that much of the necessary physical infrastructure is simply missing.

A major constraint to physical investments on the part of the government is the austerity program implemented to bring government finances into balance. As noted above, financing of deficits through money creation led to a situation of repressed inflation which undermined attempts to restore price incentives. The need to bring expenditures into line with revenues means that the government is severely constrained in the investment programs it can initiate. Given the large percentage of expenditures financed by foreign donors, spending priorities were inevitably influenced by donor priorities. As can be seen in Table 8.10, development assistance amounted to nearly 60 percent of GNP in 1989, a figure that is by far the highest in the world.

In spite of debt relief granted in support of structural adjustment, Mozambique's debt burden remains high (see Table 8.13). Although the debt service ratio has been reduced, it remains extremely high (debt service equaled 178 percent of export revenue in 1989), and total debt continues to grow both in absolute terms and as a percentage of GDP. The government remains heavily dependent on continued debt relief and aid to finance a fiscal deficit equal to well over 10 percent of GDP in 1989. This problem remains intractable even though expenditures decreased at an annual average rate of 2.7 percent during the 1980s (World Bank 1991e).

Lack of donor support for physical investments to support the government's rural strategy of development has been cited as an ongoing problem in the foreign aid program.[8] The government has struggled to ensure that a minimum of 40 percent of foreign aid is directed toward rural areas in the face of donor preference for urban or large showcase projects such as hydroelectric facilities.

In summary, both donor preferences and war conditions have contributed to a heavy emphasis on price reform as the principal vehicle for government policy reform. The extreme nature of the price reforms, with increases of between 500 and 1,000 percent for most agricultural products

[8] Interview, Ministry of Finance, August 1989.

Table 8.13. Debt ratios and net resource flow, Mozambique

	Total debt (million $)	Total debt/ GDP	Total debt/ exports	Debt service/ GDP	Debt service/ exports	Net resource flow (million $)
1980	—	—	—	6.0	32.1	—
1984	—	—	—	16.3	195.9	—
1985	2,863	87.4	1,558.3	11.6	215.5	538
1986	3,525	88.1	1,835.7	11.7	274.5	653
1987	4,261	337.1	1,820.8	36.7	227.5	757
1988	4,418	437.3	1,698.1	39.8	190.6	701
1989	4,737	474.2	1,842.4	37.3	178.4	264

Source: Data from World Bank 1991e.

and a devaluation of the exchange rate of more than 3,000 percent in dollar terms over the three-year period 1986–89, makes Mozambique an interesting case study in the possibilities for and limits to price reform in structural adjustment programs.

Conclusions

Mozambique's experience with structural adjustment over the period 1986–89 is an interesting case study in the efficacy and limits of policy because of the extreme difficulty of pursuing complementary policies of investment in physical infrastructure and inputs. In addition, the overall context of war, dislocation, and active attempts to sabotage economic activity make the country the most adverse environment imaginable for economic reform or, in fact, economic activity of any description.

That a program of structural adjustment can have any positive results at all under such circumstances is rather remarkable. The implementation of the ERP resulted in a reversal of the falling economic activity in all sectors which had characterized the previous five years, but so far it has generated only moderate rates of growth, in the area of 4–5 percent in the aggregate. Normally, much higher rates of growth would be expected in a country in which production remains far short of installed capacity. Two reasons underlie this somewhat disappointing performance. First, obsolete or destroyed capital stock means that installed capacity is not in fact available to be used. This is true both because of the war and because much of the vintage equipment left at independence is in need of replacement. Second, even where capacity is available, foreign exchange for raw materials is not. Bureaucratic allocation of foreign exchange together with donor tying of counterpart funds contributes to this problem, which basically is a result of export levels less than one-fifth of imports. Progress under the structural adjustment program to limit administrative allocation of foreign exchange in favor of auctions has help redirect needed hard currency toward more productive activities.

Agricultural supply response is constrained far more by the massive dislocation of agricultural population and the war conditions affecting producing areas than by lack of installed capacity. The most important economic constraint to increased production aside from war is the inability of peasants to convert marketed surpluses into consumption goods or agricultural inputs. This inability is a result of both the breakdown of the rural distribution system and more recently of AGRICOM's problems with timely payment. The experience of the 1990/91 season, when poor weather contributed to a disappointing harvest, emphasizes the fragility of Mozambique's recovery and its vulnerability to external shocks.

Given the extremely low incomes of much of the population, the effects of the massive price increases on consumption and nutrition have been severe for those not covered by the ration system in Maputo and Beira or by free food aid distributions. Though marketed food is generally available in the cities, much of the urban population, especially refugees, lack sufficient income. Even for those with access to the rationing system, these official distributions cover less than half of estimated minimum food requirements.[9]

The experience with the urban rationing system shows that much can be done to cushion the effects of the inevitable transition problems on the poorest, at least when they are easily targeted by virtue of being in or near large cities. Even so, such programs can have unexpected effects and costs which, though hidden, can grow to quite substantial size in a highly distorted and inflationary economy.

The problem of malnutrition in the countryside is generally due to an absolute lack of supplies in many areas rather than to insufficient cash income. This observation does not imply that these people are not poor; rather, it means that aid in the form of cash income supports will not alleviate the hunger problem. Given the virtual collapse of markets for food and for cash crops, schemes that must rely on functioning markets in order to be successful will do more to raise prices than to alleviate hunger.

The most important lesson from the Mozambican case is that progress will take time—longer than the planning horizons allowed for in planning reform programs. Even without Mozambique's problems of incessant and widespread warfare, the building of physical and institutional infrastructure will constrain the pace of any reform program. As it is, it is clear that a

---

[9] Interview, Ministry of Health, August 1989.

resolution to the armed conflict is a prerequisite for substantial progress to be made.

A second lesson is that price reforms "work" in the sense that increases can help promote marketed surpluses, but that there are limits to the response that can be expected on the basis of price incentives alone. Once physical or institutional constraints become binding, additional price increases do more to push the price of available supplies out of reach of the poorest than they do to promote additional production in the short to medium run.

The experience of the NSA supports the view that many of the poorest in the large cities are unable to afford even small amounts of food. In addition, much of the subsidization of food provides benefits to intermediaries on the marketing chain rather than to consumers. Although the riskiness of trading and marketing activities in Mozambican conditions surely accounts for some if not most of the seemingly excessive marketing margins, it is clear that price subsidization alone cannot make sufficient quantities of food accessible to all.

A third lesson is that a desire for market-based solutions should not be pursued in situations where markets are nonexistent or poorly functioning. In particular, income supports for malnutrition cannot address the base problem of nonfunctional food markets in rural areas. Maputo and Beira clearly do have functioning markets, as any visitor can attest, but the situation in the countryside remains problematic. It is important to bear in mind the limits of market-based policies when assessing projects with the attributes of a public good. Such solutions are unlikely to be effective in areas involving infrastructure such as roads or water systems.

Finally, the provision of foreign aid can at times conflict with an emphasis on rural development and smallholder agriculture, even apart from donor tendencies to fund large projects. The mere fact of large centrally administered foreign aid flows together with the large number of associated foreign "experts" can result in an unintended urban bias, since most of these advisers stay in the capital, where they are willing to pay high prices for manufactures and urban amenities.[10] This problem is further exacerbated by the large refugee populations in and around urban areas.

Overall, it is clear that the Mozambican case demonstrates that reform packages based on price incentives and exchange rate realignments can have positive effects, even in extremely adverse conditions. It is equally clear, however, that they cannot by themselves create the necessary conditions for growth over the medium to long run.

[10] As of autumn 1989, there were approximately 10,000 foreign *cooperantes* in Maputo, a large number compared to the number of counterparts in the government with whom they are working.

9

From Forced Modernization to Perestroika: Crisis and Adjustment in Tanzania

Alexander H. Sarris and Rogier Van den Brink

A severe economic crisis hit the Tanzanian economy in the early 1980s. The country's import capacity had deteriorated to such an extent that a serious and generalized shortage of consumer and intermediate goods arose. Given the particular institutional setting of Tanzania at the time, official markets effectively ceased to function, and even parallel markets had to resort in many cases to barter trade.

Recognizing the impossibility of dealing with the crisis alone, the government initially sought external assistance form multilateral donors such as the IMF in 1979. The conditionalities formulated by the IMF were at first strongly resisted on the grounds that they would undo all the social and distributional gains achieved by Tanzania since independence. But attempts by the government to implement its own stabilization and adjustment program in 1982–84 largely failed. The 1984 Tanzanian budget incorporated several policies advocated earlier by the IMF, and in 1986 Tanzania agreed on a three-year package of measures and policy reforms supported by an IMF structural adjustment facility. Support from other multilateral and bilateral donors followed. This initial program, called the Economic Recovery Program 1986–89, has been succeeded by the Economic and Social Actions Program of 1989–92, which has received only lukewarm support from the IMF.

Some controversies surround the origins of the Tanzanian crisis, which have both ideological and policy implications. The Tanzania experiment in socialization has been unique in many ways and has thus invited both supporters and critics, who have tried to view the crisis from their own perspective. Sympathizers of the Tanzanian experiment have tended to

260

blame the crisis on factors external or exogenous to domestic policies, whereas critics have tended to blame the crisis largely on domestic policies.[1]

However the blame is apportioned, there is a general consensus that the Tanzanian experiment proved to be unsustainable. Looking back at the crisis of the 1980s, it appears that the experiment escaped neither the common maladies of post-independence sub-Saharan Africa nor the universal failure of authoritarian socialism to "deliver the goods." Thus it has a lot in common with several other sub-Saharan economies. Tanzania's uniqueness lies in the fact that the economic interventionism and government control over the economy were uncommonly direct and comprehensive and had a profound impact on the economy. This was in the end both a blessing and a curse. A blessing, because Tanzania did make a credible and well-meaning attempt at institutional change; a curse, because the design and implementation of these policies were seriously flawed. They thus affected more rural and urban lives than they would have elsewhere in sub-Saharan Africa, where the government's sphere of influence often literally stops at the gates of the capital.

To understand both the unique and common factors of Tanzania's crisis, we necessarily have to take a relatively long-term perspective. Merely investigating the events of the 1980s does not suffice. To understand an economy's adjustment to crisis, one first has to understand its structure, and the path that led to the crisis. Unfortunately, we cannot analyze the dynamics of this structure with much empirical evidence at our disposal. Much of our story—and that of other observers—is based on rather weak evidence. Whereas statistical time series on key economic variables are commonly absent in sub-Saharan Africa, the socialist, "planned" economy of Tanzania has instead produced a plethora of data. The quality of much of this information is, however, highly questionable. The result of this situation is that one has to seriously question even the broad directions of certain trends as they are indicated by the official data.

Genesis of the Economic Crisis

The pre-independence British trustee government had kept Tanganyika a cheap-labor agrarian economy without a significant settler population. In 1962 the post-independence government did not inherit any industrial

---

[1] Prominent among such exogenous factors are droughts, the collapse of the East African Community in 1977, the war Tanzania fought to liberate Uganda from Idi Amin in 1979–80, the second oil crisis of 1979, a decline in commodity prices on international markets, and a rise in international real interest rates.

sector to speak of, but there was a relatively good road and railway infrastructure and an agricultural sector that held considerable promise. Endowed with a variety of agroclimates, the country had the potential to grow a wide range of commercial crops, and during the 1950s the country had shown impressive agricultural growth.

The First Five-Year Plan, 1964–69, did not change the basic structure of the free enterprise economy inherited at independence, although the state was supposed to have a more direct role in the development of the economy. Agriculture was to be the leading sector, evolving on the one hand by "improvement"—namely, by promoting government-controlled extension, marketing, and credit—and by "transformation"—namely, by implementing some pilot large-scale village settlement schemes designed to promote modern capital-intensive cultivation methods. The internal demand generated by agricultural growth would facilitate expansion of import-substituting manufacturing production, and some import restrictions were imposed. The plan emphasized Africanization, by promoting investments in education, and economic independence, by cutting the rigid links of the money supply with the sterling balance and establishing a central bank in 1966.

The economic performance in these early post-independence years was quite adequate, with average GDP annual growth more than 6 percent and agricultural growth at about 4 percent. Exports grew rapidly at about 8 percent annually and the manufacturing sector showed a similar fast expansion. No major institutional changes occurred until 1967, and the pre-independence farmer-controlled cooperative movement in agriculture continued to evolve (Coulson 1982).

All the same, President Nyerere's Arusha declaration of 1967 deemed this liberal development strategy a failure. The declaration heralded the beginning of what was to become know as the Tanzanian experiment. The principles espoused were conceptualized as *ujamaa,* which literally means "familyhood" but in political terms means socialism or, rather, Tanzanian socialism. The economic interpretation of the socialist guidelines involved a prominent emphasis on egalitarianism, basic needs, and self-reliance.

In marked contrast with a more liberal view of "induced institutional change," the new strategy outlined in the Arusha declaration and specified in more detail in the Second Five-Year Plan of 1969–74 was based on the basic premise that institutional change necessarily *preceded* economic growth. Such institutional change would be orchestrated by the government, which would use the government as its executive branch.[2] Apart

---

[2] Tanzania had officially become a one-party state in 1965, and the supremacy of the party and government was officially incorporated in the constitution of 1975.

from an increase of controls on economic transactions in general, direct asset redistribution from the private to the public sector was seen as an essential element of the strategy. Consequently, the government set out to comprehensively nationalize the economy. Nationalization initially affected industries, banks, insurance agencies, wholesale trading companies, and private estates. Later rentable private property (1971), village retail shops (1976), and private medical care (1976) would be caught in the government's fervor to take over private sector activities.

Economic practices that were perceived to be "kulak," or exploitative, were outlawed, including simple rural wage labor. In particular, the Asian and Arab commercial classes, who have been favored by the German and British governments and had obtained certain administrative and economic perquisites, seemed to be targeted. Earlier, one of the main thrusts of the indigenous cooperative movement had been to establish a countervailing power to such Asian and Arab trading blocks. After independence it was felt—and still is—that these commercial classes would always attempt to make huge short-term profits that would not be invested domestically. Certain nationalizations and economic controls could be interpreted as direct and indirect attempts at striking at the heart of these "foreign" commercial interests.

Other politically influential economic groups were, however, also subjected to "nationalization." In the same year the Arusha declaration was made, workers' right to strike and to negotiate labor contracts was curtailed. During 1967–73 "decentralization" of the public administration took place. Responsibility for planning and running of public services was transferred to the regions, and civil servants were transferred from central headquarters to the regions. But decentralization was in fact a direct assertion of the center at the expense of local government: district councils, which had been the primary elected centers of local power, were eliminated.

Economic performance slowed after the Arusha declaration. According to the official figures for the period 1967–71, GDP growth slowed to about 3 percent annually, less than half its pre-1967 rate. Growth in manufacturing, agriculture, and exports significantly decreased, whereas the volume of imports increased substantially. By 1970 a previously balanced external trade amount showed for the first time a significant deficit. Still, the external deficit was easily financed by long-term capital inflows, and the overall current and capital account balance remained positive. Investment rose quite fast, reaching about 20 percent of GDP by 1973. Public investment, mainly by parastatals, grew to reach 66 percent of total gross fixed capital formation by 1973. On average, about 30 percent of total expenditure was spent on basic needs (education, health, housing, water and electricity, and social security). In 1973 foreign borrowing made up more than half the financing of the government deficit and domestic inflation rates were kept low.

The implementation of the Tanzanian experiment was accelerated in the beginning of the 1970s. In 1973 a national price commission was set up to implement a national price policy. The commission was to set prices so that they would be "compatible with and conform to the principles of socialism and the political and social aspirations of the people." The commission initially targeted around a thousand items for price control, and this number would rise to approximately three thousand by 1978 (Mbelle 1990). In agriculture, a pan-territorial pricing system went into effect. By the mid-1970s, 80 percent of medium- and large-scale economic activity was located in the public sector, a figure that exceeded that of Soviet bloc states at a comparable time period (Young 1986). All throughout the 1970s, then, the parastatal sector expanded rapidly and, given the widespread adoption of the principle of "cost plus pricing," this expansion took place without clear reference to international standards of efficiency.

In 1973 several shocks hit the Tanzanian economy. The oil crisis of 1973 quadrupled world oil prices and necessitated sharp increases in import expenditure. Furthermore, the 1973/74 marketing year exposed the weakness of the pan-territorial pricing system. Drought conditions had driven up food prices in the parallel markets, and a major drop in domestic officially marketed food supplies resulted. Massive cereal imports, unprecedented for a country with an agricultural resource base as rich and varied as Tanzania's, were needed to feed the urban population. Cereal imports accounted for 22.1 percent of total imports in 1974. Although good weather returned in the next year, cereal imports continued at a high pace in 1975 because of a continued shortfall in domestic marketed supplies to the official channels.

The policy response of the government was to adopt a stabilization package that included measures such as substantial increases in the official producer pan-territorial food crop prices, a 40 percent increase in the minimum wage and progressively smaller wage increases for higher-echelon civil servants, sharp retail price increases of basic foods, price increases in the main consumer products, import allocation to essential goods, moderate curtailment of public infrastructure projects, mobilization of external finance to cover external deficits, and tax increases (Weaver and Anderson 1981).

At another level, however, the government reacted to the steep rise in the country's import bill by drawing on the first of the two tranches of the IMF quota and by increasing its domestic borrowing. As a consequence, the money supply increased by 22 percent, and inflation, as measured by the national CPI, jumped to 19.5 percent in 1974 compared to an annual average of 6.5 percent for the four previous years. Import expenditures (in U.S. dollars), despite the government's pledge to the opposite, increased by over 50 percent in 1974 and a further 5 percent in 1975, whereas export earnings stagnated. The current account deficit almost tripled from Tsh 755 million in 1973 to Tsh 2,037 million in 1974. The government increased long-term external borrowing and also received help from the IMF oil facility and the IMF compensatory finance facility, both relatively low-conditionality loans. Nevertheless, public expenditures sharply increased by 46 percent despite the fact that revenues only increased by 31 percent. The financing of the sharply increased domestic deficit (Tsh 1,862.5 million in 1975 vs. Tsh 851 million in 1974) came from a 99 percent increase in domestic borrowing and a 169 percent increase in foreign borrowing. Inflation continued at a 26 percent pace in 1975, and in October 1975 the government devalued the shilling by 11 percent.

Some of Tanzania's domestic problems in food and export crop production were related to the radical institutional transformations. Under the much-debated villagization program, the normally scattered African homesteads and plots were to be centralized into new socialist villages and block farms. It was an old idea dating from the British period and had been temporarily revived—with negative results—in the early 1960s. In the 1970s, the idea had once again become popular, even though the earlier experiences had been negative. In 1973 and 1974, impatience with the slow pace of socialist transformation of the peasantry resulted in villagization drives forced on the rural population by intimidation and physical force. The presumed increase in agricultural productivity in the rural areas never materialized. On the contrary, the net effect of villagization on agriculture was probably negative.

In 1976 the government compounded its institutional problems in the rural areas by abolishing the cooperative system, a system that had evolved in certain regions since the beginning of the century. The cooperative unions to which the primary cooperatives had marketed a certain region's main crops were replaced by crop-specific parastatals, the so-called crop authorities.

In the years 1976–77, the external constraint eased significantly as the world price increases in coffee and tea led to sharp increases in export earnings. The weather was normal in Tanzania and this, coupled with producer food price increases, led to increased domestic officially marketed supplies of cereals and reduced the need for cereal imports. Import expenditure in 1976 was thus sharply reduced by 22 percent, while export earnings increased by 49 percent, leading to an almost balanced trade account. The favorable export situation continued in 1977, although imports increased again considerably. Still, the continued inflow of long-term capital meant that the external account was quite healthy, and in fact by the end of 1977 the government had accumulated large foreign reserves. Within the framework of the Third Five-Year Plan (1976–81), the government implemented its basic industries strategy, which largely emphasized import substitution with respect to basic consumer goods and the intermediate and capital goods required to produce those consumer goods domestically. Given the relative abundance of foreign exchange, then, the government embarked on several import-intensive industrial and infrastructural projects (e.g., the Tanzania-Zambia railroad project).

In 1978 the government used much of the accumulated foreign exchange reserves to liberalize imports, at the recommendation of the IMF and the World Bank. The resulting high import expenditures in 1978 (up by 43 percent) were accompanied by a fall in exports of 18 percent as the coffee crisis eased and world prices declined. The result was a sharply increased trade deficit, which depleted the foreign exchange reserves, and external accounts that could not be balanced by capital inflows. Many of the official loans were transformed into grants. The situation became much worse, however, when war began with Uganda. From the end of 1978 through most of 1979, the government had to import arms and war-related supplies that are estimated to have cost about US$300 million, equivalent to more than half the 1977 export earnings.

The government's response to the trade deficit seemed inadequate. Whereas the 1978 and 1979 high import expenditures can be rationalized in terms of the spending of accumulated foreign exchange reserves in 1977, the Uganda war, and the 1979 oil price hike, it is hard to see why import expenditures stayed at a continued high level until 1981. In fact, apart from a small decline in 1981, it was not until 1982 that import expenditures were seriously curtailed. Moreover, examination of the volume and value indices of merchandise trade reveals that, through the barter terms of trade turned against Tanzania's trade between the "noncrisis," "nonboom" year 1975 and the early 1980s, it was the decline in the volume of exports that seems to have been a key element in the decline in export earnings. Svedberg (1991) estimates that 79 percent of Tanzania's decline in export earnings over the period 1970–85 can be explained by a decrease in export volumes, namely, officially marketed export crops. Only 21 percent of the decrease was associated with a decline in the barter terms of trade.

The well-documented decline in total and certainly per capita export crop production in the 1970s has been largely attributed to domestic policies. The major causes of poor export crop performance were the declining official producer prices of export crops both in real terms and relative to those of food crops (especially in the parallel uncontrolled markets), the declining efficiency and hence rising marketing margins of the official marketing system which resulted in large transfers from peasants to the state, and the nonavailability of incentive consumer goods in rural areas because of official rationing (Ellis 1983; Odegaard 1985; Bevan et al. 1989).

In early 1979 the government again resorted to the IMF and drew on its first tranche for balance-of-payment support. It also appealed for external aid. But aid was not easily forthcoming this time in the aftermath of the war. It is at this stage that major internal controversies first arose regarding policy reforms requested by the IMF (Biermann and Campbell 1989). Negotiations broke off in November 1979. In 1979 the government expanded the money supply considerably (by 47 percent) as public expenditure rose by 34 percent while revenues declined.

There have been wide debates concerning the causes of the crisis that led to the first IMF negotiations in 1979–80 and subsequent adjustment efforts. On the one hand, some have argued that the major causes were mainly external, such as the oil shocks, the drought, and the Uganda war (Green, Rwegasira, and van Arkadie 1980). Others have suggested that the causes were mostly internal and due to economic mismanagement (Sharpley 1985; Lofchie 1988, 1989). If we look at the Tanzanian balance of payments in U.S. dollars for 1970–88 as reported by the IMF (Table 9.1), a clearer picture of the evolution of the crisis emerges. The indicator of interest is the basic balance, namely, the sum of the balance of trade in goods and services and net long-term capital inflows. Table 9.1 shows that indeed in 1974 and 1978 Tanzania suffered the first major external crises.

The response of the Tanzanian government to the major shocks of the 1970s was inadequate. Current spending continued at a high pace throughout the 1970s. Moreover, whereas during the period 1976–80 current expenditures grew by an average of 13.9 percent annually, in the next three years they grew by an average of 23.7 percent annually, despite the foreign exchange crisis and the fact that recurrent revenues grew by only 18.7 percent annually. Development expenditures also continued at a high pace as a share of GDP. It thus appears that the government attempted to keep up the demand in the face of a supply shock.

The result was a sharp increase in inflation. As measured by the national CPI, prices that had increased by 6.9, 11.6, 6.6, and 12.9 percent in 1976, 1977, 1978, and 1979, respectively, jumped by 30.3 percent in 1980 and kept growing at a rate higher than 25 percent for the next several years. At the same time, the exchange rate kept appreciating because the government steadfastly refused to make more than token adjustments in the nominal exchange rate. The result was a sharp increase in the parallel market premium on foreign exchange, and hence further evasion of the official markets.

In conclusion, whereas external shocks were instrumental in initiating

Table 9.1. Balance of payments, Tanzania (million US$)

	Exports, Goods/ services (1)	Total transfers (2)	Total receipts, exports/transfs. (1+2) (3)	Imports, goods/ services (4)	Current account surplus (3+4) (5)	Long-term capital inflows (net) (6)	Ba bala (5 (
1970	321.8	12.8	334.6	−370.2	−35.6	71.6	
1971	349.7	5.8	355.5	−455.2	−99.7	137.7	
1972	411.7	−4.1	407.6	−473.3	−65.7	108.3	
1973	455.8	4.9	460.7	−568.2	−107.5	155.3	
1974	488.4	49.2	537.6	−822.9	−285.3	117.6	−1
1975	391.3	102.3	493.6	−823.6	−330.0	170.5	−1
1976	633.2	54.6	687.8	−722.3	−34.5	102.4	
1977	656.4	114.7	771.1	−843.5	−72.4	100.7	
1978	624.2	164.0	789.2	−1,262.6	−473.4	136.0	−3
1979	697.2	174.8	872.0	−1,218.5	−346.5	225.4	−1
1980	686.7	128.7	815.4	−1,249.0	−433.6	166.3	−2
1981	884.8	130.3	1,015.1	−1,187.0	−171.9	204.5	
1982	530.2	119.2	649.4	−1,030.4	−381.0	167.9	−2
1983	486.3	103.4	589.7	−785.7	−196.0	177.7	−
1984	480.3	158.6	638.9	−853.4	−214.5	89.7	−1
1985	432.8	456.3	889.7	−1,036.5	−147.4	−39.5	−1
1986	456.6	474.1	930.7	−1,105.0	−174.3	−24.7	−1
1987	447.1	707.0	1,154.1	−1,419.4	−265.3	−36.5	−3
1988	499.4	722.0	1,221.4	−1,479.8	−258.4	31.8	−2

Source: IMF, various years (b).
Note: Inflows are positive, outflows are negative.

the crisis, it was largely the weak internal adjustment efforts, coupled perhaps with optimism concerning the continuation of foreign capital and aid inflows, that eventually led to a largely uncontrollable situation in the early 1980s.

Crisis and Policy Reforms in the 1980s

In the beginning of the 1980s the collapse of the entire formal economic structure became apparent. Later, revised national accounts would show that real per capita GDP had declined in every year since 1975, and manufacturing output was contracting around 20 percent a year in the first half of the 1980s (Bevan et al. 1987). Statistically, the crisis was most apparent from the balance-of-payments situation: in 1980 official import expenditures were about US$1,249 million (27 percent of GDP) while export receipts stood at only $687 million (15 percent of GDP) (see Table 9.1).

The seriousness of the situation led to the first attempt at internal adjustment with the National Economic Survival Programme (NESP) in 1981. The basic tenet of the NESP was redistribution. It entailed a cut in salaries and social services, although minimum wages were raised in 1980 and 1981. It also entailed an increase in taxation of salaried workers and an increase in official producer prices.

Adjustment was, however, only partial: official prices did not reach the parallel market rates. Consequently, officially marketed production did not increase as inflation rose strongly due to monetary expansion. Large food imports of maize, which had not been necessary from 1976 until 1979, rose again from 21,000 metric tons in 1979 to 268,000 metric tons in 1980—nearly 20 percent of estimated national production. Such imports were less linked to national weather patterns than to the inefficiencies of the official marketing system. They continued at levels around 150,000 tons until 1986, when they dropped to 52,000 tons.

The government's "price scissors" policies proved unsustainable. Low real prices for maize in the official urban markets had been obtained through an implicit inflation tax on producers. But the existence of a substantial parallel food market had reduced the government's ability to assure the flow of maize. Faced with countervailing parallel market forces, the government had to either raise both producer and consumer prices or adopt an explicit subsidy. In 1980 it had chosen to adopt a subsidy, which resulted in the bizarre situation that maize flour (sembe) became cheaper than unmilled grain. This policy could not be sustained, given that 75 percent of the subsidy was financed directly through government expenditures and only 25 percent from a tax. At the same time, the pan-territorial pricing system also proved to be financially unsustainable and was abolished in 1981. The sembe subsidy would be abandoned only as late as 1984.[3] It had cost the government Tsh 1,233 million, or 2.2 percent of total government expenditures.

The shortage of foreign exchange affected the whole economy, but in particular the manufacturing sector. Ironically, capacity in the formal manufacturing sector had actually grown by 12.4 percent annually over 1973–81. The fact that such capacity was increasingly underutilized is often explained by pointing to a lack of infrastructure and of foreign exchange to acquire inputs (e.g., Wangwe 1983; Skarstein and Wangwe 1986). But such explanations themselves need explanation. Infrastructure had also been minimal before the 1970s; nonetheless, the industrial sector had grown at a comfortable rate. Moreover, throughout the entire post-independence period there had been by African standards a large influx of foreign capital, which was supposed to have built such infrastructure. The more fundamental explanation lies in the fact that the ineffective system of bureaucrat-

---

[3] The cost of the scheme was expected to increase to Tsh 800 million in 1984/85. Consequently, it was abandoned.

ically controlled industry had been stifling productivity ever since the late 1960s. At the same time it had become "addicted" to continuous injections of foreign capital, it had experienced a massive and protracted decline in productivity of both labor and capital (Morris 1985). The relative scarcity of foreign exchange of the late 1970s and early 1980s quickly brought the sector to its knees. In 1979–81 the growth rate of value added actually became negative relative to 1978 and the share of manufacturing declined to 7.4 percent of GDP, compared to 11.4 percent in 1972.

The short-lived NESP gave rise in 1982 to a structural adjustment program, which attempted to implement the recommendations of an impartial technical advisory group funded by the World Bank. The aims of the new program were to limit the public deficit by cutting recurrent expenditures without cutting social and economic services and to implement efficiency- and productivity-enhancing measures for parastatals while maintaining employment. Implementation of the program was, however, only partial and did not fundamentally address the structural problems of the formal economy. Simultaneously, the government undertook some small devaluations (10 percent in March 1982, 20 percent in June 1983) but kept tight import controls.

The government anticipated increased IMF and World Bank financing with its program, but this did not come because some of the basic preconditions for IMF agreement were not met. Those conditions consisted of a large devaluation; reductions of government expenditure, public sector borrowing, money supply, and the number of parastatals; increases in interest rates and producer prices for export crops; and liberalization of external and internal trade. This package of conditions was considered unrealistic by the Tanzanian government and consequently rejected. In the absence of an IMF agreement, most bilateral donors, with the exception of the Scandinavian countries, curtailed aid and thus aggravated the situation. In terms of total ODA flows, however, concessional funding continued at high levels. From 1980 to 1987, in every year Tanzania was still in the top three in absolute terms among all countries in sub-Saharan Africa. ODA funds developed from $668 million in 1980 to $888 million in 1987 (World Bank 1989a, tab. 7-1:185).

The crisis forced people to operate in parallel markets, and the foreign exchange premium (i.e., the excess of parallel over official exchange rate) rose to above 300 percent. In response, the government tried to ameliorate things with calls for legal behavior. A crackdown on parallel market activities followed, known as the "war on economic saboteurs" of 1983. Instead of viewing the emergence of parallel economic institutions as a sign that the official economy was not performing its basic functions, the government viewed the parallel economy as part of the problem. Given that the official economy continued to function poorly, the effect of the crackdown seems to have been merely to increase the riskiness of parallel market operations. Such increased transaction costs only drove up parallel market prices of food and basic consumer goods, thereby increasing the cost of living. Such a crackdown might also have amplified inequalities between socioeconomic groups through increased differentiation of access to parallel markets.

The government's exhortations for social and economic discipline became irrelevant as the formal economy was unable to supply even the most basic goods (e.g., sugar, soap, kerosene). Just to survive, people were forced to operate in the "second" economy.[4] The country quickly spun into a downward spiral of overvaluation, decreasing exports, deficit financing, more inflation, more overvaluation, and so on.

On the political front, the hard-liners, those who stressed development as a purely institutional process, were starting to lose ground to the technocrats, those who saw development more as a technical process.[5] The budget of 1984 represented a more technocratic approach to development. The crisis had caught up with the government, and it now fundamentally altered course. A 26 percent devaluation of the shilling, increases in producer prices, and removal of the sembe subsidy were adopted. The socialist distribution system was slowly dismantled: some user fees were raised (e.g., passports, secondary school fees, certain taxes), and the number of items controlled by the price commission was brought down from about two thousand to seventy-five.

An "own-funded" import scheme, announced in the 1984 budget and implemented in 1985, heralded the first real attempt at economic reform. The scheme basically allowed individuals to open a foreign exchange account with no questions asked about the origins of the foreign currency. Such foreign currency could then be used to officially import goods. The economic reaction was immediate. Shortages of fuel and spare parts disappeared quickly, and consumer goods reappeared in the shops. Apparently a massive, but hidden, import capacity stepped into the open. Official statistics showed a near doubling of the ratio of imports to GDP from 15.6 percent in 1985 to 28.4 percent in 1986. This ratio even grew to 40.4 percent in 1987. In fact, such growth was probably initially merely a "stepping out" of the parallel economy into the official sphere, and not a "real" increase. But the flow of own-funded imports continued unabated,

[4] In the minds of its people, Tanzania had now become "Bongoland" (*bongo* meaning "brain"). It was the country in which one had to use one's brain (i.e., find ways around the official system) in order to survive.

[5] In 1982 the constitution was once again amended to restore the separation of the powers of state and party. At the same time, elected district councils, which had been abolished in 1972, were reintroduced.

pointing to a substantial parallel export capacity in the Tanzanian economy.

In Table 9.2 we exhibit the official and parallel market rates as well as estimates of the real exchange rate (column 7) computed by a simplification of the IMF method and based on relative consumer prices (Sarris and Van den Brink 1993). This approach assumes that the national CPI in Tanzania is a reasonable proxy for the domestic price level, but that index is dominated by official prices and underestimates true prices. Hence the estimated overvaluation is smaller than the true one. It is quite evident from the table that the parallel market premium (column 6)—that is, the percentage by which the parallel market rate exceeds the official rate—jumped to more than 150 percent in 1980 and continued at levels above 200 percent until 1987. The computations of the real exchange rate show a significant appreciation of the shilling starting in 1981 and continuing until 1986. As noted above, this appreciation is an underestimate of the true appreciation.

In 1985, President Nyerere resigned and Mwinyi assumed the presidency. After this transition, the pace and extent of economic reforms picked up. The crisis and the lack of foreign exchange forced the government to abandon its efforts at dealing with the crisis alone, and it again started discussions with the IMF. The government seemed willing to reconsider its socialist paradigm but wanted change to occur slowly. This "gradual approach" to reform was based on the logic that current distortions should be alleviated slowly in order to lower adjustment costs and produce less political opposition to reform (Ndulu 1986).

By this time, however, distortions in the official system had rendered official markets largely inoperative and inefficient. In particular after 1984, many parallel market activities had become de facto legal and had compensated for or directly replaced formal economic institutions. Thus, the abolition of the sembe subsidy had not stirred up the riots that had been predicted. Similar informal adjustment was observed in the health sector, which saw a proliferation of private hospitals and clinics, and in the urban transport sector. In other words, large parts of the economy were already adjusting autonomously and operated under a market system of sorts. Government "reform," then, merely made official de jure what was already there de facto. Such structural adjustment was relatively costless.

In August 1986 an agreement with the IMF was reached based on the adoption of a three-year economic recovery program (ERP), the main elements of which were quite orthodox: (1) a very large exchange rate devaluation (from Tsh 17 per U.S. dollar in early 1986 to Tsh 190 per dollar in March 1990); (2) establishment of an open general licensing facility for foreign exchange allocation; (3) positive real interest rates; (4) removal of

Table 9.2. Official parallel and real exchange rates, Tanzania (Tsh per US$)

	Official rate		Parallel rate		Ratio		Real exchange rate index (1969=100) (7)
	End of period (1)	Average (2)	End of period (3)	Average (4)	End of period (3)/(1) (5)	Average (4)/(2) (6)	
1965	7.143	7.143	8.52	NA	1.193	NA	149.41
1966	7.143	7.143	8.64	8.80	1.210	1.218	141.65
1967	7.143	7.143	8.68	8.70	1.215	1.218	122.30
1968	7.143	7.143	8.25	8.50	1.155	1.190	109.93
1969	7.143	7.143	9.10	8.70	1.274	1.218	99.99
1970	7.143	7.143	10.45	10.10	1.463	1.414	102.93
1971	7.143	7.143	15.00	11.60	2.100	1.624	112.91
1972	7.143	7.143	15.40	15.20	2.156	2.128	109.79
1973	6.900	7.021	13.45	14.53	1.949	2.070	111.26
1974	7.143	7.135	14.00	13.46	1.960	1.886	112.28
1975	8.264	7.367	25.00	20.58	3.025	2.794	109.24
1976	8.324	8.377	20.40	21.93	2.451	2.618	111.10
1977	7.960	8.289	15.05	21.47	1.891	2.590	116.67
1978	7.415	7.712	11.75	13.07	1.585	1.695	116.31
1979	8.221	8.217	13.50	11.98	1.642	1.458	121.50
1980	8.182	8.197	26.50	21.02	3.239	2.564	105.59
1981	8.322	8.284	24.35	26.57	2.926	3.207	80.21
1982	9.567	9.283	29.15	32.60	3.047	3.512	69.15
1983	12.457	11.143	50.00	39.62	4.014	3.556	67.87
1984	18.105	15.292	70.00	57.08	3.866	3.733	66.06
1985	16.499	17.472	150.00	100.80	9.091	5.769	57.08
1986	51.719	32.698	180.00	165.00	3.480	5.046	160.07
1987	83.717	64.260	190.00	180.00	2.270	2.801	250.39
1988	125.000	99.292	230.00	210.00	1.840	2.115	276.43
1989	192.300	143.377	300.00	250.00	1.560	1.744	337.01

Sources: Cowett, 1985; Pick's Currency Yearbook, various years; IMF, various years (b); Bagachwa et al, 1990; column (7) computed by the authors.
Note: NA, not available.

price controls (from four hundred to twelve categories of goods); (5) increases in official producer prices for export crops; (6) agricultural marketing reforms; and (7) establishment of fiscal and monetary targets. The ERP was supported by an IMF standby loan and finalized in October 1987 by a three-year structural adjustment facility of SDR 64.2 million. The adoption of the program and agreement with the IMF led to substantial foreign inflows by the World Bank and by other donors. In November 1987 an International Development Association (IDA) multisectoral rehabilitation credit ($130 million) from the World Bank was accorded with supplemental financing approved in 1988. Bilateral donors also supported the ERP and joined the rescue effort: the Paris Club rescheduled Tanzania's debt in September 1986.

Still, the long-term trend of falling export earnings continued. The main problem was in agricultural exports, which constituted 80 percent of export earnings. Export crop production of such crops as cotton, sisal, and tobacco continued to fall. Only the coffee sector seemed resilient, probably because it had a strong indigenous cooperative system and operated under an auction system that ensured that world prices were partly passed on to the producer. With respect to the export sector as a whole, however, no fundamental restructuring of the parastatal-dominated export sector was undertaken. Moreover, Tanzania's fledgling export sector received a further downward shock when world market coffee prices collapsed in 1987 and coffee export revenues fell by 41 percent.

Gradual reform may well have been counterproductive. The fact that quasi-adjustment measures are often double-edged is best illustrated by the developments in the financial sector. In 1987, as part of the ERP, the nominal savings interest rate was raised from 10 percent in 1986 to 21.5 percent. In 1988 the interest rate was further raised to 29 percent. These increases seemed substantial, but in fact, given the high domestic inflation rates (30 percent in 1987 and 31 percent in 1988), the real interest rates on savings remained negative. The inflation was directly linked to the financing of parastatals by the nationalized Tanzanian banking system. In 1970 the share of total lending of the banking sector to official entities had been 50 percent, but by 1985 it stood at approximately 90 percent. The increases in nominal interest rates implemented under the ERP accelerated inflation, because interest payments on domestic debt automatically increased also (Collier and Gunning 1991, 538).[6] Additionally, the average recovery rate on loans from the National Bank of Commerce over the period 1984–90 was only 27.1 percent.

From 1984 to 1988, then, gradual reform only reduced the number of

---

[6] In April 1991 a parliamentary act was passed which allowed private banking.

parastatals from 425 to 400 (Mbelle 1990). In 1988 the public sector claimed 78 percent of total domestic credit, whereas 41 parastatals received 97 percent of commercial credit against 3 percent by nine private enterprises. The process continued for the rest of the decade: from June 1989 to June 1990, 77 percent of lending of commercial banks went to the marketing of agricultural produce, and parastatals accounted for over 86 percent of outstanding bank credit in 1990.

The socialist price system, however, had been dismantled: price controls were lifted from over three thousand items in 1978, to four hundred in the beginning of the 1980s, to only nine items in 1991. And government crackdown on the parallel markets witnessed a revival. Just as the late premier Sokoine had led the "war against economic saboteurs" in 1983, during which enforcement of official institutions rose dramatically, at the end of 1990 Minister of Home Affairs Mrema started a similar war. In line with the times, "economic incentives" were not attached to the exposure of illegal practices. The finder was entitled to 10 percent of the value of the fraud. Without a fundamental reform of the official economy, it was again difficult to gauge whether such crackdowns attained anything more than a mere increase in transaction costs in the parallel markets.

Such parallel markets were, however, still necessary to compensate for the inefficiencies of formal markets. The minimum wage, for instance, was Tsh 2,500 per month in 1991, and even top civil servant wages were close to the poverty level (Sarris and Van den Brink 1993). Such below-subsistence wages induced widespread parallel market activities.[7] Induced corruption, then, became generalized across all strata of the population, fundamentally eroding formal social and economic institutions.

In 1989 the government adopted the new three-year Economic and Social Action Programme (ESAP), which stresses social dimensions of adjustment and poverty alleviation while continuing the pace of economic reforms. The plan did not receive immediate support from the IMF, since Tanzania was still seen to be slow on the issues of foreign exchange rate adjustment and a restructuring of the financial sector. An official IMF agreement was to be reached by mid-1991 in the form of an enhanced structural adjustment facility worth about $100 million in each of the three years.

By the end of the 1980s, Tanzania was in certain ways back where it had started in the 1960s. Foreign exchange reserves in 1989 were $54.2 mil-

---

[7] Because the Arusha declaration entailed a leadership code that precluded party members from having any source of wage or business income other than their official wage, the economic situation forced even party members to engage in illegal activities. In this respect, party members in the rural areas are said to be lucky, since they can at least grow their own food.

lion. They were $61.3 million in 1966. In the process, however, Tanzania has incurred a total foreign debt of $6 billion. The slow pace of adjustment kept crucial sectors of the economy stationary. In agriculture, although the cooperative system had been restored in 1984, the government was still controlling the cooperative unions and marketing boards. As of June 30, 1990, the total debt of the cooperative unions and the marketing board was Tsh 53.667 billion, or 15.3 percent of 1989 GDP at current prices (Mbelle 1990). Still, Tanzanian glasnost and perestroika fueled democratic "independence" movements of primary cooperative societies in an attempt to form their own unions.

The World Bank extended several loans in 1990, among which was a $200 million IDA agricultural adjustment program. Agreements were also reached for IDA projects in the health sector ($70 million) and education ($67.5 million). Moreover, the World Bank took the lead in an integrated road project of a total of $871 million in which the Bank itself participated with $180 million. The project will attempt to ameliorate the appalling state of the national road network. It is unclear, however, whether the "absorption capacity" is there to handle such a large infrastructural project. In more than one way, then, Tanzania is back where it started in the 1960s: it still needs to build its basic infrastructure and it is still uncertain whether it has the management capacity to do so.

Economic Performance during the 1980s

Growth and Macro Performance under Adjustment

In Table 9.3 we summarize some of the sectoral and macro developments over the period 1976–89 as exhibited by official statistics. After relatively stagnant GDP over the period 1979–83, the significant growth observed since 1983 is due mostly to the growth in agricultural production; manufacturing output has stagnated. The output of the trade sector has not kept up with that of agriculture, and the output of public administration, after a brief decline in the first two years of the ERP, grew in 1989 to levels close to its pre-ERP level. The most interesting development seems to be occurring in gross fixed capital formation: private sector capital has grown very fast since 1983, whereas public sector capital declined and has remained stagnant at its already low pre-ERP levels.

Examining the implicit deflators for the various sectors, one can see that the deflator for agriculture has grown faster than all others, except the one for trade. Notice the very small rise in the deflator for public administration, which reflects the small rise in public sector wages in the face of inflation.

Table 9.3. Recent macroeconomic developments, Tanzania

	1976	1977	1978	1979	1980	1981	1982	1983	1984	1985	1986	1987	1988	1989
GDP real index	100.0	100.4	102.5	105.5	108.2	107.6	108.3	105.7	109.3	112.1	115.5	120.0	124.9	130.6
GDP agri. total index	100.0	101.0	99.5	100.2	104.1	105.1	106.6	109.6	114.0	120.8	127.8	133.4	139.4	145.7
GDP manuf. real index	100.0	94.0	97.1	100.4	95.4	84.7	82.0	74.8	76.8	73.8	73.8	73.8	77.8	81.8
GDP constr. real index	100.0	103.5	88.6	99.4	105.4	100.7	105.2	62.1	74.7	68.0	79.8	81.6	85.6	92.9
GDP trade real index	100.0	98.0	98.5	100.0	100.0	96.0	94.0	92.0	93.0	93.8	104.2	109.6	114.0	119.0
GDP public admin. real index	100.0	106.6	128.0	139.0	136.1	151.6	151.8	151.3	151.5	154.4	137.7	138.5	142.7	147.0
GDP fixed cap. form. real index	100.0	110.7	110.7	124.1	108.0	112.5	117.3	78.3	114.2	140.0	135.8	134.3	129.8	NA
GFCF[a] public real index	100.0	114.0	104.2	110.9	107.2	108.0	112.9	66.0	76.5	98.1	94.1	66.6	60.7	NA
GFCF private real index	100.0	106.7	119.4	141.7	111.0	118.7	123.2	95.0	164.8	196.0	191.9	225.1	222.4	NA
GFCF buildings real index	100.0	115.3	102.0	117.1	119.1	141.2	125.3	84.3	98.8	95.7	105.3	137.9	126.4	NA
GFCF other works real index	100.0	96.7	79.7	88.5	97.0	76.2	91.8	45.8	52.2	48.8	62.3	49.2	78.6	NA
GFCF machin. eq. real index	100.0	118.1	133.3	148.9	112.4	124.7	130.3	96.5	158.6	213.3	193.1	186.2	163.0	NA
GDP deflators (1976=100)														
GDP at factor cost total	100.0	118.2	128.7	141.4	159.9	188.4	224.2	273.6	330.3	445.2	563.0	743.0	967.7	1,242.0
Agriculture	100.0	121.7	139.0	162.5	176.6	213.8	274.4	330.2	400.5	560.2	728.2	1,002.3	1,413.7	1,571.0
Manufacturing	100.0	124.5	141.4	137.1	152.7	189.0	189.3	231.5	274.8	321.2	429.5	712.9	1,118.1	1,320.3
Construction	100.0	121.4	134.4	139.8	160.7	181.4	200.3	228.1	251.7	342.9	444.1	480.9	601.1	719.0
Trade	100.0	122.5	139.0	153.0	166.0	201.1	255.4	311.9	395.7	533.2	658.4	834.3	1,353.5	1,585.9
Public admin.	100.0	104.0	95.9	102.7	124.2	133.3	153.2	208.1	242.7	296.9	314.4	398.0	481.2	651.9
Terms of trade (TOT) agri./manuf.	100.0	97.7	98.3	118.5	115.7	113.2	145.0	142.6	145.8	174.4	169.5	140.6	126.4	119.0
TOT agri./constr.	100.0	100.2	103.4	116.2	109.9	117.9	137.0	144.8	159.1	163.3	164.0	208.4	235.2	218.5
TOT agri./trade	100.0	99.3	100.0	106.2	106.4	106.4	107.4	105.9	101.2	105.9	110.6	120.1	104.4	99.1
TOT manuf./constr.	100.0	102.5	105.2	98.1	95.0	104.2	94.5	101.5	109.2	93.7	96.7	148.2	186.0	183.6

Source: Computed from the national accounts and Bank of Tanzania yearbooks.
Note: NA, not available.
ᵃGFCF = gross fixed capital formation.

In terms of relative prices, it can be clearly seen that the terms of trade of agriculture after 1983 (these reflect official prices) initially grew vis-à-vis manufacturing but since 1986 have declined. But the terms of trade of agriculture relative to construction, a largely nontraded activity, have increased substantially. The same is true of the terms of trade of manufacturing (which is largely an import substitute) versus construction. It is not clear to what extent these trends are influenced by the use of official prices in the national account statistics.

One of the major problems facing the analyst of the Tanzanian economy is that the numbers that presumably describe the evolution of aggregate variables suffer from inaccuracies due to incomplete coverage as well as inaccurate estimates of the activities covered. In Tanzania a large segment of the formerly active private sector either stopped producing or went underground (i.e., kept producing but with the higher cost of evading the various government controls). Given official policy, these underground activities were not recognized and hence no effort was made to estimate them. The resulting "hidden" or "second" or "parallel" or "underground" economy in Tanzania is thought to have gradually grown as the public controls became more binding, especially as shortages of various goods became widespread with the post-1979 foreign exchange crisis. A major question of macroeconomic relevance is whether the hidden economy has followed the fluctuations of the officially observed economy, or, if not, whether it has tended to compensate or accentuate the observed secular decline in economic activity.

This problem has been analyzed by Maliyamkono and Bagachwa (1990) and by Sarris and Van den Brink (1993). Table 9.4 presents the results of some estimations of the evolution of total GDP, including the second economy, over the period 1967–88. What is reported is the total real GDP (the sum of the official real GDP and the second economy real GDP), the yearly changes in total and official real GDP, and the size of the second economy as a share of official and total GDP. It is interesting to note that a very different picture of total real GDP obtains from this table compared to the official figures. Years of decline in total real GDP appear to be 1973, 1979, 1982, 1983, and 1986. Growth appears to have been quite strong in 1984, the first year of liberalization, and has continued to be strong since then. The first year of the ERP, 1986, appears to have been marked by a fall in total real GDP, though the official figures indicate a rise. Since 1987 growth appears to have occurred at rates faster than officially reported.

Of interest also are the last two columns, which indicate the share of the second economy in the total (official plus unobserved) and official GDP. Although the share of the second economy in official GDP appears to have fluctuated considerably between 1967 and 1988, with peaks in 1981 and

Table 9.4. Size of the second economy, Tanzania (million 1976 Tsh)

	Official real GDP (1)	Real second GDP (2)	Total real GDP (1+2) (3)	Yearly change of total GDP (%) (4)	Yearly change of official real GDP (%) (5)	Second GDP as % of total GDP (6)	Real second GDP as % of real official GDP (7)
1967	14,438	3,996	18,434	NA	NA	21.68	27.7
1968	15,186	4,017	19,203	4.17	5.18	20.92	26.5
1969	15,465	3,971	19,436	1.21	1.84	20.43	25.7
1970	16,362	6,633	22,995	18.31	5.80	28.85	40.5
1971	17,046	7,445	24,491	6.51	4.18	30.40	43.7
1972	18,192	9,558	27,780	13.43	6.72	34.51	52.7
1973	18,748	7,169	25,917	-6.71	3.06	27.66	38.2
1974	19,217	8,078	27,295	5.32	2.50	29.60	42.0
1975	20,352	8,320	28,672	5.04	5.91	29.02	40.9
1976	21,652	8,797	30,449	6.20	6.39	28.89	40.6
1977	21,739	9,401	31,140	2.27	0.40	30.19	43.2
1978	22,202	11,546	33,748	8.38	2.13	34.21	52.0
1979	22,849	9,816	32,665	-3.21	2.91	30.05	43.0
1980	23,419	9,705	33,124	1.41	2.49	29.30	41.4
1981	23,301	11,634	34,935	5.47	-0.50	33.30	49.9
1982	23,439	9,042	32,481	-7.02	0.59	27.84	38.6
1983	22,882	8,124	31,006	-4.54	-2.38	26.20	35.5
1984	23,656	11,179	34,835	12.35	3.38	32.09	47.3
1985	24,278	12,381	36,659	5.24	2.63	33.77	51.0
1986	25,008	9,983	34,991	-4.55	3.01	28.53	39.9
1987	25,972	10,716	36,688	4.85	3.85	29.21	41.3
1988	27,039	11,314	38,353	4.54	4.11	29.50	41.8

Source: Sarris and Van den Brink 1993, courtesy of New York University Press.
Note: NA, not available.

1985, it seems that with minor fluctuations its share in total GDP has stayed remarkably constant, around 28–32 percent throughout the two decades tracked. It must, of course, be noted that all these conclusions depend strongly on the method of estimating the second economy GDP.

It must also be realized that a currency-based approach to estimating second economy GDP (which underlies the calculations above) neglects barter transactions or those involving foreign exchange. These in fact could have been sizable in Tanzania all throughout the 1970s and 1980s, for many of the agricultural export crops and mining sites are close to border areas. Barter transactions or transactions involving foreign currency would not be captured by domestic currency statistics, and in fact most of the parallel export trade occurs in convertible currencies. This factor would tend to underestimate the domestic currency-based estimates of second economy GDP.

If the numbers of Table 9.4 represent a more reliable picture of the evolution of the Tanzanian economy, then it appears that, apart from the external crisis years (e.g., 1973 and 1979), years of extreme government controls (1982 and 1983), and the first adjustment plan year (1986), growth occurred at reasonable rates, and in most years at rates above population growth in contrast to the official statistics.

The history of adjustment efforts in Tanzania over the 1980s reveals efforts at expenditure control but consistently excessive optimism in terms of economic targets. Ndulu (1988) presented tables contrasting structural adjustment targets and actual performance of several macro indicators over the years of the first adjustment efforts, 1982–85. In terms of fiscal targets, the performance during that period was better than planned. But the expected increases in external resources, both in terms of export earnings and as external loans and grants, fell far short of planned targets. The resulting serious external gap pushed the government to resort to domestic inflationary finance, and money supply growth targets were surpassed in all but the first year of the adjustment plan.

The targets set for the 1986/87–88/89 ERP were also quite ambitious: an average rate of economic growth of 4–5 percent annually, a progressive reduction in the rate of inflation to less than 10 percent in 1988/89 from over 30 percent in 1985, improvements in the external position through faster export growth, increases in utilization rates in manufacturing from 20–30 percent in 1985 to 60–70 percent by 1989, and increases in export earnings of 16 percent annually.

In terms of official GDP growth, the performance since 1985 has indeed been one of revival, with agriculture leading the way. All sectors, with the exception of mining and quarrying, have improved their 1980–85 performance. In 1989 first estimates indicated a real GDP growth rate of 4.4

percent with all sectors growing, agriculture in particular growing at 4.6 percent and manufacturing at 5.1 percent. Despite this growth, however, only agriculture, electricity and water, commerce, finance, and public administration exhibited real total product level higher in 1989 than in 1980 (again measured by official statistics).

On the external account front, the target of 16 percent growth in export earnings was not attained. Total export earnings from the six major agricultural exports were at their lowest level of the decade in 1989 at US$180.4 million, after a brief revival in 1986 due to a small coffee boom, compared to $264 million in 1980 and $196.1 million in 1985 at the depth of the crisis. Only cotton appears to have exhibited a strong export volume increase, generally because very little marketing liberalization in agricultural export crops took place (see below). But nontraditional exports, especially manufactured products, staged a strong recovery. From a continuous decline between 1980 and 1985 from US$241.5 million to $90.5 million, they recovered to $214.8 million in 1989. This recovery was undoubtedly aided by the generous export retention scheme instituted under the ERP. Imports, on the other hand, surpassed their 1980 level in 1987, and because of donor support as well as the own-fund import scheme they stayed quite close to their targeted levels.

On the inflation front, the performance has not lived up to the targets. The change in the national CPI for the years 1986, 1987, 1988, and 1989 has been 32.4, 30.0, 31.2, and 25.9 percent, respectively. This is no different from the performance during 1980–85, when annual inflation ranged between 26 and 36 percent.

The ambitious targets on industrial production also did not materialize. The doubling or tripling of capacity utilization envisioned in 1986 implied a corresponding increase in output. But of thirty-one industries whose output for the period 1980–88 is reported in the 1989 Bank of Tanzania Economic and Operations Report, only fifteen exhibited an increase in production between 1985 and 1988, and of these only ten experienced a more than 50 percent total increase over the three-year period—this despite the fact that the bulk of imports over the period were intermediate goods and machinery.

Deficits and Monetary Adjustments

Table 9.5 presents a summary of central government revenues and expenditures from the period 1975/76 to the most recent available year. Since the fiscal year 1975 there has been a public sector deficit amounting to as much as 14.5 percent and never less than 7.5 percent of GDP at market prices. In the composition of public expenditure, current expenditure has

Table 9.5. Summary of central government operations, Tanzania

	1975/76	1976/77	1977/78	1978/79	1979/?
Current revenue (million Tsh)	3,848.4	5,200.2	6,005.6	6,435.7	7,360.
Current expenditure (million Tsh)	3,933.2	5,014.8	5,598.9	7,227.9	7,419.
Surplus/(deficit) (million Tsh)	(84.8)	185.4	406.7	(792.2)	(58.
Development expenditure (million Tsh)	2,253.0	3,244.0	3,331.0	4,750.0	5,184.
Total surplus/(deficit) (Million Tsh)	(2,337.8)	(3,058.6)	(2,924.3)	(5,542.2)	(5,242.
Deficit as % of total expenditure	37.8	37.0	32.7	46.3	41.
Share of GDP at market prices (%)					
Current revenue	18.53	19.41	19.34	19.01	18.
Current expenditure	18.32	18.38	19.94	20.18	20.
Surplus/(deficit)	0.21	1.03	(0.60)	(1.17)	(1.
Development expenditure	11.26	11.39	12.56	13.69	11.
Total surplus/(deficit)	(11.05)	(10.36)	(13.16)	(14.86)	(13.

Source: World Bank 1989l; authors' computations.
Note: For each calendar year, the average of the two adjacent fiscal years was used in the numerator of relevant fractions.

been quite high, whereas development expenditure since 1982 has declined to about half its pre-1980 share of GDP. A large share of total public expenditure is for internal transfers such as interest on public debt and subsidies to parastatals. As a share of official GDP, the total deficit, after declining significantly between 1982/83 and 1984/85, increased again to the levels of the boom years 1975–1977 but stayed below the very high levels of the early 1980s.

Table 9.6 exhibits the composition of public revenues and expenditure from 1975/76 to 1986/87. Trade taxes (import and export duties) constitute a small portion of total public revenue, which from a high share of about 21 percent in 1978/79 declined to only 6 percent in 1982/83, to recover at 13 percent in 1986/87 and to an estimated 15 percent in 1988/89. It is quite noticeable that, in contrast to many primary exporting countries, export taxes since 1980 are almost negligible. Export producers are, however, heavily taxed implicitly, with the implicit revenue being largely absorbed as parastatal marketing cost. The bulk of public revenue comes from consumption and excise taxes, as well as from personal and income taxes.

The most noticeable trends in the composition of public expenditure since 1975 are the increase in the share of general public services (mainly administration), the sharp decline in the shares of education, health, and economic services to about half their 1975/76 shares, and the quadrupling of the share of servicing of public debt.

Education in most countries, and Tanzania is no exception, tends to be strongly associated with higher income. In Tanzania the emphasis on

)80/81	1981/82	1982/83	1983/84	1984/85	1985/86	1986/87
528.7	9,777.4	12,313.9	15,467.3	18,252.7	20,632.0	31,102.0
531.6	13,040.0	14,055.0	15,272.3	21,370.6	26,381.0	42,793.6
)02.9)	(3,262.6)	(1,741.1)	195.0	(3,117.9)	(5,749.0)	(11,690.8)
759.0	5,196.0	4,380.0	4,712.0	5,949.0	5,817.0	11,832.0
761.9)	(8,458.6)	(6,121.1)	(4,517.0)	(9,066.9)	(11,566.0)	(23,322.8)
40.3	46.4	33.2	22.6	33.2	35.9	43.1
18.64	18.97	19.70	18.97	16.12	15.98	18.05
22.98	23.27	20.80	20.61	19.79	21.36	22.57
(4.34)	(4.30)	(1.10)	(1.64)	(3.68)	(5.39)	(5.55)
10.14	8.22	6.45	6.00	4.88	5.45	6.38
(14.48)	(12.52)	(7.54)	(7.64)	(8.55)	(10.78)	(11.93)

broad-based literacy and primary education in the 1970's is regarded as having produced one of the most literate populations in Africa. The International Labour Office (1982) estimates that from 1970 to 1980 the illiteracy was reduced from 70 to 21 percent, especially after the introduction of universal primary education in 1977. In 1981 primary school enrollment was 8.5 times higher than in 1981, 3.54 million against 0.42 million, and over the same period secondary education increased 4.9 times from 14,000 to 68,000. It is uncertain, however, to what extent the quality of the educational services offered by the public school system has been able to keep up with the dramatic quantitative increases. In the recent period, primary and secondary school enrollment rates have declined, whereas enrollment in private secondary schools has increased much more than in public schools.

In Table 9.7, as an illustration of the general crisis in public services, we exhibit some statistics on the growing crisis in education. It can be clearly seen that, whereas in the late 1970s and early 1980s slightly more than 90 percent of all primary school–aged children were enrolled in primary schools (almost totally public), by 1986 that proportion had dropped to less than 80 percent. Whether this is a trend cannot be ascertained with the few years of data available. But if the adjustment program reductions in education spending (see Table 9.6) result in lower enrollment rates, this will have obvious significant long-term consequences for national human capital.

The secondary school enrollment data illustrate that only a very small proportion of eligible students (around 3.2 percent) are enrolled, and even

Table 9.6. Composition of revenue and expenditure, Tanzania

	1975/76	1976/77	1977/78	1978/79	1979/80	1980/81	1981/82	1982/83	1983/84	1984/85	1985/86	1986/87
Share of revenues (%)												
Import duties	11.7	6.2	10.1	13.9	11.0	7.7	6.9	5.9	6.2	8.4	7.5	13.0
Export duties	4.1	16.1	8.9	7.0	6.3	2.5	0.2	0.1	0.1	0.1	0.0	0.0
Consumption/excise duties	42.4	33.6	36.4	40.6	40.2	51.3	52.5	49.5	51.6	55.9	52.1	53.4
Income/personal taxes	28.4	26.8	23.1	29.5	33.0	32.2	33.2	30.8	26.5	25.8	30.9	23.0
Other taxes/income sources	13.4	17.2	21.6	9.0	9.5	6.2	7.3	13.7	15.6	9.9	9.5	10.6
Share of expenditures (%)												
General public service	15.8	17.4	16.0	16.0	14.7	16.4	17.9	17.1	22.0	29.9	28.9	25.5
Defense	12.2	12.3	13.5	24.6	7.7	11.0	12.5	13.3	12.8	13.9	10.4	14.6
Education	14.1	13.6	13.3	11.3	11.2	11.8	12.5	13.2	11.7	7.3	7.3	6.4
Health	7.1	7.1	6.7	5.3	5.0	5.4	5.4	5.1	5.5	5.0	4.3	3.7
Social security/welfare	0.4	0.2	0.2	0.3	0.4	0.3	0.3	0.3	0.3	0.5	0.2	0.3
Housing/community amenities	1.8	1.2	0.9	0.8	1.0	1.2	1.0	1.1	1.0	1.0	0.6	0.4
Other community/social services	2.4	2.3	1.7	1.8	1.9	2.1	2.1	2.0	2.0	2.2	0.5	0.5
Economic services	36.9	38.1	34.1	35.1	36.1	34.8	29.8	27.0	26.0	24.2	22.8	16.5
Other purposes	9.2	7.9	13.6	4.8	22.0	16.9	18.5	21.0	18.8	16.1	25.0	32.1
Public debt	7.3	5.9	7.0	3.9	7.5	11.4	17.8	20.2	18.1	15.4	24.2	31.4

Source: Computed from data in World Bank 1989l.

Table 9.7. Primary and secondary education statistics, Tanzania

	1978	1979	1980	1981	1982	1983	1984	1985	1986
Primary school pupils (PSP)									
TOTAL (thousands)	2,994	3,211	3,367	3,539	3,513	3,561	3,494	3,170	3,159
% in public schools	97.3	99.6	99.8	99.8	99.7	99.8	99.7	99.7	99.9
Population in 6–12 age group (thousands)	3,461	3,624	3,792	3,834	3,874	3,914	3,955	3,996	4,037
PSP as % of 6–12 age group	86.5	88.6	88.8	92.3	90.7	91.0	88.3	79.3	78.3
Secondary school students (SSS)									
TOTAL (thousands)	64.2	68.3	67.3	67.6	69.1	71.2	74.2	83.1	91.6
% in public schools	65.1	59.0	57.7	56.7	56.4	55.8	54.7	50.9	47.4
Population in 13–17 age group (thousands)	1,785	1,843	1,903	2,034	2,174	2,323	2,483	2,653	2,835
SSS as % of 13–17 age group	3.4	3.7	3.5	3.3	3.2	3.1	3.0	3.1	3.2

Note: The population in the given age group was interpolated from 5-year intervals reported in World Bank (1989) from their population, constant fertility scenario, according to the growth rates inferred from those data.

this proportion seems to have declined since the late 1970s. The low overall proportion is explained by the official bias against secondary school education of the government. Of interest, however, is the diminishing share of secondary school students that attend public schools. In essence, between 1978 and 1986 the total number of secondary students enrolled in public schools has stayed about constant, while the population in the relevant age group has increased by about 59 percent. It is quite possible that the decline in public education expenditures leads to a growing share of children's education expenses being born by parents, with the consequence that children of wealthier parents receive proportionally more education and hence future access to high-paying jobs. This would tend to accentuate income inequality in the future.

Table 9.8 presents the public financing requirements on a fiscal year basis from 1976 to 1989 and the method of financing them. All throughout the period, the major sources of finance have been foreign grants and loans and import support and counterpart funds (columns 6, 7, and 8). During the crisis years 1980–86, this share declined considerably and the main method of finance became domestic borrowing (column 9), of which bank borrowing became quite important (column 10). Since 1986 bank borrowing (financing through money creation) appears to have declined dramatically, and in fact in fiscal year 1988/89 it appears that some of the domestic public bank debt was redeemed.

If this picture truly reflects the mode of financing, then the money supply (M1 or M2) must not have expanded by very much since 1986, in accordance with ERP guidelines. Analysis of the money supply components shows, however, that both currency in circulation and money supply (M2) have grown since 1986 by more than 30 percent annually, much larger than that justified by reported domestic bank borrowing.

The explanation of this apparent paradox, as Collier and Gunning (1991) argue, is that in Tanzania there is no commercial banking system in the traditional sense and the main function of the existing commercial banking system, which is basically only one bank (the National Bank of Commerce), is to extend loans to the parastatals. In fact, 70–90 percent of all lending goes to official entities. A large part of that is to cover current parastatal deficits, especially those of official agricultural marketing authorities (the chief being NMC), and cannot be repaid. It should therefore be treated very much like money creation.

Between 1982 and 1988, 50 percent of the increase in total commercial lending is accounted for by increases in lending to marketing parastatals. For 1985–88 the ratio is 49 percent. If we consider the changes from 1985 to 1988 in the money supply (M2) and the total commercial bank lending to official entities, then the latter makes up 80.1 percent of the total change

Table 9.8. Financing of the public deficit, Tanzania (million Tsh)

	Public financing requirement (1)	Foreign grants and loans (2)	Import support and counterpart funds (3)	Domestic borrowing (4)	Nonbank borrowing (5)	Share (%)				
						(2)/(1) (6)	(3)/(1) (7)	(2)+(3)/(1) (8)	(4)/(1) (9)	(4−5)/(1) (10)
1976	2,461	1,033	0	1,422	NA	41.97	0.00	41.97	57.78	NA
1977	1,604	1,402	0	202	NA	87.41	0.00	87.41	12.59	NA
1978	2,836	1,529	0	1,320	NA	53.91	0.00	53.91	46.54	NA
1979	3,891	1,930	0	1,940	NA	49.60	0.00	49.60	49.86	NA
1980	6,668	2,268	802	3,511	NA	34.01	12.03	46.04	52.65	NA
1981	6,380	1,845	875	3,686	1,374	28.92	13.71	42.63	57.77	36.24
1982	4,018	1,838	1,022	1,112	−1,499	45.74	25.44	71.18	27.68	64.98
1983	4,837	1,858	751	2,215	−1,231	38.41	15.53	53.94	45.79	71.24
1984	7,995	1,895	611	5,486	1,913	23.70	7.64	31.31	68.62	44.69
1985	7,931	2,658	1,236	4,016	2,144	33.51	15.58	49.10	50.64	23.60
1986	9,709	2,045	1,489	6,182	1,257	21.06	15.34	36.40	63.67	50.73
1987	18,712	6,235	8,001	4,514	2,858	33.32	42.76	76.08	24.12	8.85
1988	27,719	9,881	12,909	4,938	2,355	35.65	46.57	82.22	17.81	9.32
1989	29,739	10,619	20,889	−1,774	2,686	35.71	70.24	105.95	−5.97	−15.00

Source: Computed from data in Bank of Tanzania, Economic and Operations Report, June 1989 and June 1982.
Note: NA, not available or unable to compute.
[a]Fiscal year starting the previous year; e.g., 1976 is the 1975/76 fiscal year.

Table 9.9. Financing of the current account deficit, Tanzania (million Tsh)

	1983	1984	1985	1986	1987	1988
Current account	−3,394	−5,491	−6,500	−10,633	−28,655	−37,323
Capital acct. (MLT) net	2,038	−1,133	−323	−688	−450	3,932
Supplier's credit (net)	1,134	1,679	−559	−1,816	321	477
Imprt. sup./exc. finance	675	751	1,048	2,720	2,744	9,998
Errors/omissions	−664	1,768	−560	−2,154	7,982	−4,217
Overall balance	−211	−2,426	−6,894	−12,571	−18,058	−27,133
As % of GDMP[a]	−4.8	−6.2	−5.4	−6.7	−13.1	−12.
Overall balance at % of GDPMP	−0.3	−2.7	−5.7	−7.9	−8.2	−8
Financing						
IMF (net)	−333	−421	−248	429	2,506	−5,563
Reserve decrease (− increase)	−168	47	175	−816	−643	4,448
Arrears (+ increase)	713	2,800	6,967	−27,635	−64	15,379
Debt rescheduling	0	0	0	37,479	12,081	12,870
Others	−1	0	0	3,114	4,178	−1

Source: Bank of Tanzania, Economic and Operations Report for the year ending June 30, 1989.
[a]GDPMP = GDP at market prices.

in M2. It therefore appears that the change in domestic money supply is indeed closely associated with bank credit to official and especially agricultural marketing parastatals. This obviously makes agricultural marketing reform an item with important macroeconomic consequences.

Table 9.9 exhibits the mode of financing of the external deficit during 1983–88. It is quite obvious that accumulation of arrears and debt rescheduling have been the main methods of financing a growing current account deficit since the onset of the ERP. The current account deficit has grown from −5.4 percent of GDP at market prices in 1985 to −12.0 percent in 1988, and capital flows have not ameliorated the situation by much. It must be recalled that own-funded imports are counterbalanced by an equivalent current transfer item on the credit side, so they do not affect the overall current account balance. The figures manifest the continuing unsustainability of the external deficit and its reliance on debt rescheduling.

Exchange Rate and Trade Policy

One of the subjects of major debate in Tanzania in the early and mid-1980s was that of currency depreciation as a tool of balance-of-payment adjustment. In fact, it was mostly over the reluctance of the government to devalue that negotiations with the IMF broke off several times. There has been extensive literature on the reasons for the government of Tanzania's resistance to devaluation despite the fact that it had accepted several other orthodox stabilization measures (e.g., van Arkadie 1983; Jamal 1986;

Loxley 1989; Ndulu 1988; Singh 1986). In fact, almost all the measures recommended in the report of the mostly neutral technical advisory group were implemented in the first structural adjustment program, except the suggestion of a mild devaluation.

The basic arguments of those opposing devaluation were that devaluation would not lead to an improvement in the balance of payments, that it would generate inflation and hence tension over income shares, and that it would not lead to an improvement in the real exchange rate (the ratio of traded to nontraded goods prices) because of inflation, thus necessitating further devaluations and an inflationary spiral.

The basic reason devaluation would not improve the balance of payments, according to the critics, was elasticity pessimism. Supply response of export crops to price increases was thought to be very small given that many of them were perennial. Smuggling would not decline, according to the critics, because the main motivation of smuggling was the exchange with consumer and luxury goods not available in Tanzania and the desire to hold foreign exchange as the real value of domestic currency declined. On the demand side, most of the rationed imports were of intermediate and capital goods, which were already quite compressed. Hence very little reduction of imports was possible. Inflation was seen as inevitable, as those whose incomes would be most affected (mainly urban salary earners) would resist and demand wage adjustments, thus creating an inflationary spiral and frustrating efforts to change relative prices. The conflicting claims on income could generate a contraction along the lines of the well-known argument by Krugman and Taylor (1978).

The basic counterproposal put forth by critics of devaluation, including the government, was that the severe foreign exchange constraint made an initial injection of foreign exchange a prerequisite for any recovery. This increase in foreign exchange inflows would improve the availability of incentive consumer goods and motivate export crop supply response in addition to mobilizing domestic idle manufacturing capacity. This was also one of the suggestions of the advisory group. But that group believed that the currency had become so overvalued that some devaluation was necessary, and that the arguments of the devaluation critics were exaggerated. The inflationary impact would not be that severe because official price controls had already broken down and most prices, especially those of food, followed parallel market rates. Furthermore, the strength of urban workers was overestimated, and the argument of those favoring devaluation was that the lower-paid workers could be protected by increases in the minimum wage.

The government's line of reasoning was supported by influential economic observers. In their publication *East African Lessons on Economic*

Liberalization, Bevan, Bigsten, Collier, and Gunning (1987) derived several important policy implications from what is known as the consumer rationing story. Along the lines of the "leisure effect" in labor supply, Bevan et al. claimed that the raising of producer prices in a rationed economy (which would presumably follow a devaluation) creates a perverse supply response. In other words, when people receive higher prices for their output in a context where there is nothing of interest to buy, they in fact decide to work less. To avoid this leisure effect, the government has to make sure the shelves are filled with "incentive goods" and hence a *prior* injection of foreign exchange is needed. The government therefore asked the donors to subsidize consumer goods imports.

Perusing the debate in retrospect, it appears that what led to the acrimonious arguments between the government and the IMF was not so much the principle of devaluation but the magnitude and pace of devaluation and adjustment. The IMF argued for a shock treatment with more than 100 percent devaluation, abolition of the subsidy on sembe and hence a sevenfold increase in its price, substantial increases in producer prices of export crops, and liberalization of all price controls. Apart from the speed of adjustment, the Tanzanian government's other major concern was about the incidence of the burden of adjustment. The government was worried about the equitable sharing of this burden, something the IMF was not so concerned with.

According to the counterargument to the reasoning of the government and its supporters, based on more conventional economic thinking, liberalization of markets basically implies a reduction of risk premia, and hence prices, on trading in the parallel markets. The emergence of better functioning markets for output as well as simultaneous improved availability of incentive goods then ensures that any increase in real income will cause increased demand. On this argument, it was the combination of ineffective official markets and high risks on parallel market trading in the early 1980s that had created consumer rationing in the first place. The sudden increase in imports under the own-funded import scheme in 1985 seems to confirm this argument: the foreign exchange that was hidden in the parallel economy up to that point was immediately available without risks for imports and automatically "flushed" the system, without the necessity of government-mediated consumer goods imports.

It is also interesting to note in this context that despite the enormous (by Tanzanian standards) official devaluation between 1985 and 1989 (on average more than 80 percent per annum), and the substantial narrowing of the parallel premium (see Table 9.2), inflation during 1986–89 (26–32 percent) was on average not much different than it had been during the period 1980–85 (26–32 percent). Most consumers made their purchases at prices reflecting parallel rates throughout the decade of the 1980s.

More important than the economic reasoning behind the "gradual reform and devaluation" argument seems to have been the political rationale of piecemeal adjustment, combined with more resources to dispose of. The government probably felt that it did not have the political legitimacy to enforce austerity on its bureaucracy, and in particular to dismantle the parastatal system and lay off large numbers of government employees, without a prior injection of foreign exchange to supply the official markets, on which many official employees depended, with consumer goods.

Turning to trade policy, it was already noted from Table 9.6 that export duties have had a negligible contribution to public revenue, whereas import duties declined substantially between 1980 and 1986 and started increasing after the onset of the structural adjustment program. The structure of protection in Tanzania has been aimed mostly at generating revenue and, given the import substitution strategy of the 1970s, has tended to protect proportionately more industries with domestic manufacturing.

Despite the relatively high tariff rates indicated in Table 9.10, one of the major problems in the 1980s has been the low collection rates of tariff revenue. Ndulu et al. (1987) estimate that in 1986 only 42.5 percent of the scheduled tariff revenue on imports of consumer goods was collected. For intermediate goods the collection rate was 45.8 percent, for capital goods only 23.1 percent. The reason for this widespread phenomenon was the general deterioration of public services and incomes, which encouraged rent seeking by public officials and encouraged general tax evasion. The "war on economic saboteurs" in 1983–84 briefly improved tariff collection, but after 1985 the phenomenon of tax evasion reappeared.

Table 9.10. Nominal and effective rates of protection, Tanzania, 1986

Sector	Nominal protection (%)	Effective protection (%)
Beverages and tobacco	65.6	83.8
Textiles and apparel	43.8	55.4
Food products	29.8	65.1
Tanneries and leather	28.7	41.3
Plastics and pharmaceuticals	26.8	45.4
Iron, steel, and metal products	24.1	28.1
Agriculture	23.8	24.0
Machinery and transport equipment	22.3	25.0
Rubber, glass, wood paper, printing, and cement	19.8	27.9
Chemicals and fertilizers	8.2	1.6

Source: Ndulu et al. 1987.

Table 9.11. Import duties for selected commodities, Tanzania (percentage) ·

	1980	1981	1983	1985	1987	1988	1990
Skimmed milk	25	25	25	25	60	60	60
Meat and edible offal of animals	25	25	25	25	60	60	60
Milk and cream not concentrated/sweetened	10	10	25	25	60	60	60
Beet/cane sugar	25	25	25	25	25	25	30
Cigarettes	135	135	135	135	100	100	60
Common salt	30	30	30	30	40	40	60
Asbestos	10	10	10	10	40	20	30
Soap	40	40	40	40	40	40	40
Silk woven fabrics	45	25	25	120	69	100	60
Yarn							
Counts 40+	30	30	30	30	60	60	60
For fishnet twines	30	30	30	30	25	25	0
Other	30	30	30	30	60	60	60
Woven fabrics of manmade fibers							
Grey unbleached	10	20	30	60	20	20	20
Other	45	60	60	60	60	60	60
Bed sheets/towels, cotton							
Grey unbleached	45	45	45	75	60	60	60
Other	25	25	25	75	60	60	60
Footwear	40	40	25	25	25	40	40
Umbrellas and sun shades	30	30	30	30	30	60	60
Lavatory basins							
Lavatory cisterns	30	30	30	30	25	25	30
Other	15	15	15	15	25	25	30
Sheets, plain or corrugated	30	30	30	20	25	25	30
Cycles	30	30	30	30	15	15	20
Pocket watches/wrist watches	40	40	40	25	40	40	40

Source: Data from Mbelle 1991.

This might not be unrelated to the fact that scheduled tariff rates for many items seem to have increased after the onset of structural adjustment, contrary to what one would expect. Table 9.11 exhibits scheduled tariff rates for several items over the period 1980–90. Although a small list, it does indicate a tendency toward increased scheduled tariffs. This tendency might be related to the own-funds import scheme, under which imports expanded considerably and hence increased the potential tax base of the government.

Agriculture and Food Markets

Tanzanian government intervention in the organization of production and marketing of agricultural products has been subjected to a range of radical policy shifts. During the 1950s the country had exploited its rich agricultural resource base at a fast pace and had shown impressive agricultural growth. Moreover, Tanganyika's farming community had been allowed to develop in a relatively autonomous way. In several regions of the country (e.g., Kagera, Arusha, and Kilimanjaro) export crop–producing smallholders had created a strong and indigenously grown cooperative sector.[8] After the Arusha declaration, however, the pace of radical institutional change picked up, and it would be dictated from above.

The trends in agricultural production in Tanzania seem to be closely associated with such institutional changes. At independence, agricultural marketing was organized along what was known as the "three-tier single-channel marketing system." At the apex level, crop-specific marketing boards were responsible for the final sale of the agricultural product. The marketing boards bought product from the regional cooperative unions, who in turn procured them from primary cooperatives or directly by private producers. Crops falling under this regime had their prices set by the government and were called "scheduled" crops. They included maize, paddy, wheat, oilseeds, cashew nuts, and cotton. All but cotton were marketed by the National Agricultural Products Board. Final marketing of all other crops (i.e., cotton, tobacco, coffee, pyrethrum, sisal, and tea) took place through crop-specific marketing boards that derived producer prices from actual export sales. Under the three-tier single-channel marketing system, significant geographic differences existed due to the normal variation of marketing costs by region.

By 1973 the organizational makeup of this marketing system consisted of about 2,300 primary coops (which usually combined several villages and marketed several crops), twenty regional cooperative unions, and eleven marketing boards for coffee, oilseeds, cashew nuts, cotton, pyrethrum, sisal, sugar, tea, tobacco, and cereals. In 1973, however, the government replaced the marketing boards with parastatal crop authorities. In the food sector, the National Milling Corporation (NMC)—until then only involved in milling—took over from the Products Board. All marketing functions of the cooperative system were transferred to the crop authorities, who received considerable funding from Western donors.

Whereas the unions had specialized geographically, the crop authorities were specialized by crop nationwide. In the 1974/75 crop season the "into-store price," which had resulted in differential pricing at the regional level, was replaced by a "producer price," which became fixed for the whole nation. This was called the "pan-territorial price." Additionally, the list of scheduled crops was extended to include sorghum, bulrush millet, finger millet, and cassava—the so-called drought crops. Various pulses were add-

---

[8] The cooperative sector's history goes back as early as 1925, the year the Kilimanjaro Native Planters' Association was registered. By 1926 this association had over 10,000 members (Coulson 1982, 61).

ed to the list later in the 1970s. The resulting system proved inefficient: whereas marketings of the NMC, for instance, rose by only 49 percent between 1972/73 and 1977/78, its operating costs increased by 672 per cent (MDB 1980). The same sort of cost inflation would affect the other crop authorities.

The massive villagization campaigns of the early and mid-1970s attempted but failed to change the mode of production radically. The main presumed advantages of the socialist villages were the following: economies of scale of modernized agricultural production could be captured by consolidated communal farms; delivery of social services would be less costly if people lived in centralized villages rather than in dispersed homesteads; and the threat of an increase in the influence of large, commercial farmers—called "kulaks" in rather un-African jargon—would be thwarted.

Achievement of economies of scale rested on the assumption or hope that there was an agricultural technology "on the shelf" which would enable farmers to intensify or mechanize production in the typical agroclimates and soil conditions of sub-Saharan Africa. But villagized agriculture never received such yield-improving or labor-saving technology. Moreover, plot and farm scattering in the extensive farming system as it is practiced in the majority of regions in Tanzania constitutes an appropriate adaptation to the ecosystem given the available low-input technology. Ironically, then, the results of the villagization program were ecological degradation and declining yields.

The second presumed advantage of villagization rested on the belief that the state would in effect be capable of sustained delivery of a range of social services in the rural areas of Tanzania. This belief proved to be equally unfounded. Especially during the 1980s, social services eroded as quickly as they had been set up in the 1970s. We refrain from commenting on the third presumed advantage, relating to the "kulak" issue. Suffice it to say that is was deeply embedded in Marxist economic ideology concerning the "peasant question."

The Village Development Act passed in 1975 made the village a legal agent able to enter into contracts with other legal entities. With the village the preferred instrument of government policy, it was felt that cooperatives, which often cut across several villages, were now superfluous. Consequently, the cooperative system was dismantled in 1976. The political rationale for abolishing the coops was that their leadership was corrupt and that they were dominated by "kulaks"—the rural economic class targeted by the government's policies. Most private retail shops in rural areas were closed under "Operation Maduka," and regional trading corporations became responsible for the retail distribution of food and consumer goods in the regions. Due to the general inefficiency of the new system, consumer goods rationing soon followed.

In the beginning of the 1980s, the crisis of the official marketing systems of food as well as of export crops forced the government to reconsider its stance on the cooperative system. The reintroduction of the three-tier cooperative system went into effect in 1984, although the "top down" approach that had characterized the marketing bureaucracy of the 1970s seemed to have been carried over to the new cooperative system.

In 1984 a first set of liberalization measures took effect. For instance, the limit on privately traded quantities of foodgrains was raised substantially and the controls at roadblocks were removed. Whereas the reinstated cooperatives de jure continued to hold a marketing monopoly, the government de facto allowed private traders to step into the market. The cooperatives continued however, to purchase maize at pan-territorial producer prices.

All remaining quantity restrictions on interregional private grain trade were lifted in March 1987. Grain trade liberalization, combined with favorable weather, made Tanzania a net exporter of maize as of the 1987/88 season, ending its status as a substantial foodgrain importer since the early 1970s. This is perhaps the most symbolic positive development that seems to be directly linked to the economic recovery programs.

The desire for gradual reform in the official marketing system continued to compound the government's financial problems. The government remained particularly hesitant to privatize certain crucial links in the marketing and processing system (e.g., cotton gins) to noncooperative actors. In general, there was still a fear of rural capitalists and of a renewed monopolistic entry of Indian and Arab traders into the rural economy. Thus no liberalization in export crop marketing was undertaken, and the pace of liberalization of cereal crop marketing was very slow. There, the cooperative unions were still under an obligation to purchase all quantities offered for sale at the official prices. As a result of the increases in open market quantities and favorable weather conditions over the 1985–87 period, official producer prices in the more remote maize-producing regions were above the open market producer price. During the 1984–87 period, the NMC stocks of maize reached record levels of up to 200,000 tons, with average increases in official purchases of 40 percent per year. In an effort to decrease its accumulated stocks, the NMC opened over one hundred retail shops in December 1988 which sold maize flour below open market prices.

In June 1988 the NMC officially lost its monopoly on grain trade. Still, the private sector did not seem very interested in purchasing from the unions given the price situation in the open markets, and the unions continued to rely on the NMC as the buyer of last resort. Both institutions suffered considerable losses on their marketing transactions in the 1988/89 and 1989/90 seasons. In December 1988 the government was forced to assume the responsibility for the overdrafts of the NMC, which had an inflationary impact. By 1990 foodgrain stocks of approximately 250,000 tons were stranded predominantly in Arusha, Rukwa, and Shinyanga. NMC operations in 1990 virtually came to a halt after the government prohibited it from undertaking nonprofitable operations. So-called gradual reform had created a situation in which the NMC's and other parastatal's financial losses had risen to such levels that they were posing a threat to the Tanzanian banking system: only in September 1990 were traders allowed, for the first time since 1967, to go into the villages and buy up crops from individual farmers. The 1991/92 marketing season was thus the first season during which grain marketing was completely liberalized at all levels.

The tumultuous changes in agricultural marketing manifested themselves in marked patterns of prices. Figure 9.1 exhibits indices of real official producer prices of food and export crops from 1965 to 1988. The indices of weighted nominal official prices are deflated by the nonfood national CPI. The patterns are quite clear. The period before 1973 was one of declining real producer prices, as agricultural taxation was used to create investible surplus. Between 1974 and 1978, real official export prices increased substantially with the coffee boom, and food prices also increased in response to the reduced official marketings. From 1978 to 1983, substantial reductions in real official producer prices of both foods and export crops are observed. These were the years of increased farmer taxation and growing inefficiencies of the official marketing channels. After 1983 real official food prices increased in response to liberalization and competition from private trades. But, after a small rise during the coffee miniboom of 1984, real official export prices stagnated and even declined in 1988. This pattern was related to the much slower pace of marketing liberalization in the export crop sector, which continued to be dominated by publicly owned parastatals.

In the food crop sector, the inefficiency of public food marketing quickly gave rise to parallel food markets. Figure 9.2 presents real indices of official and open market prices over the period 1969–88 of the main staple food products. The deflator used is the nonfood national CPI. What becomes apparent is that during both the period of villagization (1974–76) and the crisis (1980–84) open market food prices appreciated considerably. The official food marketing system functioned erratically, whereas parallel markets for food were usually localized and diffuse due to strict government controls on interregional trade. After the onset of food marketing liberalization, open market food prices quickly came down.

Whereas the overall decline of export crop production is well documented, the dramatic increase of per capita food production as it appears

Figure 9.1. Export and food crops, official real prices, Tanzania

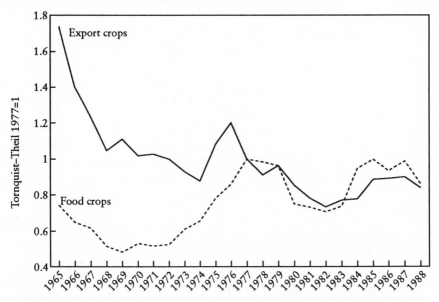

Source: Sarris and Van den Brink 1993, courtesy of New York University Press.
Notes: Nominal prices are deflated by the national CPI. Export crop series is composed of coffee, cotton, cashew, tea, tobacco, and pyrethrum. Food crop series is composed of maize, rice, beans, sorghum, millet, and cassava. The Tornquist-Theil index is a discrete approximation of a Divisia index. The index was chained and based to 1977.

in the official data since the early 1970s seems to be exaggerated (Sarris and Van den Brink 1993). There are several reasons for this. First, "agriculture is politics"; official estimates may have been routinely adjusted upward during the 1970s. Second, survey estimates of national production are consistently below the official series. Third, the official estimates point to a considerable "calorie overhang," which is difficult to square with the massive foodgrain imports in the 1970s and 1980s and with the reported widespread child malnutrition. Fourth, a comparison between the 1971/72 agricultural census and the 1986/87 agricultural sample survey shows no apparent increase in cultivated area under food crops and no substantive technological change in agriculture. Fifth, real consumer prices of food in rural markets appreciated throughout the Tanzanian experiment and the consequent crisis, but open market food prices fell dramatically after liberalization in 1984. These trends point to an increasing scarcity of marketed food until the mid-1980's and a substantial supply response of marketed surplus afterward. More probable, then, is the suggestion that export crop

Figure 9.2. Real official and parallel food prices, Tanzania

Figure 9.2

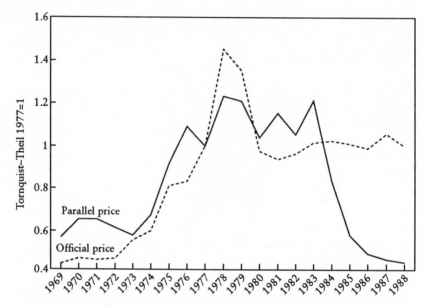

Source: Sarris and Van den Brink 1993, courtesy of New York University Press.
Notes: Nominal prices are deflated by the nonfood part of the national CPI. Official price series is composed of maize, rice, beans, sorghum, millet, and cassava. Parallel price series is composed of maize, rice, and beans only. The Tornquist-Theil index is a discrete approximation of a Divisia index. The index was chained and based to 1977.

production decreased in those areas where parallel exports were logistically constrained, but that food crop production remained approximately stable, keeping up with population growth.

Parallel exports of coffee were particularly important in Arusha and Kilimanjaro, where a lively trade with Kenya existed. Such trade developed mainly after the coffee boom of 1976, which ensured high relative prices in the parallel markets. According to Moshi traders, as of 1977 parallel exports of coffee increased significantly, and high levels of parallel exports were reached by 1980. This increase implies that some of the observed decline in export crop production should be accounted for by a shift from official to parallel exports. Little information exists about parallel coffee trade in other regions, however.

The slow pace of liberalization of export marketing is one of the main reasons for the stagnating output of export crops after the adoption of the

ERP. Consider again coffee, which is the most important export crop. The 1980s saw a reduction in explicit taxation when the export tax was abolished in 1982, but explicit taxation was replaced by implicit taxation. First, the government has regularly engaged in barter contracts in nonquota markets which were not in the interest of the producer. It is estimated that around 60 percent of Tanzania's coffee is traded directly for oil from Germany (Kristjanson et al. 1990). Such noncompetitive contracts tended to create a significant difference between the Tanzanian selling price and the world market price.

Second, increased inflation has implicitly taxed the producer price. Coffee marketing functions under a residual pricing system. This means that the producer price is calculated as a residual of the price realized at the auctions where the Tanzanian Coffee Marketing Board markets the coffee after it subtracts its marketing cost. The residual pricing system implies that the farmer receives an advance, an interim, and a final payment. Whereas the residual pricing system and its direct link to the world market seem to be one of the main reasons coffee producers receive relatively higher shares of the world market price compared to the producers of other export crops, long time delays exist between first, interim (6 months), and final payments (18 months). Such delays are obviously costly in an inflationary environment.

Third, as inefficiencies of the coffee marketing board increased, the cost of marketing increased, resulting in a price to producers that became a smaller share of the world price. Table 9.12 exhibits the average export and producer prices for the two main coffee varieties over the period 1981/82–87/88. It is clear that the devaluation-induced increases in average export prices realized by the marketing board since 1985 were not passed to the producers, whose share of the export price declined considerably since the early 1980s.

Conclusions

The Tanzanian experiment of 1967–84 can be summed up as an attempt at institutional change without microeconomic foundations. Moreover, the resource requirements of the institutional superstructure that was erected preempted and stifled economic growth. Thus the socialist edifice was left suspended in midair, and ready to crash. The crash occurred in 1979, triggered by exogenous shocks.

On the one hand, then, the institutional changes in the economy of Tanzania stifled economic development. For instance, the mostly public manufacturing sector was brought to near-zero activity levels in the early

Table 9.12. Export and producer prices for arabica and robusta coffee, Tanzania (Tsh per kilogram)

	1981/82	1982/83	1983/84	1984/85	1985/86	1986/87	1987
Arabica							
Average export price	20.97	26.06	39.16	55.61	72.36	150.00	171
Producer price	17.38	18.96	28.59	37.10	57.25	75.94	82
(advance + interim + final)							
Producer share of export price (%)	82.88	72.76	73.01	66.71	79.12	50.63	47
Robusta							
Average export price	16.38	22.22	40.38	46.67	52.78	126.90	131
Producer price	11.14	21.10	32.70	30.14	36.60	65.00	75
(advance + interim + final)							
Producer share of export price (%)	68.01	94.96	80.98	64.58	69.34	51.22	57

Source: Sarris and Van den Brink 1993, courtesy of New York University Press.

1980s. The indigenous cooperative marketing system was disrupted. Villagization interventions in the rural areas had the same negative impacts as they have had elsewhere in sub-Saharan Africa (e.g., Ethiopia in the early 1980s). The private estate sector was erased in many parts of the country, except in those regions where it could fend off the effective implementation of the socialist experiment (e.g., Arusha and Kilimanjaro).

On the other hand, parallel institutions emerged during the experiment. These institutions had to compensate continuously for the unstable walk of the formal economy, and at one point they even had to regress to a barter system in order to keep the economy operating. These parallel institutions, which kept the economy growing at a modest pace for most of the 1970s and 1980s, are currently "stepping out" after the implementation of liberalization measures, and they have unveiled a basic economic structure that has remained remarkably stable over time.

There is no single type of institutional framework that can make an a priori claim to efficiency and equity. Given the current state of Tanzania's infrastructure, both in physical and in institutional terms, getting rid of government is clearly not the answer. Given the past record of government policies, however, it would seem unwise to revive institutional experimenting again. On the contrary, the resilience of the Tanzanian people throughout the crisis can be attributed only to the effective functioning of what has been termed "the second economy." It seems only fair to finally acknowledge and build on the effectiveness of this economy rather than discard it as backward and corrupt.

Moreover, the prevailing attitude in the country seems to be that the Tanzanian government has lost the legitimacy to decide for its citizens what they should sell, at what price, and to whom. The government has not been able to create a functioning formal economy, in spite of the donor generosity that has accompanied Tanzania's quest for a socialist paradise. At this time, then, it might be advisable for public institutions to start earning their place in the economy in competition with private institutions. Government's legitimate concerns regarding the equity of liberalization policies seem best directed toward the creation of a stronger private sector accompanied by a more effective, but indirect, redistribution system.

10

Structural Adjustment in a Small, Open Economy: The Case of Gambia

Cathy Jabara

Gambia is one of Africa's smallest and poorest countries. It forms a small enclave in Senegal on the west coast of Africa, and its per capita income was about US$350 in 1989/90. Total population was estimated at 832,000 in 1989, giving a population density of 73 per square kilometer, the fourth highest in Africa. About two-thirds of the population is engaged in agriculture: subsistence farming, livestock raising, and cultivation of groundnuts for export. With the exception of approximately 2,400 hectares of irrigated rice (less than 2 percent of cultivated land), agriculture in Gambia is entirely rainfed. Agricultural technology is still rudimentary and extremely labor intensive.

Gambia's financial crisis emerged from the acceleration of its investment effort in the mid-1970s under the First Development Plan (1975/76–80/81). This plan was designed to increase investment in basic economic and social infrastructure through the use of funds largely borrowed from donors. Before this period, Gambia had followed fairly prudent financial policies which, aided by favorable external circumstances, maintained broad equilibrium in public sector operations and in the balance of payments. From 1970 to 1979, the Gambian economy experienced an average real growth in GDP of 5.5 percent per year, a rate well above population growth. From 1979 to 1986, however, average real GDP growth fell to 2.6 percent per year. The latter period was characterized by drought, adverse declines in the external terms of trade, and chronic imbalances in the balance of payments and in the government accounts.

Gambia's small size, undiversified production base, and trade openness made its economy particularly sensitive to shortfalls in agricultural production and to changes in its terms of trade. The country is dependent on trade for up to one-half its food requirements, and it imports most manufactured goods and all fertilizers, fuel, and equipment. Exports of groundnuts and groundnut products accounted for about 12 percent of foreign exchange revenues from exports of goods and nonfactor services in 1988/89, the remainder being made up from the tourist trade and from Gambia's activity in the reexport trade as an entrepôt.

The government of Gambia adopted a far-reaching economic recovery program (ERP) in consultation with the IMF, World Bank, USAID, and other donors in July 1985. The ERP has generally been cited by the donor community as a significant contributing factor to the real economic growth that has occurred in Gambia since 1986/87, although more favorable external factors have also been important (McPherson and Radelet 1989; Radelet 1990). The ERP concentrated on reducing government expenditures, increasing the prices of imports, and raising the economic returns to Gambia's principal export activities—groundnut production and the reexport trade. In the first year of the ERP, Gambia did not receive any support from the donor community in terms of structural adjustment loans or grants. The ERP was, however, extended by the government for two years, 1986/87–87/88, with financial support from the donor community. This was followed by a second three-year ERP phase from 1988/89 to 1990/91.

The analysis of Gambia's economic recovery program from 1985/86 to 1988/89 reveals some important insights into the structural adjustment process. For instance, the Gambian example illustrates that a structural adjustment program that includes liberalization of the exchange rate need not result in permanent and chronically high rates of inflation. At the same time, however, economic analyses of exchange rate changes on particular sectors should take account of secondary effects, such as restrictions on credit and changes in interest rates that may adversely affect export sectors. For instance, groundnut producers in Gambia no doubt benefited from exchange rate depreciation, but they were adversely affected by the government's restrictions on access to seasonal credit and crop finance which were needed to support a depreciation in the real exchange rate. Real public expenditures declined only in the first year of the ERP, but an important shift occurred in the mix of expenditure that favored consumption over investment priorities. Finally, the Gambian case illustrates that donors should pay attention to the role foreign assistance projects play in hindering efforts to privatize certain operations, as illustrated by example of fertilizer marketing.

Causes and Manifestations of Economic Disequilibrium

The Origin of the Financial Crisis

Gambia's First Five Year Plan, which started in the 1975/76 fiscal year, channeled development expenditure into transport and communications, public utilities, schools, agricultural extension stations, and health clinics. Gambia initially relied on highly concessional foreign loans and grants to finance 70–75 percent of its investment program. The remainder was financed from its current budgetary surplus, limited domestic borrowing, and grants.

It has been argued that the public investment program implemented under the First Plan contributed to the economic deterioration of the early 1980s because of its low average returns and its failure to place the economy on a higher growth path (World Bank 1985a). More important, however, was the failure of government officials and donors to foresee that a rapid expansion of development expenditure would lead to a structural problem of higher public recurrent expenditure and imports. For example, civil servants were rapidly recruited under the investment program to furnish new services, with the result that established posts in the government doubled and recurrent expenditure increased by 20 percent per year. In addition, an estimated 40 percent of public investment was channeled to publicly owned enterprises created in an attempt to develop new industries. This resulted in a doubling of the number of parastatals between 1975 and 1981. The rapid increase in public consumption and development expenditure led to a "structural deficit" in Gambia's external trade due to the high import consumption of the fast-growing urban population, especially public sector wage earners, and to the high import component (over 60 percent) of general government expenditures and development projects. Domestic exports, even in a good year, were increasingly unable to generate foreign exchange sufficient to cover the cost of domestic imports.

The rapid expansion of public expenditure for largely nonagricultural uses limited the government's ability to increase production of groundnuts, the country's primary commodity export, through producer price incentives. The Gambia Produce Marketing Board (GPMB), which had the sole right to purchase groundnuts at established prices, was called to use its accumulated financial reserves to capitalize other parastatals through share purchases or loans and to provide large direct transfers to the government in support of development expenditure. The GPMB was also responsible for importation of rice and fertilizer and for local sales of groundnut oil. The price structures established for sales of these products to wholesalers and distributors rarely covered the GPMB's operating costs. The need to divert GPMB surpluses to nonagricultural uses, combined with financial losses incurred on commodity sales, limited the ability of the GPMB to raise producer prices for groundnuts during the late 1970s, and groundnut production began a downward trend.

Gambia's balance of payments went into deficit in 1976, into surplus in 1977, and then into deficit thereafter. The external current account deficit increased from an average level of about 4.0 percent of GDP in the early 1970s to 20.8 percent of GDP in 1979/80, whereas the public sector deficit increased from about 5.0 percent of GDP in the same period to 11.0 percent in 1979/80. Over 90 percent of this deficit was financed with foreign loans and grants, but the overextension of the public sector and the drawdown of foreign and domestic financial reserves made it increasingly difficult for the government to meet its remaining obligations. More important, by 1979 the public sector was so overextended that it could not act as a buffer against the climatic factors and exogenous shocks that soon followed.

Exogenous Shocks, Capital Flight, and Foreign Aid

Drought and the deterioration in the terms of trade severely reduced real GDP growth in Gambia over the two-year period 1979/80–80/81 and export earnings growth during 1979/80–81/82.[1] Gambia's terms of trade fell by 32 percent in 1979/80 and by 31.4 percent in 1981/82, contributing to the long-term decline noted in Table 10.1. Real GDP growth in 1979/80 was limited to 1.4 percent, and it fell by 8.3 percent in 1980/81. Exceptionally high levels of official foreign aid and foreign borrowing during 1980/81 and 1981/82 enabled the economy to avert severe economic adjustment. The overall public sector deficit increased to 20 percent of GDP in those years, but highly concessional foreign loans and grants, which more than doubled during this period, continued to finance more than 90 percent of this deficit. Foreign financing allowed the country to maintain annual investment plans and to increase government consumption. The

---

[1] The original estimates of real GDP are used throughout this essay, rather than an updated version that was made available in 1991. Although the estimates of the revised data seem reasonable for the later years, I have more confidence in the percentage changes in GDP obtained from the original data. For example, a comparison of the percentage change data for 1985/86 reveals a sharp difference: the revised data would have us believe that Gambia experienced an 8.3 percent increase in real GDP, its highest annual increase ever, in the first year of the ERP; given the cutbacks in government spending and in foreign aid, this result is highly implausible. Most likely the analysts who made the revised estimates did not take into account cutbacks in government spending in that year which were not recorded in the government's annual published budget estimates. Similarly, the revised estimates contain increases in sector value added that may have taken place during the period 1983/84–89/90, but the specific years to which some dramatic increases were assigned by the analysts need to be more closely examined.

Table 10.1. Real GDP growth and real and nominal exchange rate indices, Gambia

	GDP (% change)	Exchange rate indices (1980 = 100)	
		Real	Nominal
1979/80	1.4	92.3	103.3
1980/81	−8.3	93.7	108.6
1981/82	9.7	92.6	104.1
1982/83	13.4	93.4	102.8
1983/84	−7.2	94.6	98.9
1984/85	1.6	89.7	88.8
1985/86	−0.3	88.6	72.8
1986/87	5.5	72.2	41.0
1987/88	5.5	77.6	43.0
1988/89	4.5	80.5	46.6
1989/90	7.9	NA	NA

Source: Jabara 1990a, courtesy of Cornell Food and Nutrition Policy Program.
Note: NA, not available.

government also borrowed heavily on a commercial basis: credit extended from suppliers and financial institutions increased from 5.5 percent of medium- and long-term debt in 1979 to 22.5 percent in 1981.

Under an IMF standby program adopted in February 1982, the Gambian government attempted to lower the public sector deficit through reductions in consumer subsidies and selective tax increases and to reduce its external current account imbalance through increased producer prices for groundnuts and rice. It also increased petroleum prices, raised electricity tariffs, and introduced petroleum-related taxes to discourage consumption and increase fiscal revenues. The standby program also included a significant rise in the interest rates paid on commercial bank deposits, and it removed interest rate ceilings on loans and overdrafts.

With the return of foreign aid inflows to more normal levels in 1982/83, the government was forced to implement abrupt expenditure cuts and to curtail imports. It raised producer prices for groundnuts in the 1982/83 season under the advice of the IMF. With improved weather conditions, however, the area planted in groundnuts rose to its highest level ever. Increased production, coupled with a decline in the world groundnut price in that year, plunged the GPMB deeply into debt. The government was forced to grant the GPMB direct access to central bank financing at the government's lending rate, producing domestic credit growth of 45 percent in a twelve-month period.

By 1982/83 the growing inability of the central bank to convert dalasi deposits into foreign exchange led to a rapid loss of confidence in the banking system by the trading community. The private short-term capital account in the balance of payments became highly negative. Foreign exchange from the reexport trade became increasingly unavailable to the domestic banking system as traders began transferring it directly to foreign banks or into the active parallel foreign exchange market that developed about this time. The premium received for CFA francs on the parallel market also made cross-border exports of groundnuts to Senegal more attractive.

The economic situation continued to deteriorate in 1983/84 as the economy stalled with unfavorable weather conditions. Under a second standby arrangement negotiated with the IMF in early 1984, the dalasi, which had been pegged at D 4.00 per pound sterling since 1974, was devalued by 25 percent. Subsidies on rice, fertilizer, and public transport were also reduced.

Gambia's economic and financial difficulties continued to worsen in 1984/85. The external current account deficit, which had declined during the two years ending 1982/83, widened from 19 percent of GDP in 1983/84 to 26 percent in 1984/85. The scope of the parallel foreign exchange market continued to expand, and the foreign exchange receipts accruing to the banking system declined. External payment arrears rose to about SDR 55 million at end June 1985, equivalent to over two and one-half times 1984/85 domestic exports. Shortages of essential imports, such as petroleum and rice, became widespread.

The 1984 devaluation of the dalasi resulted in a significant real compression of domestic imports, which declined by an estimated 27 percent from 1983/84 to 1984/85. Crop prices were increased in June 1984. The 11 percent increase in the groundnut price, however, was considerably less than the amount of the devaluation, and it was not competitive with the price offered in Senegal, particularly in light of the black market premium on the CFA franc. The GPMB further increased the groundnut price by 24 percent in late January 1985, but by that time a substantial part of the crop had already been sold. The price increases and lower-than-expected groundnut purchases, reflecting a poor harvest, resulted in the GPMB showing an overall deficit for 1984/85, which had to be financed from bank credit.

The government's development expenditure rose by 57 percent under a second development plan in 1984/85. Due to the increase in net foreign assets of the commercial banks from foreign financing of these expenditures, broad money increased by 34 percent and contributed to an acceleration of domestic inflation to 22 percent for the fiscal year. As a result, by August 1985 the real effective exchange rate had appreciated significantly and the parallel exchange market discount quickly reemerged.

The Economic Recovery Program

Gambia's ERP was introduced by the government at the start of the 1985/86 fiscal year. Its objectives were to reduce Gambia's external current account and internal fiscal deficits through pricing reform, exchange rate liberalization, demand management, and reduction in the size and role of the public sector. More specifically, liberalization of the exchange rate, liberalization of key trading activities, such as rice and fertilizer, promotion of the reexport trade, increases in the prices of traded goods, particularly for groundnuts, and elimination of government subsidies were the most important components to the ERP.

The exchange rate liberalization was required to restore to the banking system the large sums of foreign currency that were believed to be held outside the system due to the exchange rate overvaluation. This liberalization, combined with reduced tariffs on products important to the reexport trade, would also increase foreign exchange earnings and government revenues from the reexport trade. At the same time, privatization would encourage private traders to use their foreign exchange to finance the country's imports and thus reduce government expenditure. Agricultural policy reforms under the ERP focused on raising producer prices for groundnuts and other agricultural commodities (in the first two years of the ERP), on reducing subsidies for fertilizers and credit, on restoring the financial viability of the farm credit system, and on improving the groundnut marketing system (in the latest years of the ERP).

In the first year of the ERP, and government received little financial support from the donor community. Net official foreign loans to Gambia in 1985/86 actually declined from SDR 33.8 million (21 percent of GDP) to SDR 26.4 million (16 percent of GDP). Although reaction to the ERP was favorable, donors were reluctant to provide initial support because Gambia was in arrears to the IMF and did not have a standby arrangement.

Financial support from the donor community and a standby arrangement with the IMF followed in 1986/87 once these arrears were cleared. In addition to drawings of SDR 5.13 million under the standby agreement in that year, the ERP was supported by a loan of SDR 3.4 million under the first annual arrangement under the IMF's structural adjustment facility, a three-year $37 million World Bank structural adjustment credit with cofinancing, and $7.3 million in commodity aid from STABEX grants. Numerous bilateral donors also provided technical assistance and balance-of-payment support. In 1986/87 net official foreign loans and grants increased to SDR 61 million, or over 50 percent of GDP. These arrangements were followed by a second annual structural adjustment facility arrangement in 1987/88, approval of a three-year arrangement under the IMF's enhanced structural adjustment facility in 1988/89, and a second three-year World Bank Credit in 1989/90.

Price Reforms

Price reforms in the first year of the ERP were designed to reduce domestic absorption and to raise the prices of traded relative to nontraded goods. An increase in the general import tax and higher item-specific import duties increased the prices of most imported consumer goods. The government of Gambia also raised fertilizer prices, liberalized the importation of fertilizer in January 1986, and announced that fertilizer subsidies would be phased out by December 1986. The export duty on fish (previously 17 percent) was eliminated to encourage exports. Controls on rice prices were abolished and all restrictions on private sector participation in rice trade and marketing were eliminated. The new rice policy not only permitted consumer rice prices to rise, thereby reducing import demand, but also reduced the public sector's demand for foreign exchange.

Government controls on prices for all foodstuffs (such as groundnut oil, meat, eggs, and bread) were eliminated. Administered prices for petroleum products, public transport, telecommunications, water, and electricity were raised at various times to ensure pass-through of higher import tariffs and exchange rate adjustments. To reduce subsidies on health care services provided by government hospitals and clinics, the fees on hospital services were raised by 25–400 percent.

In the initial year of the ERP, the government raised the producer price for groundnuts by 58 percent, from D 620 to D 980 per ton effective at the start of the planting season. The price was later raised to D 1,100 per ton at the start of the buying season in November and to D 1,260 per ton in January 1986 in order to bring the Gambian price more in line with the Senegalese price (which was higher than the equivalent world price) at the prevailing parallel market exchange rate. But the new producer prices were still not sufficiently competitive to prevent the occurrence of significant cross-border outflows in light of the depreciation of the parallel exchange rate during much of the buying season.

To meet the conditions of the 1986/87 IMF standby arrangement, the government agreed to a further increase in the farmgate price for groundnuts—to D 1,800 per ton, a level 43 percent above the level of January 1986 and some 10 percent above the export price for shelled groundnuts. This price was chosen in collaboration with the IMF in order to assure that the GPMB would procure and export at least 60,000 tons of groundnuts. GPMB losses on its groundnut account were covered by a budgetary transfer financed partly from external assistance and partly by fiscal measures, including an increase in the import duty on rice from 26 to 30 percent and higher import taxes on petroleum products. Due to much larger groundnut purchases, however, the GPMB losses were greater than anticipated, which necessitated additional financing through bank credit to the GPMB above the programmed level. The producer price increase, combined with good weather, resulted in a 46 percent increase in groundnut production, while food crop production declined by 16 percent.

The 1986/87 IMF standby arrangement also committed Gambia to promote the reexport trade. Import duties were reduced on items such as textiles, tomato paste, radios, batteries, corrugated sheet metal, and cement, all of which are important in that trade. Lower import duties, combined with improved access to foreign exchange through the interbank market (see below), helped to encourage this trade.

Producer prices for groundnuts were lowered during the 1987/88 and 1988/89 marketing seasons in order to phase out the groundnut price subsidy. New groundnut pricing and marketing rules introduced in the 1989/90 season eliminated the groundnut subsidy altogether and allowed anyone—farmers, traders, or Gambia Cooperative Union (GCU) agents— to sell groundnuts directly to the GPMB at its depots. To cushion the impact on farmers, the government suspended the export tax on groundnuts. Marketing restrictions were entirely eliminated in the 1990/91 season, and farmers and traders were allowed to sell groundnuts to anyone willing to buy.

Exchange Rate Reforms

Among the policies implemented by the government to correct for the distortions in the prices of traded to nontraded goods, the boldest and most effective was the liberalization of the foreign exchange market in January 1986. This move was deemed necessary to open the large parallel market for foreign exchange and to reduce the capital flight associated with that market. The system adopted in 1986 involved a floating exchange rate and "interbank" market in which the commercial banks and the central bank set foreign exchange rates weekly by bidding for the available foreign exchange.[2] During the first month of operation, the official exchange rate was set at D 7.45 per pound sterling, representing a dalasi depreciation of 49 percent. By March 1986 the official exchange rate had depreciated further to D 10 per pound sterling, thereby reducing the

---

[2] Three commercial banks along with the central bank participate in the interbank trading sessions every week. Foreign exchange is sold by these banks at market rates after these weekly sessions.

premium on the parallel market rate to only about 5 percent; by mid-October 1986 the parallel market and interbank market rates were practically identical.

To strengthen the liberalized system, the government removed existing restrictions on payments and transfers for current international transactions, established foreign exchange surrender requirements for public enterprises, and set limits on the foreign exchange working balances of the commercial banks.[3] Despite these measures, an initial lack of confidence in the system resulted in increased private short-term capital outflows in 1985/86. In 1986/87, however, higher interest rates and improved confidence in the interbank system reversed the outflow of private short-term capital.

Demand Management

To support the liberalized exchange rate system and to reduce capital outflows, the government allowed the banks to raise the interest rates paid on savings deposits from 8 to 15 percent. All interest rate ceilings were abolished in February 1986, and minimum deposit rates that ranged from 15 to 17 percent were introduced. The interest rate on treasury bills and on central bank advances was also increased from 8 to 15 percent.

In July 1986 the government introduced a flexible, market-oriented system of interest rate determination. Under the new system, key interest rates were determined on the basis of a biweekly tender for treasury bills. Central bank rediscount rates and the minimum rate on 3-month savings deposits were linked to this treasury rate, and other deposit and lending rates became freely determined. Under this new system, interest rates on treasury bills fluctuated between 16 and 20 percent in 1986/87, as compared with the previous fixed rate of 15 percent.

To reduce inflationary pressures, the government initially intended to limit (1) the growth of total domestic credit to 9 percent, (2) the increase in net credit to government to zero, and (3) the increase in credit to public enterprises to 10 percent. It believed that the rate of inflation could be held to around 20 percent, notwithstanding the possible inflationary impact of movements in the exchange rate.

Increased foreign financial inflows in 1986/87 enabled the government to retire its debt to the banking system and to establish a net creditor position. The growth in money supply (44 percent) during that year derived entirely from an increase in net foreign assets of the banking system.

---

[3] Commercial banks must sell end-week foreign exchange balances in excess of established reserves to the central bank.

Gambia's balance of payments registered a small surplus in 1986/87 despite the external current account deficit widening to 32 percent of GDP. The government also reached agreement with Paris Club creditors to reschedule payments pertaining to publicly guaranteed or insured commercial credits and official loans contracted before July 1, 1986.

In 1987/88, as part of financial restructuring, the government wrote off accumulated debts owed to the Gambia Commercial Development Bank and the Central Bank of Gambia by two parastatals—the GPMB and the Gambia Utilities Corporation. To improve the financial viability of the agricultural credit system, government instituted a program for recovery of outstanding debt owed by farmers to the GCU and new credit eligibility rules for cooperative members. It also made provisions to write off accumulated debts owed by the GCU to the GCDB in 1989/90.

Public Sector Policies

To limit upward drift in domestic prices of nontraded goods, the government froze public sector wages and salaries in 1985/86, sharply curtailed its development expenditures, and constrained the growth of noninterest current expenditure. The government workforce was reduced by more than 24 percent in 1985/86 through layoffs of 2,300 temporary and daily laborers and 460 civil servants in established posts (Jabara 1990a). The government also instituted a ban on the creation of publicly owned enterprises and commenced a program to rationalize and privatize parastatal corporations. Actions taken to tighten customs revenues collection resulted in both higher revenues and increased prices for imported goods.

Foreign loans and grants accommodated a 77 percent increase in development expenditures in 1986/87. Although the general freeze on government wages and salaries continued, recurrent public expenditure rose by 66 percent to accommodate higher costs for debt service and imported material supplies and a budgetary transfer of D 83 million (about 8.8 percent of GDP) to the GPMB. Government staff levels were reduced by an additional 17 percent with the termination of 750 civil servants in established posts, firing of 340 temporary and daily paid laborers, and elimination of 750 vacant posts. The government also adopted a plan for rationalizing the public enterprise sector which involved the phased divestiture of government holdings and negotiations of performance contracts with selected parastatal enterprises.

To improve both cost recovery and efficiency in the health system, a user fee system for health services was instituted in all government hospitals and clinics in August 1988. The revenues from these fees were linked to the establishment of a revolving fund that would be used for financing recurrent costs of drugs and medical supplies and for overhauling the health services sector.

An income tax reform was approved in February 1988 which greatly simplified the income tax system, lowered marginal tax rates, and increased taxable income thresholds. At the time it was implemented, the new tax system effectively exempted most civil servants from income taxes. A 10 percent national sales tax at the importer/manufacturer level was introduced in July 1988 to replace the 6 percent general import duty and other excise taxes. The national sales tax was broadened to include domestic services, such as hotels/restaurants, telecommunications, insurance services, night clubs, and casinos in July 1989. In January 1989 a long-delayed general wage and salary increase for government employees was introduced which raised civil servant salaries by 55 percent, on average, based on substantially restructured pay and grading scales.

Economic Performance under the ERP

Economic Growth

Gambia's economy experienced a decline in real GDP in the first year of the ERP (Table 10.1). This decline could be attributed to exogenous factors, such as poor weather or the decline in external aid flows of 1985/86, or to the policies implemented under the ERP. Real GDP rose by about 5.5 percent per year during the 1986/87–87/88 period and by an estimated 4.5 percent in 1988/89. Again it is not possible to attribute these increases in real output to the adjustment program or to the improvement in exogenous factors such as adjustment-induced foreign financial inflows, improved weather conditions, and the halt in the deterioration of the terms of trade in 1986/87. It should be noted that the large inflows of structural adjustment-induced foreign assistance allowed the government to stabilize its exchange rate at lower interest rates, and at a higher level of real income, than would have been possible if the current account deficit had to be eliminated.

Despite the 1985/86 price adjustment for groundnuts, poor rainfall distribution contributed to a decline in groundnut production in that year whereas cereal production rose by 34 percent. Improved weather combined with the 1986/87 groundnut producer price adjustment resulted in a 46 percent increase in groundnut production and a decline in cereal production. Groundnut production increased further to 120,000 tons in 1987/88, despite the decline in the real price in that year. It fell in 1988/89 and then rose to 130,000 tons in the 1989/90 season, the highest level since 1982/83.

Despite the ERP's emphasis on transferring income to the agricultural sector through the groundnut subsidy, real output growth in the industrial sector appears to have outstripped that of agriculture during the 1986/87–87/88 period. Although groundnut prices remained high from 1986/87 to 1987/88, increased fertilizer prices, the restructured farm credit system, and more stringent credit requirements may have all contributed to the relatively slower growth experienced by this sector. In the industrial sector, value added from groundnut processing rose in both 1986/87 and 1987/88. Construction activity benefited from increased public infrastructure projects starting in 1986/87 as well as from private sector initiatives in commercial and residential construction.

The service sector also expanded during the 1986/87–88/89 period. Higher output from groundnut marketing, as well as other trading, government encouragement of tourism, and increased public sector expenditures for transport and communications contributed to this growth. In addition, the creation of the interbank market, introduction of a flexible interest rate policy, and other financial policies appear to have encouraged the business and financial services sector after 1986/87.

The available information suggests that the ERP was successful in promoting increased domestic savings and investment and in reducing consumption. In particular, consumption fell from 83 percent of GDP in 1983/84 to 76 percent in 1985/86 and 72 percent in 1988/89. The evidence also suggests that much of this investment was in the form of increased residential and business construction. Value added in the construction sector (in constant prices) increased at an average annual rate of 13.5 percent from 1985/86 to 1989/90.

Because of the temporary nature of the groundnut price increases, the extent to which the ERP promoted agricultural investment is not clear. The 1986/87 groundnut price of D 1,800 per ton was announced in advance of the planting season, but the extent to which farmers would have perceived such a high price to be permanent—and therefore used it as a basis for long-term investment decisions—is not well understood. Moreover, as agricultural implements were imported, primarily from Senegal, their prices also rose with the depreciation of the dalasi.[4] Farm implement price data from the GCU indicate price increases of 70–100 percent after the start of the ERP. Von Braun et al. (1990) have suggested that farmers invested more in draught animals than in new machinery between 1985 and 1987.

---

[4] To the extent that farmers purchased implements with CFA francs earned through groundnut sales on the parallel market, the price increase for implements may not have been so large.

Inflation

Being an open economy, the rate of inflation in Gambia is highly dependent on changes in the exchange rate, international prices, and changes in domestic bank credit. The annual average rate of inflation in Gambia ranged from 5 to 10 percent during 1979/80–82/83. The average annual inflation rate rose to 16 percent in 1983/84 in response to the higher level of bank financing required to support the GPMB and other government initiatives, and to 22 percent in 1984/85 in response to the devaluation of the dalasi.

The exchange rate liberalization resulted in the value of the dalasi declining by over 50 percent from its prefloat level in terms of major currencies by the end of 1986/87 (Figure 10.1). In both 1985/86 and 1986/87 inflation (on a period average basis) rose sharply in response to the dalasi's depreciation and to the increase in bank credit required to support GPMB operations. During 1987/88 the value of the dalasis stabilized, and it appreciated slightly. Inflation also dropped sharply in 1987/88.

Gambia's experience with a liberalized exchange rate system demonstrated that introduction of a system need not be accompanied by lasting and uncontrolled inflation. Inflation (on an end-period basis) actually started to decline in 1986/87. This decline reflected the government's commitment to stabilize the exchange rate with high interest rates and restrictive monetary policy and by freezing government wages and salaries. The tight fiscal and monetary environment limited the extent of the depreciation of dalasi after 1986 by restraining the demand for and increasing the supply of foreign exchange. In contrast to previous years when large inflows of foreign assistance resulted in large increases in the money supply, the government sterilized a portion of the massive increase in foreign aid that occurred in 1986/87. Some of the assistance was used to effect a 53 percent reduction in domestic credit, and thus its effects on the money supply were moderated.

Consumer Prices

Consumer prices more than doubled during the ERP. This would be expected, since the dalasi depreciation resulted in a permanent increase in the prices of traded goods, and most items in the CPI are either imported or exported. Government pricing policies under the ERP also promoted differential increases in the prices of selected commodities and commodity groups. For instance, the liberalization of rice marketing and elimination of government controls on the prices of foodstuffs such as groundnut oil, beef,

Figure 10.1. Variations in CPI and nominal exchange rate, Gambia

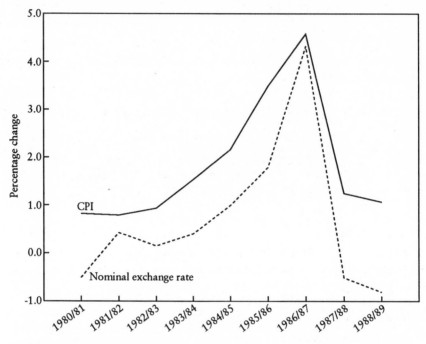

Sources: IMF 1987a, 1988a, 1989b.
Note: Nominal exchange rate is the negative of change in value of dalasi per unit of foreign exchange.

and eggs allowed consumer prices to rise with market forces. To increase revenues, the government also raised import tariffs on petroleum products and made adjustments in the prices of government-supplied services such as health, electricity, water, and public transport.[5]

Price movements since January 1985 for major consumption items are shown in Figure 10.2. Food prices rose slightly more, on average, than nonfood prices during the 1985–88 period. Energy prices, which more than tripled, exhibited the largest increase. The relatively large jump in the price of energy products, fuel and light, reflects continued increases in import tariffs on petroleum products ranging from 50 to 73 percent as well as the pass-through of the dalasi depreciation and tariff changes to

[5] In addition, with the liberalization of the dalasi all import duties were changed over to their ad valorem tariff equivalents. This ensured that all international price movements would be fully passed through to domestic prices.

government-administered prices, such as the price of electricity. In addition to electricity, government-determined prices for water rose by 60 percent (July 1986–January 1988) and bus fares increased by over 100 percent (April 1986–October 1988).

The Real Exchange Rate and the Current Account Trade Deficit

Gambia's real exchange rate, which measures the real value of the dalasi in terms of foreign currencies, remained virtually unchanged from 1979/80 to 1985/86. An approximate 20 percent decline in the real exchange rate was achieved in 1986/87. After 1986/87 the real exchange rate appreciated, but it still remained slightly below its pre-ERP level in 1988/89 (Table 10.1).

The cushioning effect of higher adjustment-induced foreign aid flows

Figure 10.2. CPI by quarter for selected products, Gambia

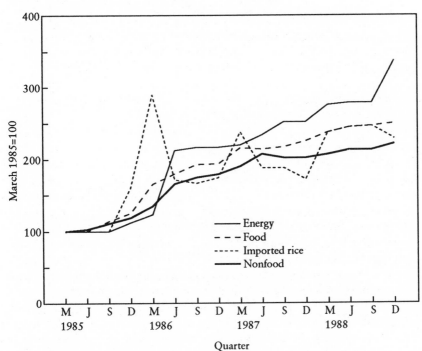

Source: Gambia Central Statistics Department, unpublished data.

since 1986/87 is shown clearly in Figure 10.3. The decline in the real value of the dalasi was not nearly enough to eliminate Gambia's current account deficit (excluding transfers), which remained at roughly 20 percent of GDP in 1988/89. At the same time, the depreciation of the dalasi contributed to an increase in the import share of GDP from around 50 percent pre-ERP to roughly 60 percent. Gambia's balance-of-payment surplus that was achieved in 1988/89 was the result of foreign transfers and highly concessional foreign loans.

Real Incomes and the Urban-Rural Terms of Trade

The government's primary goal for income distribution under the ERP was to transfer income to the factors of production engaged in tradable activities relative to the income of factors engaged in nontradable activities. In the Gambian economy, most agricultural activities can be regarded as tradables, whereas activities in hotels and restaurants and in the reexport trade are important tradables in the urban areas. One of the primary instruments used under the ERP to transfer resources into tradable activities was to raise the official producer price for groundnuts. Because the formal labor market is quite homogeneous in Gambia, the government also froze government wages and salaries. The reexport trade was promoted with reduced tariffs on imports important to the reexport trade, which primarily affected the entrepreneurial incomes of the informal sector, and by providing a favorable environment for domestic and foreign investment in hotels and restaurants.

As shown in Table 10.2, the 1986/87 increase in the official farmgate price for groundnuts raised it to over three times the level of 1982/83. Cotton prices were similarly raised. The higher minimum price for paddy offered in 1985/86 reflects the increase in market prices for rice after liberalization of this market in early 1985. The temporary increases in farmgate groundnut prices substantially raised the incentives to produce groundnuts relative to cereals from 1985/86 to 1987/88.

The 1985/86 groundnut price increase resulted in a 46 percent subsidy to farmers based on prevailing export prices and GPMB handling costs (Table 10.3). In 1986/87 the announced groundnut producer price resulted in a 42 percent subsidy. The GPMB purchased over 70,000 tons of groundnuts in that year, an amount that easily exceeded the purchase target of 60,000 tons set by the IMF. The subsidy increased in 1987/88 to 46 percent despite the sharp drop in the producer price to D 1,500 per ton. The groundnut producer price was lowered in 1988/89 to D 1,100 per ton—a price approximately equal to the GPMB's breakeven cost.

Figure 10.3. Fiscal account as a percentage of GDP, Gambia

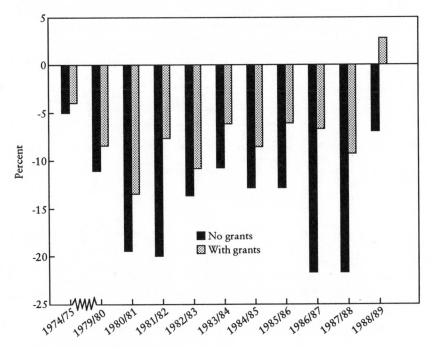

Sources: IMF 1987a, 1988a, 1989b.
Note: Negative percentages denote fiscal deficit.

Despite the price inflation of 1986/87, real per capital income from agricultural crop production (groundnuts, cereals, and cotton) rose substantially in 1986/87 to a level 50 percent above that of 1983/84 (Table 10.4). Most of this increase was due to higher prices and production of groundnuts, although real income from other crops also increased. The groundnut subsidies appear to have fairly insulated the agricultural sector from the most adverse effects of the ERP during both 1986/87 and 1987/88.

Information on real wages and salaries paid in the formal sector from 1979 to 1987, in contrast, indicates substantial declines in real earnings of both salaried (established) and daily paid (regular) workers (Table 10.4); these workers live primarily in urban areas. Most of this decline occurred between June 1983 and December 1986. Real daily earnings of established workers fell by about 9 percent from 1979 to 1983, and earnings of regular

Table 10.2. Official producer and market prices for agricultural commodities, Gambia (dalasis per ton)

	1982/83	1983/84	1984/85	1985/86	1986/87	1987/88	1988/8
Cotton	560	610	700	1,282	1,346	1,800	2,200
Groundnuts	520	450	620	1,260	1,800	1,500	1,100
Palm kernels[a]	314	314	314	410	NA	NA	NA
Rice (paddy)[b]	510	510	600	900	945	1,000	900
Maize[c]	390	390	390	600	800	1,000	1,250
Millet[d]	460	460	460	650	800	1,000	1,250

Source: Jabara 1990a, courtesy of Cornell Food and Nutrition Policy Program.
Note: NA, not available.
[a]With effect from the 1986/87 crop season, no official minimum producer prices are set for palm kernel
[b]Minimum price.
[c]With effect from the 1986/87 crop season, no official minimum producer prices are set for maize.
[d]Market price only.

workers fell by about 10 percent. In contrast to the earlier period, established workers' real daily earnings fell by 37 percent from 1983 to 1986, and those of regular workers fell by 45 percent. It is doubtful, however, that real incomes could have been maintained at their 1983 levels even without the ERP, given the government's precarious financial situation at the start of the ERP. By 1987 real incomes of established workers appear to have stabilized, but those of regular workers fell by an additional 13 percent.

The decline in the earnings of the daily paid workers as compared to salaried workers probably reflects the latter's greater ability to protect their real incomes in times of economic belt tightening and the relatively greater demand created by the ERP for higher-skilled labor. Since the government retrenchment released a much greater number of regular and temporary workers, their absorption into the private sector may have had a relatively more adverse effect on earnings of this particular type of worker.

The extent to which the ERP was successful in turning the rural-urban terms of trade in favor of the agricultural sector is illustrated by the groundnut/daily wage price ratio, also shown in Table 10.4. By 1986/87 the return to groundnut production relative to urban employment had more than doubled from the base level of 1978/79. Policy changes implemented in 1988/89, however, should reverse much of the decline in real formal sector incomes since 1979.[6] In November 1988 a new civil service grade structure was introduced, and in January 1989 wages and salaries for the established posts were raised, on average, by 55 percent. The minimum

---

[6] In addition to the 1988/89 wage and salary changes, urban workers in 1988 benefited from new income tax measures, which lowered taxable incomes for almost all civil servants.

Table 10.3. Cost/price structure for decorticated groundnuts, Gambia (dalasis per ton)

	1979/80	1980/81	1981/82	1982/83	1983/84	1984/85[a]	1985/86	1986/87	1987/88
Producer price	421	460	500	520	450	620	1,100	1,800	1,500
Board costs	211	262	152	153	186	187	275	243	250
Cost/ton unshelled	632	723	652	673	636	807	1,375	2,919	1,750
Cost/ton shelled	903	1,032	931	961	909	1,153	1,964	2,846	2,500
GPMB overhead/interest paid	75	87	173	144	121	121	809	512	327
Cost ex-Banjul before tax	978	1,120	1,104	1,105	1,030	1,274	2,785	3,409	2,881
Export duty (10–12% f.o.b. Banjul price)	102	108	89	78	180	231	246	41	346
Realized f.o.b. Banjul price	800	1,129	920	826	1,844	1,929	2,046	2,346	1,896
GPMB net trading profit (loss)	(280)	(99)	(273)	(357)	634	424	(971)	(1,473)	(1,331)
Memorandum items									
Real producer price[b]	341	341	343	327	245	277	364	408	302
Implied producer tax (subsidy) (%)[c]	(29.6)	0	(25.8)	(37.6)	126.7	73.9	(46.3)	(42.2)	(46.0)

Source: Jabara 1990a, courtesy of Cornell Food and Nutrition Policy Program.
[a]Estimate.
[b]Constant 1976/77 prices. Deflated by the CPI.
[c]Calculated as a percentage of the producer price, shelled basis.

Table 10.4. Earnings of urban formal workers and agricultural producers, Gambia (dalasis)[a]

Earnings/employee type[b]	June 1979	June 1983	December 1986	December 1987	Janu 198
Real urban daily earnings					
Established	10.9	9.9	6.2	6.2	9
Regular	6.1	5.5	3.0	2.6	4
Real urban monthly earnings[c]	NA	195.1	118.8	125.3	200
Real agricultural per capita earnings[e]					
Total crops	253.2	157.7	238.1	203.9	153
Groundnuts	193.3	105.1	168.4	134.2	72
Other	59.9	52.6	69.6	69.7	81
Groundnut/daily wage[f]	100.0	83.9	242.0	197.6	90

Source: Jabara 1990a, courtesy of Cornell Food and Nutrition Policy Program.
Note: NA, not available.
[a]1980=100. Deflated by the CPI.
[b]Established workers are salaried; regular workers are paid on a daily basis.
[c]All employees. Includes earnings of expatriate employees but excludes earnings of temporary employ
[d]Estimated on the assumption of a 60 percent increase in real wages over 1987.
[e]Income from production of groundnuts, cereals, and cotton less expenses for fertilizer and seeds.
[f]1978/79=100. Ratio of groundnut producer price to average daily earnings of regular workers employe
the formal sector.
[g]Estimated on the basis of a groundnut producer price of D1,100 per ton and a 60 percent increase in
average daily wage.

daily wage scale was also revised in 1989. This new wage scale increased the minimum wage for the highest-paid general workers (foremen and artisans) by 63–75 percent (from D 13.3 to D 21.7 and from D 7.8 to D 37.7, respectively) and increased the minimum wage for lower-paid laborers by 63 percent (from D 5.5 to D 9.0). The extent to which these higher minimum wages actually increased incomes is not clear, however, because some wages paid in the formal sector in 1987 for daily rated workers were at the same level or higher than the new minimums. Assuming, however, that these reforms resulted in a real 60 percent increase in formal sector incomes over the 1987 level, the real average monthly income of urban workers was approximately at the level of June 1983 in 1988/89.

Reduced groundnut prices and upward adjustments in urban wages and salaries appear to have eliminated most of the decline in the urban-rural terms of trade which occurred in the early years of the ERP, as shown by the decline in the estimated groundnut/urban wage price ratio for 1988/89 (Table 10.4). Rural incomes increased in 1989/90 due to higher crop prices and a larger groundnut crop, but the rural-urban income disparity still approximates that of the pre-ERP level.

The Fiscal Deficit and Public Expenditure

Total real public expenditure in Gambia fell sharply in the first year of the ERP, 1985/86. The foreign assistance programs implemented in the second year of the ERP allowed a 33 percent increase in real public expenditure in 1986/87 and a 22 percent increase in 1987/88. On average, real public expenditure during 1986/87–88/89 averaged 96 percent of the average level during the immediate pre-ERP period (1979/80–84/85) (Tables 10.5 and 10.6).[7] Increased real public expenditures in 1986/87 and 1987/88 were financed by domestic credit creation, higher tax rates, broader tax coverage, and more effective tax enforcement, as well as by foreign loans and grants. The government maintained a fiscal deficit (excluding grants) of about 22 percent of GDP in both 1986/87 and 1987/88. In 1988/89 it reduced its fiscal deficit to only 7 percent of GDP.

Although real public expenditure was maintained at its approximate pre-ERP level from 1986/87 to 1988/89, significant changes were made in the composition of public expenditure during the ERP. Public expenditure shifted sharply in favor of government consumption at the expense of public investment. Increased consumption was in the form of higher debt service, transfers to parastatals for temporary agricultural subsidies and to undo past mismanagement, and enforcement of the reforms instituted under the ERP. Development expenditure averaged only 80 percent of the level of the pre-ERP period, whereas real recurrent expenditure during the same period was actually higher.

The change in the mix of services provided through the recurrent budget during the ERP can also be observed in Table 10.6. Spending on transfer payments to pay operating costs and accumulated debt of parastatals increased from around 9 percent of recurrent expenditures during 1979/80–84/85 to over 40 percent in 1987/88. Debt service increased from 8 percent of recurrent expenditure in the same period to over 19 percent in 1987/88.[8] If these two items are excluded from the recurrent budget, the remainder of the recurrent budget averaged only 60 percent of the pre-ERP average from 1986/87 to 1988/89. Expenditures on social services, agriculture, and public works were the most significantly cut, and in real terms they were about half of their pre-ERP levels.

---

[7] Real expenditures are defined as nominal expenditures deflated by the CPI. As such, changes in real expenditure levels cannot be equated with an equivalent change in the level of services supplied by the expenditures. Rather, they represent the level of expenditure in constant monetary terms.

[8] As reported by the Gambian government, debt service includes both interest and amortization.

Table 10.5. Real public expenditure, Gambia (constant thousand dalasis)[a]

	Average 1979/80–84/85	1985/86	1986/87	1987/88[b]	1988
Recurrent expenditures					
Agriculture	8,906.3	4,607.8	2,759.6	3,468.7	3,1ᵉ
Public works	11,300.0	4,826.9	3,010.0	3,929.1	3,4
Education/youth	15,093.0	10,391.7	7,301.6	7,428.7	9,1.
Health/social welfare	9,898.3	6,302.8	5,269.0	5,422.5	5,6
Debt service[d]	8,385.0	8,577.3	20,129.5	28,750.3	35,7:
Parastatals[e]	239.0	6,988.8	26,922.9	35,197.1	3,3
Other	45,756.5	21,851.4	27,418.2	34,452.9	43,7
TOTAL	99,578.1	64,754.3	92,810.8	118,649.2	104,1
Development expenditures					
Agriculture	6,515.6	8,634.0	17,736.6	11,782.2	6,8:
Public utilities	3,514.3	5,670.1	4,811.6	1,327.6	7,9.
Industry	2,517.7	0.0	2,722.5	0.0	
Transport/commun.	25,074.3	12,843.6	13,266.9	15,957.6	11,4
Tourism/trade	857.0	0.0	144.8	233.5	4
Social welfare	13,555.3	10,352.2	4,715.9	17,420.1	8,1
Education	(9,431.1)	(644.3)	(817.5)	(457.5)	(1,5
Health	(1,958.4)	(859.1)	(688.9)	(11,075.8)	(3,0
Housing/comm. dev.	(2,165.8)	(8,848.8)	(3,209.5)	(5,886.8)	(3,5.
General	3,465.1	2,921.0	1,709.5	2,478.7	2,9
TOTAL	55,405.5	39,132.3	45,105.3	49,199.5	37,8.
TOTAL EXPENDITURES	154,983.6	103,886.6	137,916.1	167,848.7	142,0

Source: Jabara 1990a, courtesy of Cornell Food and Nutrition Policy Program.
Note: Gambian fiscal year, July–June.
[a]1980 = 100. Deflated by the CPI.
[b]1987/88 data do not include an unplanned transfer to the GUC to repay its debt to the Central Bank ⟨ Gambia.
[c]Estimate.
[d]Debt service includes both interest and amortization on internal and foreign debts.
[e]Budgetary transfers for operating costs and for retiring debts of parastatals.

To improve the productivity of public investment, the government introduced a "rolling" three-year investment program linked to the annual preparation of the recurrent budget in 1986/87. Stringent selection criteria were established for new projects, including the requirement of a minimum 15 percent rate of return. Analysis of the development budget reveals the priorities alternatively given to public utilities, agriculture, tourism and trade, and social welfare, particularly health and housing, during the ERP.

Financial and Credit Policies

Financial and credit policies were important in Gambia's success in stabilizing the value of the dalasi on the interbank market and in reducing *le 10.6.* Sectoral share of real public expenditures, Gambia (percentage)

	Average 1979/80–84/85	1985/86	1986/87	1987/88[a]	1988/89[b]
urrent expenditures					
Agriculture	9.2	7.1	3.0	2.9	3.0
Public works	11.3	7.4	3.2	3.3	3.3
Education	15.5	16.0	8.0	6.3	8.8
Health	10.1	9.7	5.7	4.6	5.4
Debt service[c]	8.4	13.2	21.7	24.2	34.3
Parastatals[d]	0.2	10.8	29.0	29.7	3.2
Other	45.2	33.7	29.5	29.0	41.9
TOTAL	100.0	100.0	100.0	100.0	100.0
velopment expenditures					
Agriculture	11.6	22.1	39.3	23.9	18.1
Public utilities	6.6	14.5	10.7	2.7	21.0
ndustry	4.5	0.0	6.0	0.0	0.2
Transport/commun.	45.4	32.8	29.4	32.4	30.3
ourism/trade	1.6	0.0	0.5	0.5	1.2
ocial welfare	24.3	26.4	10.4	35.4	21.4
Education	(16.5)	(1.6)	(1.8)	(9.3)	(4.1)
Health	(3.9)	(2.2)	(1.5)	(22.5)	(8.0)
Housing	(4.0)	(22.6)	(7.1)	(12.0)	(9.3)
;eneral	6.0	7.5	3.8	5.0	7.8
TOTAL	100.0	100.0	100.0	100.0	100.0

ource: Jabara 1990a, courtesy of Cornell Food and Nutrition Policy Program.
lote: Gambian fiscal year, July–June.
1987/88 data do not include an unplanned transfer to the GUC to repay its debt to the Central Bank of the
nbia.
Estimate.
Debt service includes both interest and amortization on internal and foreign debts.
Budgetary transfers for operating costs and for retiring debts of parastatals.

the rate of inflation. Before the ERP, all deposit and lending rates in Gambia were fixed by the central bank, often resulting in negative real interest rates. A flexible and market-influenced interest rate policy was adopted under the ERP in order to provide more rational credit allocation, to mobilize savings, and to support the value of the dalasi under the interbank system. Ceilings on the banking system and on net credit to government with the goal of reducing inflationary pressures, while at the same time increasing the flow of credit to the private sector, were also instituted under the advice of the IMF.

As shown in Figure 10.4, real interest rates in Gambia declined steadily from 1979/80 to 1986/87. But the introduction of the flexible interest rate system in 1986/87, under which key interest rates are determined on the basis of a biweekly tender for treasury bills, and lower consumer price inflation resulted in a sharp increase in real interest rates to positive levels in 1987/88. The government has maintained positive real interest rate levels on both bank deposits and commercial loans since 1987/88, and currently these rates are among the highest in Africa. Farmers were somewhat cushioned from the high interest rate policies: the rates charged for seasonal credit supplied to farmers by the GCU remained slightly below the commercial rates charge by banks for loans to individuals in other sectors during the ERP. The lower agricultural credit rate was due to the availability of the no-interest loan to the GCU under the second multidonor-sponsored agricultural development project (ADP II).

Total bank credit sharply contracted from 1986 to 1989 under IMF guidelines. At the same time, private sector credit increased from 34 percent of total bank credit extended in June 1984 to over 90 percent in June 1989 (Table 10.7). The nominal data in Table 10.7 mask the decline in real credit available to the economy under the ERP. Although domestic credit

Figure 10.4. Trends in real interest rates, Gambia

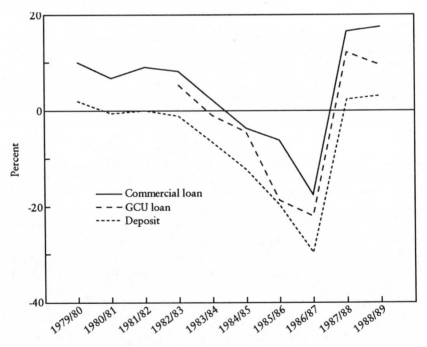

Sources: IMF 1987a, 1988a, 1989b; World Bank 1985a, tab. 21.

 10.7. Distribution of domestic bank credit, Gambia (million dalasis)

	1984	1985	1986	1987	1988	1989
estic credit[a]						
overnment (net)	80.5	86.8	100.3	(64.7)	(37.3)	(115.2)
blic entities	160.5	166.6	217.4	216.7	101.4	132.5
of which GPMB	(111.3)	(95.4)	(132.8)	(156.0)	(56.0)	(75.1)
ivate sector	143.0	159.0	185.6	174.2	192.3	224.5
AL CREDIT	384.0	412.0	501.3	326.0	256.4	241.9

urce: Jabara 1990a, courtesy of Cornell Food and Nutrition Policy Program.
 ata are for June in the year shown.

extended to the private sector increased by 56 percent from 1984 to 1989, in real terms it declined by 55 percent.

Credit extended to farmers through the GCU, on the other hand, declined by 94 percent in real terms.[9] New credit eligibility criteria, which were instituted under a structural adjustment credit with the World Bank in 1987/88, reduced access to seasonal credit by farmers. The new eligibility criteria excluded past loan defaulters from new loans and placed limits on the amount of credit extended to individuals. Restricted access to seasonal credit, along with increases in fertilizer prices, contributed to a sharp decline in fertilizer use under the ERP (see below). In addition, credit ceilings on bank lending for groundnut buying were instituted in 1987/88 to prevent extensive default of crop financing loans. These limits were primarily directed at the GCU, which had incurred significant financial losses in groundnut marketing in the past.

Marketing Reforms

Reform of public enterprises was a key component of Gambia's ERP. For some public enterprises these reforms involved the government's write-off of debts owed to the central bank or the Gambia Commercial Development Bank and the implementation of programs to improve operational efficiency. For other enterprises, reforms involved privatization of the entire enterprise or privatization of certain activities the enterprise had undertaken before the ERP. The most notable reforms, however, involved marketing of agricultural products (groundnuts and rice) and fertilizer. The implementation of policy reforms that allowed private traders to partici-

[9] GCU credit in real terms is computed by deflating the nominal amount issued by the price of SSP fertilizer. Real credit to the private sector is obtained by using the CPI as a deflator.

pate in the marketing of these products was central to the ERP because the two agencies that previously had the marketing monopolies, the GPMB and the GCU, were heavily in debt at the start of the ERP.

Government efforts to promote private sector participation in groundnut and rice marketing were relatively successful, but efforts to promote the private sector in the fertilizer market were hindered by the availability of large supplies of fertilizer that had been supplied by donors at the start of the ERP at no cost. The provision of the grant fertilizer also hindered government efforts to establish an "economic cost" for the fertilizer.

In the case of groundnuts, marketing reforms were phased into the ERP after the elimination of the groundnut subsidies. Thus reforms implemented in the 1989/90 season allowed the GPMB to establish its own procurement price (ex-depot) and to purchase groundnuts at its depots from anyone—farmers, private traders, or the GCU. Previously the GPMB had purchased groundnuts only from the GCU or other authorized agents. Roughly 50 percent of GPMB purchases came from private agents, including farmers, in that year. Additional reforms implemented in the 1990/91 season eliminated the GPMB's monopoly on groundnut purchases and sales as farmers were allowed to sell groundnuts to any willing buyer. The results from the 1990/91 season are not yet available.

The government's pre-ERP policy was to stabilize the retail price of rice, and rice importation and wholesale distribution were under the exclusive monopoly of the GPMB. GPMB involvement in rice marketing was effectively ended in mid-1985, and private traders were encouraged to take over this trade. Retail rice prices rose by 50 percent from September to December 1985 and then by an additional 100 percent from December 1985 to March 1986. The December–March increase also included the price effects from liberalization of the foreign exchange market. It should be noted that in late March and April 1986 the government sold some of its rice stocks at below-market prices in order to moderate the inflation in the rice market. In early 1986, however, international rice prices fell, and retail rice prices stabilized in late 1986.

Fertilizer importation was decontrolled in January 1986, and the GPMB was stripped of its fertilizer handling activities. Fertilizer subsidies were also reduced. As shown in Table 10.8, fertilizer prices were raised each year from 1985/86 to 1987/88. Despite the announced change in fertilizer handling, however, private importers expressed little interest in fertilizer importation, because at the time of the announcement large stocks of fertilizer that had been supplied by ADP II were still on hand. In both 1985/86 and 1986/87, the GCU was appointed to be the sole distributor of this fertilizer, which was supplied to the government on a grant basis. The grant fertilizer was auctioned at the port in the 1987/88 season, but the GCU made the only bid. Lack of private sector participation was largely attributable to the GCU's preferential financing arrangements under ADP II and to the private sector's lack of knowledge about fertilizer marketing.

Consumption of both SSP (used on groundnuts) and NPK compound (used on cereals and cotton) declined steadily during the ERP (Figure 10.5). It is believed that some of this reduced consumption was due to the reduction in subsidies which discouraged cross-border sales to Senegal. Nonetheless, apparent consumption of fertilizer in Gambia in the 1987/88 and 1988/89 seasons was far less than in the 1970s.

Private sector participation in fertilizer marketing became effective in the

le 10.8. Fertilizer subsidies, Gambia (dalasis/ton)

	SSP			Compound (15-15-15)			Compound (8-24-24)		
	Cost price[a]	Sale price	Subsidy (%)	Cost price	Sale price	Subsidy (%)	Cost price	Sale price	Subsidy (%)
'4	178	80	55.0	228	97	57.4	NA	NA	NA
'5	343	95	72.3	282	120	57.4	NA	NA	NA
'6	329	92	72.0	453	120	73.5	NA	NA	NA
'7	333	108	67.6	532	129	75.7	NA	NA	NA
'8	304	106	65.1	425	137	67.7	NA	NA	NA
'9	417	93	77.7	440	134	69.5	NA	NA	NA
;0	443	104	76.5	694	133	80.8	NA	NA	NA
;1	325	108	66.7	581	135	76.8	NA	NA	NA
;2[b]	495	140	71.7	837	194	76.8	NA	NA	NA
;3[c]	501	160	68.0	738	245	66.8	NA	NA	NA
'4	540	270	50.0	1,042	370	64.4	NA	NA	NA
;5[d]	590	460	22.0	1,136	600	47.2	NA	NA	NA
;6[e]	1,008	740	26.0	1,351	1,140	15.6	NA	NA	NA
'7[f]	840	840	NA	1,220	1,220	NA	NA	NA	NA
'8	840	840	NA	1,220	1,220	NA	NA	NA	NA
'9[g]	NA	NA	NA	NA	NA	NA	1,820	1,095	39.8

urce: Jabara 1990a, courtesy of Cornell Food and Nutrition Policy Program.
ote: NA, not available.
ased on financial costs to GPMB and GCU.
Jnder standby arrangement with IMF, the Gambian government agreed to gradually eliminate fertilizer idies.
ertilizer distributed free to groundnut farmers in 1982 and 1983 in exchange for retaining groundnut seed. idy would be higher if this scheme were taken into account.
irst tranche of fertilizer supplied by the Italian government under ADP 2 arrived late. Old stocks sold ng 1985/86 season. Fertilizer prices raised in June 1985 as part of the ERP.
ambian government unable to raise fertilizer prices to full economic cost due to large unused stocks of the an grant fertilizer.
econd tranche of fertilizer supplied by the Italian government sold at estimated c.i.f. import price plus lesale costs.
ertilizer prices lowered after third tranche of Italian grant fertilizer retendered under pressure from the d Bank.

Figure 10.5. Fertilizer use, Gambia

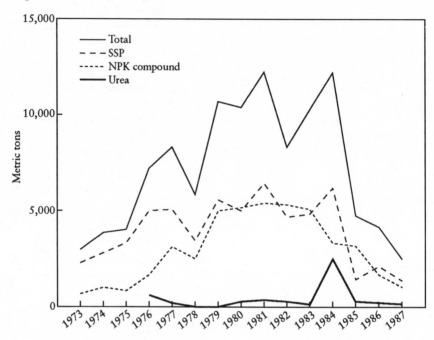

Sources: Langan 1987; Puetz and von Braun 1988.

1989/90 season. In that year, however, fertilizer was sold at an auction price that was roughly 60 percent of the estimated economic cost.[10] This change in fertilizer pricing policy was made to reflect the grant nature of the fertilizer and to put a halt to the sharp increases in fertilizer prices that had occurred during the ERP.

Conclusions

Gambia's economic recovery program is generally considered to be one of the more successful ones carried out among African countries during the 1980s. Policy reforms were designed to ease the macroeconomic imbalances of a small, open economy. The discussion of the ERP suggests some of the limitations of the structural adjustment process as a vehicle for raising the Gambian economy to a higher growth path. Despite the policy

[10] Fertilizer prices were lowered under a retender after pressure from the World Bank.

adjustment's made under the ERP, Gambia is still in need of substantial inflows of grants and concessional foreign aid to finance its current account deficit. Gambia's external debt has risen from under $100 million in 1979 to over $300 million in 1988. Practically all this debt increase was due to foreign aid from multilateral and bilateral donors.

Although Gambia's external arrears have been reduced to one-third of their peak level of June 1986, efforts to service this debt will weigh heavily on the economy over the coming decades. High debt service ratios combined with Gambia's structural merchandise trade deficit will require even larger foreign financial inflows and increased foreign exchange earnings to balance the external payments gap in the future.

11

Staggered Reforms and Limited Success: Structural Adjustment in Madagascar

Paul Dorosh and René Bernier

In the 1980s, Madagascar, an island nation of 11 million inhabitants, managed to overcome a serious economic crisis and came to be regarded as a star pupil of the IMF. Madagascar undertook a major investment program in the late 1970s in an attempt to promote rapid and equitable growth. Serious and unsustainable macroeconomic imbalances developed, however, making necessary a protracted period of stabilization and structural adjustment. Significant reductions in inflation, budget deficits, and balance-of-payments deficits were achieved by the mid-1980s, but there has been little real economic growth until recently, and GDP per capita was only $250 in 1989 (World Bank 1991c). Moreover, the early reforms were sharply criticized for their adverse effects on some of the poorest segments of the urban population.

Structure of the Economy

Madagascar is typical of many low-income countries with large agricultural and service sectors and a small industrial sector. In 1989 agriculture accounted for 29 percent of GDP, whereas industry represented 13 percent of GDP and the service sector approximately 49 percent of GDP.[1] The sectoral share of value added had changed little since 1970, when

[1] Dorosh, Bernier, and Sarris (1990) gives sector shares that show agriculture as dominating the economy, with about 43 percent of GDP, and services having a lower share at 37 percent. Those shares were based on earlier national accounts figures (compiled by the Ministry of Planning). These revised figures are from World Bank 1991.

agricultural, industry, and services accounted for 22, 15, and 53 percent of domestic production. In 1989, 84 percent of GDP was absorbed by private consumption, 13 percent of production was used in investment, and government consumption accounted for 8 percent (World Bank 1991c).

Agriculture dominates Madagascar's economy in terms of employment, accounting for about 87 percent of total employment (IMF 1988b) as well as income generation. Much of the population lives on the high plateau, which ranges from the north to the south in the center of the island. Here the population density in 1984 was 0.363 rural inhabitants per square kilometer, compared to the national figure of 0.147 (Ministère de la Production Agricole et de la Réforme Agraire [MPARA] 1987a).

Although some large farms managed by parastatals exist, the bulk of agricultural production is carried out by traditional smallholders who use few purchased inputs. Traditional farmers account for 95 percent of total cultivated area, with an average holding of 1.15 hectares.[2] Food crop production represents 75–80 percent of total agricultural production and 78 percent of cultivated area. Rice and cassava account for 43 and 15 percent of agricultural production in value terms, respectively.[3] The principal export crops (coffee, cloves, and vanilla) accounted for 8 percent of agricultural production in 1984 and are also largely produced by smallholders (Dorosh, Bernier, and Sarris 1990, tab. 6).

Madagascar's agriculture differs from that of most of sub-Saharan Africa because of the dominance of irrigated land, especially on the high plateau. Irrigated area, planted primarily with rice or cotton, accounts for 44 percent of traditional cultivated area nationwide (MPARA 1988a–f). A shortage of irrigated land is a principal constraint to increased production by rice farmers on the high plateau.[4] Unfortunately, few opportunities remain for economically profitable investments that increase the amount of irrigated land, although rehabilitation of existing systems may be profitable since inadequate water control limits double-cropping in many irrigated perimeters (Associates for International Resources and Development [AIRD] 1990; Barghouti and LeMoigne 1990).

Besides providing livelihoods for the majority of Madagascar's population, agriculture makes three important contributions to the economy. First, agricultural products are the primary source of foreign exchange

[2] Traditional farmers are defined in the 1984 agricultural census as farmers owning 10 or fewer hectares, hiring fewer than five full-time, paid workers, and not using any specialized modern equipment or machinery.

[3] These figures are derived from Ministry of Agriculture annual production statistics. Estimates of production based the 1984/85 agricultural census yield figures of 38 percent for rice and 19 percent for cassava. See Dorosh, Bernier, and Sarris 1990 for a discussion of the differences between the two data sources.

[4] See Bernier and Dorosh 1993.

earnings. Agricultural exports (mainly coffee, cloves, and vanilla) accounted for 54 percent of total exports in 1989 (Banque des Données de l'Etat [BDE] 1989).[5] A second key contribution made by agriculture is that it supplies the bulk of domestic consumption. Rice consumption represents 54 percent of total calorie consumption alone (Food and Agriculture Organization [FAO] 1984). Finally, agriculture provides many of the primary inputs into domestic industries such as textiles, sugar, oil products, and other food processing industries.

Madagascar has a relatively small industrial sector, typical of most developing countries. Cities in Madagascar are relatively small by international standards, and, in the 1984, 81 percent of the population lived in rural areas. The capital city, Antananarivo, is the only city with more than 600,000 inhabitants (World Bank 1991d). Production of the formal industrial sector is concentrated in easy import-substitution sectors such as food processing, textiles, beverages, and nontraded sectors such as water and electricity. Value added for the 355 enterprises included in the 1984 industrial census represented 50.4 percent of total industrial value added. Textiles accounted for 30 percent of value added and 19 percent of employment for the sample. Food processing enterprises accounted for 19 and 28 percent of value added and employment, respectively (BDE, undated). Limited data are available on the informal sector; based on a 1978 survey in Antananarivo covering 343 of 3,913 informal enterprises (and excluding transport, construction, and trade), the primary activities of the informal sector are textiles and clothing, basketry, woodworking, metalworking, food processing and sales, and mechanical repairs (Project PNUD/BIT 1980). Total value added for the sample firms was estimated at FMG 2.449 billion, or 0.5 percent of national GDP.

Madagascar's structure of external trade has changed gradually since the early 1980s. The major traditional exports, coffee, cloves, and vanilla represented 65.7 percent of total exports in 1980 ($263 million) (World Bank 1986b) but, due in large part to the collapse in world markets, the three products only accounted for 51.5 percent of total exports ($158 million) in the 1987–89 period (World Bank 1991c, tab. 10). Prospects for expansion of traditional exports are poor due to unfavorable conditions in international coffee markets and Madagascar's already large shares in world trade of vanilla (75 percent of world supplies) and cloves (61 percent of world supplies). Among the nontraditional products, shellfish exports have increased most rapidly, from 3.9 percent of total exports in the 1978–80 period ($16.5 million) to 10.2 percent ($31.2 million) in 1987–89, exceeding the shares of both cloves and vanilla (World Bank 1991c, tab. 10).

---

[5] Including agroindustrial exports such as cloth, preserved meats, and essences of cloves and ylang ylang, the share rises to 62 percent (IMF 1991c).

The structure of imports is characterized by a predominance of raw materials, energy, and capital goods, which represent 72.6 percent of the import bill (World Band 1991c, tab. 11). This share is only slightly below the 1975 share (74 percent), prior to the import-substitution industrialization drive (World Bank 1980b). Madagascar is wholly dependent on foreign trade to supply its oil needs. The trade data show a fairly constant level of oil imports during the 1976–78 period, at about 497,000 tons (or FMG 11.5 billion). The oil price hike of 1979/80 resulted in a drop in the volume of oil imports, falling to 299,000 tons in 1979 and to 130,000 tons in 1980. Oil imports recovered in 1981, reaching 486,000 tons, and combined with a 49.8 percent increase in the c.i.f. price of oil led to a large increase in the oil import bill which reached FMG 33 billion (BDE 1988).

The generally poor-quality data on the distribution of income and wealth in Madagascar at least suggests that inequality is a problem, particularly in rural areas, and that the situation may have deteriorated during the 1960s and 1970s. The Gini coefficient for the country as a whole was .447 in 1980, up from .391 in 1960 (Pryor 1988). Rural inequality worsened significantly. The 1962 Gini for rural areas was .290; by 1980 it had increased to .410. A more recent survey of Antananarivo in 1989 suggests a highly skewed distribution of income, with 10 percent of the urban population receiving 80 percent of total income (World Bank 1990c). Data from the 1984 agricultural census suggest that a significant proportion of rural households (38 percent) lack enough land to grow enough rice to be self-sufficient (Dorosh, Bernier, and Sarris 1990) and that 25 percent of farm households have fewer than 0.1 hectares per person. Gini coefficients measuring the distribution of rural land range from .323 in Fianarantsoa to .429 in Toliary (World Bank 1990c).

Poverty lines for rural and urban households were calculated based on food requirements and typical expenditure patterns for 1980 (Dorosh, Bernier, and Sarris 1990). This exercise yielded estimates of FMG 150,000 and FMG 132,000 for urban and rural households. Based on these poverty lines, approximately 37 percent of rural households, 26 percent of households in small urban areas, and 18 percent of households in the seven large urban areas can be classified as poor. Nationally, 34 percent of all households are poor, 90 percent of which are in rural areas (Dorosh, Bernier, and Sarris 1990). Estimates of average incomes of small farmers in 1984 ranged from FMG 102,700 per capita on the plateau to FMG 118,300 per capita in the south and west, compared to an estimated average income of FMG 271,300 per capita for large farmers and other rural higher-income households (Dorosh et al. 1991).

Expenditures and diets of the poor also provide some indication of the nature of poverty in Madagascar. Expenditures on rice account for a large portion of household budgets for all income groups, including the poor.

Rice expenditures account for between 39.6 and 47.7 percent of total expenditures for the poorest 20 percent of the population of Antananarivo. The middle class spends 26–34 percent of its income on rice (AIRD 1985). The elasticities for meat and fish are greater than unity. Poorer households consume more cassava relative to bread when compared to richer households. In most urban areas, the shares of food expenditures in total expenditures rose during the 1978–82 period while real expenditures fell. Rice expenditures grew as a share of total expenditures as well.

Rural, poorer households consume more cassava, on average, than do urban, better-off households, although rice still has the largest share in total expenditures. Combined with the fact that many rural households are not food self-sufficient, variations in the prices of these basic food items are therefore important determinants of the welfare of the poor.

Roots of the Crisis

The roots of the crisis in the early 1980s can be traced back to mounting frustration about slow growth rates and the slow rate of malgachinization of the economy during the 1960s and early 1970s.[6] From independence in 1960 through 1972, Madagascar's first government, headed by Philbert Tsirananana, adopted conservative fiscal and monetary policies and tolerated a large foreign presence in the economy: foreign-owned industries accounted for 65 percent of total industrial sales, French colonists owned large export crop estates, and trading was conducted by non-Malagasy merchants (Hugon 1987). Close ties were maintained with the French, including participation in the franc zone and the stationing of French troops in Madagascar.

Tsiranana's regime was overthrown and replaced in 1972 by a socialist military government that stressed malgachinization and centralization of the economy through state intervention and nationalizations, basic-needs satisfaction, and reduced external dependance. Madagascar severed formal economic and military ties with France and pursued a policy of nonalignment, resulting in increased aid and trade flows with the Soviet Union and North Korea.

One of the means by which the development objectives of the new regime were to be achieved was through an expansion in parastatal involvement in the economy through the nationalizations of industry, services, and shipping. The government also aimed at satisfying basic needs and reduc-

---

[6] Originally used to mean a demand for education in the national language, the term *malgachinization* gradually came to encompass the ouster of the French from Madagascar (see Covell 1989).

ing income disparities through increased government expenditures on health, education, and housing. Because the marketing system had been previously dominated by non-Malagasy, the government assumed responsibility for the marketing of agricultural and other products. This entailed state determination of producer and consumer prices and state control of agricultural marketing institutions. Some parastatals marketed products directly. In other cases, the state set maximum markups that were to be allowed in price determination. Large, often foreign-owned plantations were expropriated and converted into collective farms. Land reform was implemented, based on traditional *fokonolona* communes and distribution cooperatives (Mukonoweshuro 1990).

The 1978–80 development plan, instituted under Didier Ratsiraka, who became president after the assassination of President Richard Ratsimandrava in February 1975, sought rapid growth combined with satisfaction of basic needs through higher investment rates in key sectors, including social services. This "investing to the limit"[7] program entailed investments requiring large imports of capital and intermediate goods, which were to be financed from domestic savings facilitated by high world coffee prices. In practice, financing was achieved by foreign borrowing and domestic credit expansion.

Most of the investments were ill conceived and had little economic viability. As a result, loans came due without the increased capacity to repay them. In the short run (1978–80), though, the economy boomed as foreign and domestic borrowing spurred aggregate demand and later inflation. Real GDP growth accelerated from an annual average growth rate of −0.01 percent between 1973 and 1977 to 2.7 percent between 1978 and 1980 (Table 11.1). The greatest acceleration was in 1979, when real GDP grew by 9.86 percent. High levels of coffee export receipts based on favorable terms of trade and inflows of foreign capital provided the foreign exchange for the unsustainable surge in imports. Capital and intermediate goods imports rose from $188.9 million in 1977 (World Bank 1980b) to $505 million in 1979 (World Bank 1986b) and $539 million in 1980 (World Bank 1991c). A policy bias against agriculture, especially in the form of low fixed producer prices and subsidized consumer prices for rice, contributed to stagnating domestic production and a rapid increase in rice imports as well.

Thus the balance-of-trade deficit grew from a surplus of US$1 million in 1978 to a deficit of $328 million in 1980. The fiscal deficit grew largely because of increased capital expenditures, which quadrupled from FMG 31.2 billion in 1978 to FMG 126.8 billion in 1980. Foreign financing of

---

[7] The French term is *investir à outrance.*

Table 11.1. Macroeconomic summary, Madagascar

	1961–72	1973–77	1978–80
Real GDP (billion 1984 FMG)	1,189.00	1,712.90	1,797.60
Real GDP per capita (billion 1984 FMG)	188.30	224.30	212.10
Average GDP growth rate (%)	2.87	−0.01	2.70
Annual % change in GDP deflator			
Average	3.70	11.60	11.00
End of period	5.20	8.60	15.00
Trade deficit/GDP (%)			
Average	5.80	−4.10	−13.00
End of period	3.30	−3.60	−16.40
Budget deficit/GDP (%)[a]			
Average	0.80	−3.28	10.05
End of period	3.80	−6.28	14.51
Rice imports (thousand tons)			
Average	−11.00	86.00	161.00
End of period	13.00	95.00	176.00
Exchange rate (FMG/$)			
Average	252.00	233.00	217.00
End of period	253.00	226.00	211.00
Industrial value added (billion 1984 FMG)			
Average	211.00	252.70	267.60
End of period	237.00	262.40	265.50

Sources: Pryor 1988 for 1961–72; World Bank 1991c for 1973–80.
Note: Industrial value added 1961–72 in factor prices, all other years in market prices.
[a]Budget deficit on a commitment basis.

the deficit was significant in 1978 (58.3 percent), especially as compared to earlier years when the average deficit was 38.9 percent foreign-financed. Foreign financing fell in 1979 and 1980 to 44.2 and 37.8 percent, respectively, but loans from the domestic banking system increased. Money supply (M2) grew 20.65 percent in 1980 and 23.9 percent in 1981.

This monetization of the deficit resulted in an acceleration of inflation. The CPI inflation rate increased from an annual rate of 7 percent in 1978 to 13 percent in 1979, 17 percent in 1980, and 29 percent in 1981. Real interest rates, although rarely positive during the previous eight years, became more negative in 1979 and 1980, falling to −5.8 and −9.5 percent, respectively.

Poor and unsustainable sectoral performances were another aspect of the mounting economic crisis. Agricultural production consistently lagged behind overall GDP growth performance, averaging 1.8 percent in 1979 and 1980, whereas GDP growth averaged 5.3 percent during the two years. Industrial value added grew an impressive 9.8 percent between 1978 and 1979 but fell by 5.6 percent between 1979 and 1980 and by another 20 percent between 1980 and 1981.

By the end of 1980, serious macroeconomic imbalances had developed due mainly to the expansionary economic policies of the Malagasy government. Although external factors, especially a decline in Madagascar's external terms of trade and a tightening of international credit markets, played some role in the development of the economic crisis, they were not the principal causes of that crisis.

The decline in Madagascar's terms of trade was largely due to lower world coffee prices as the coffee boom that began in 1977 ended, reducing export receipts. Furthermore, the second oil price shock increased the import bill. But even after adjusting the trade balance to reflect constant terms of trade at the 1978 level, there was still a large gap in the trade balance that was primarily due to the investment push. Table 11.2 provides real counterfactual trade figures under the assumption that the 1978 terms of trade held for the entire period. Madagascar's terms of trade deteriorated by 30 percent between 1978 and 1981. Assuming constant terms of trade, however, the current account deficit would still have averaged US$145 million (1984) between 1979 and 1981, instead of $232 million (1984).

A contributing factor to the disastrous performance after the investment push was the rapid drying up of international credit. Commercial lenders were less willing to extend new credit to Madagascar as it became increasingly apparent that the loans could not be repaid and as arrears mounted. With little prospect of obtaining new commercial loans, Madagascar was faced with a full-blown crisis requiring assistance from the major international financial institutions and from bilateral donors. It was clear that stabilization and structural adjustment were necessary to reverse the disastrous effects of the short-lived, ambitious program which, though initially yielding notable aggregate performance, created serious growth-hindering distortions and was in the end unsustainable.

Policy Reform in the 1980s

Macroeconomic reform in Madagascar began in earnest in 1981 with the signing of the IMF standby agreement.[8] In the short run, reductions in aggregated demand were at the heart of the policy reform. Government expenditures were cut to reduce the fiscal deficit, and imports were sharply curtailed to reduce the balance-of-trade deficit. Only in the mid-1980s did

---

[8] A previous standby agreement, signed in June 1980, was cancelled nine months later with only SDR 10.0 million of the negotiated SDR 64.45 million having been disbursed, because the Malagasy government did not meet the government budget and balance-of-payments targets that were conditions of the loan (Ramahatra 1989, 169).

Table 11.2. Effects of the terms of trade on the balance of trade, Madagascar

	1970	1971	1972	1973	1974	1975	1976	19⁚
Export prices, f.o.b. (1980=100)	29	30	32	42	55	52	73	1
Import prices, c.i.f. (1980=100)	23	27	27	35	58	56	57	
Terms of Trade (1980=100)	125	111	119	120	96	93	129	1
Merchandise exports (million $)	145	147	166	200	240	320	289	3
Merchandise imports (million $)	−142	−178	−168	−178	−238	−332	−262	−3
Trade balance (million $)	3	−31	−2	23	2	−12	27	
Merchandise exports (million $ 1984)[a]	351	343	344	332	333	415	378	4
Merchandise imports (million $ 1984)[a]	−344	−415	−348	−296	−330	−431	−343	−3
Trade balance (million $ 1984)[a]	7	−72	−4	37	3	−16	35	
Merchandise exports, 1978 t.o.t. (million $ 1984)[b]	490	477	495	461	422	592	382	3
Merchandise imports, 1978 t.o.t. (million $ 1984)[c]	−480	−514	−480	−396	−320	−458	−356	−3
Trade balance 1978 t.o.t. (million $ 1984)	10	−37	15	65	102	134	25	−
Terms of trade effect, exports (million $ 1984)	139	134	151	129	89	177	3	−
Terms of trade effect, imports (million $ 1984)	−136	−99	−132	−100	10	−28	−14	
Terms of trade effect (million $ 1984)	3	35	19	29	99	150	−10	−
Dollar world price index[d]	100.0	103.9	116.8	145.8	174.5	186.7	185.1	19

Sources: World Bank 1988j; authors' calculations.
[a]Converted to 1984 prices using the dollar world price index.
[b]Values of exports at 1978 export prices, expressed in 1984 dollars, x′(t) = x(t) [px(1978)/px(t)][$\mathbb{V}$ (1984)/$WPI(1978)], where x(t) is exports in year t, px is the price index of exports, and $WPI is the do world price index.

the major emphasis of reform switch to efforts to increase aggregate supply in the economy.

Early Stabilization Policies

Cuts in government expenditure focused mainly on the investment programs that were the major cause of the budget deficit (Table 11.3). Capital expenditures were slashed from their peak of FMG 229.3 billion (1984) in 1980 to FMG 99.9 billion (1984) in 1982. Despite this large reduction in spending, real capital expenditures in 1982 were still 180 percent greater than in 1977 before the *investir à outrance* development push. Moreover, real government investment continued at about the same level through 1986.

Current expenditures were cut by 37 percent in real terms between 1980 and 1983. Salaries, which accounted for 56 percent of current budgetary

⁷8	1979	1980	1981	1982	1983	1984	1985	1986	1987	1988	1989
90	101	100	89	90	89	93	94	108	89	101	94
72	85	100	102	96	93	93	91	78	85	82	87
25	120	100	87	94	96	100	104	139	105	122	108
05	.414	437	332	327	313	335	270	329	327	284	313
04	−662	−764	−564	−464	−374	−353	−326	−331	−315	−319	−314
1	−249	−328	−214	−137	−61	−19	−56	−2	−12	−35	−1
37	384	364	300	313	308	335	269	273	239	195	215
36	−615	−636	−494	−444	−367	−353	−325	−275	−231	−219	−215
1	−230	−272	−193	−131	−60	−19	−56	−2	9	−24	−1
37	395	422	360	350	341	348	277	295	354	273	321
36	−605	−591	−415	−374	−312	−295	−278	−329	−288	−300	−278
1	−210	−169	−55	−25	29	53	−1	−34	66	−27	43
0	11	59	60	37	33	13	8	22	114	78	107
0	9	44	78	70	56	58	48	−55	−57	−81	−63
0	20	103	139	106	89	72	55	−33	58	−3	44
4.3	260.9	291.0	267.9	253.1	246.5	242.2	242.7	291.7	330.8	353.1	353.0

Values of imports at 1978 export prices, expressed in 1984 dollars, $m'(t) = m(t)[pm(1978)/pm(t)][\$WPI(1984)/\$WPI(1978)]$, where $m(t)$ is imports in year t, pm is the price index of imports, and $WPI is the dollar rld price index.
Constructed as the weighted average of wholesale price indices, expressed in dollars, of Madagascar's major ding partners.

expenditures in 1980, bore the brunt of the slowdown in spending. Although nominal wages and employment changed little, the total government wage bill fell by 30 percent in real terms during the 1980–83 period. Other categories of current expenditure were also reduced, especially subsidies, which fell by 34 percent.

Although government revenues actually fell in real terms by 24 percent between 1980 and 1983 (due mainly to a drop in import tariff revenues), the large reduction in real spending reduced the budget deficit from FMG 271.8 billion (1984) in 1980 to FMG 80.7 billion (1984) in 1983. As a share of GDP, the budget deficit was reduced from −14.9 to −4.8 percent over the same period.

Reducing the budget deficit combined with continued inflow of foreign financing to sharply decelerate the increase in money supply. Domestic financing of the budget deficit fell from FMG 78.8 billion in 1980 (equal to 63 percent of the 1979 money supply) to FMG 36.6 billion in 1982 (equal

Table 11.3. Real government expenditures, deficit, and financing, Madagascar (billion 1984 FMG)

	1971	1972	1973	1974	1975	1976	1977	19
Currency	178.5	192.2	202.8	168.6	175.7	205.0	232.5	27
Budgetary	178.5	192.2	202.8	168.6	175.7	205.0	232.5	24
Personnel	88.6	91.9	109.9	100.5	104.1	121.9	133.4	14
Other goods/services	59.7	55.8	41.6	33.7	38.8	52.9	65.5	6
Interest	11.2	10.8	21.4	13.1	11.2	10.3	11.6	
Transfers/subsidies	19.1	33.7	29.9	21.3	21.7	19.9	22.1	3
Extrabudgetary	—	—	—	—	—	—	—	2
Capital	67.3	81.7	63.7	50.7	46.6	62.5	35.7	5
Other	6.0	−8.7	−2.8	10.2	10.9	−16.4	21.1	
TOTAL	251.8	265.2	263.6	229.4	233.3	251.1	289.2	32
Real revenues	238.8	221.7	208.8	184.1	188.5	196.0	202.9	27
Real deficit	−13.0	−43.5	−54.8	−45.3	−44.8	−55.1	−86.4	−5
Foreign financing of deficit (%)	83.1	23.0	37.6	27.4	52.1	26.6	22.8	5

Sources: Pryor 1988; World Bank 1980b, 1986b, 1991c; IMF 1988b, 1991c.
Note: Deflated using GDP deflator.

to 19 percent of the 1981 money supply) (Table 11.4). Growth in the broad money supply (M2) fell from 20.7 percent in 1980 to only 8.9 percent in 1982 and −9.1 percent in 1983. Domestic inflation, in excess of 20 percent per year between 1981 and 1983 (as measured by the GDP deflator), finally slowed in 1984 to 10.3 percent per year.

Improvements in Madagascar's balance-of-payments position were achieved mainly through the tightening of import controls. In part, the reduction of imports was a direct result of slowing down the implementation or cancelling projects in the government's investment program: imports of capital goods fell by 35 percent in real 1984 dollars from 1981 to 1984. Imports of intermediate goods were also cut back, contributing to a sharp decline in industrial output. Rice imports, which accounted for 2.4 percent of imports in 1980, rose sharply to a record level of 351,000 tons, 19 percent of imports, in 1982 before being cut back along with other imports of consumer goods in subsequent years (IMF 1988b). In all, imports fell by $269 million (1984) and the current account deficit was reduced by 56 percent between 1980 and 1983 (World Bank 1991c).

Madagascar achieved this improvement in the current account deficit in spite of a reduction in export earnings. The decline in world coffee prices from their extraordinarily high levels of the late 1970s together with unstable markets for vanilla and cloves were major factors behind a 24.1 percent drop in average export earnings (measured in 1984 dollars) between 1981 and 1983 compared with the average level between 1977 and 1980.

Because the allocation of foreign exchange and the level of imports were

979	1980	1981	1982	1983	1984	1985	1986	1987	1988	1989
84.2	283.3	232.6	207.7	176.3	189.7	190.3	182.2	191.1	180.3	207.2
54.6	247.8	199.2	176.3	160.8	172.7	175.7	169.3	179.6	169.7	192.9
36.5	139.6	125.7	114.9	97.1	98.6	97.5	91.4	90.2	86.5	83.7
78.0	71.5	40.6	29.5	28.9	28.4	35.6	28.0	29.4	25.0	31.4
4.8	8.7	11.7	14.6	16.4	25.4	24.6	27.0	36.1	37.5	50.1
35.4	27.9	21.2	17.3	18.3	20.3	17.9	22.8	23.9	20.8	27.8
29.6	35.6	33.4	31.5	15.5	15.4	13.1	10.0	11.0	0.0	0.0
79.1	229.3	157.7	99.9	97.4	108.9	101.0	93.8	122.3	123.7	183.7
31.9	28.1	25.1	17.8	11.6	0.0	0.0	4.6	14.1	0.0	0.0
95.1	540.7	415.5	325.5	285.2	298.6	291.3	280.7	327.5	304.1	390.9
86.7	268.9	206.0	203.0	204.5	243.4	225.3	219.8	268.5	246.9	295.6
08.4	−271.8	−209.6	−122.4	−80.7	−55.2	−66.0	−60.8	−59.0	−57.2	−95.3
44.2	37.8	48.7	57.9	56.7	55.5	66.6	70.8	103.3	111.7	111.5

directly controlled through the government's import licensing system, exchange rate changes had little initial direct impact on Madagascar's balance-of-payments deficit. Despite a devaluation of the official exchange rate of 112 percent in nominal terms and 9.8 percent in real terms between 1981 and 1984, there remained excess demand for, and therefore rationing of, foreign exchange. Devaluation did serve two other major functions in the adjustment process, however. First, by increasing the price of imported goods, especially imported intermediate and capital goods, it was a key part in restructuring the domestic price system, helping to narrow the gap between domestic and world prices. Second, the decision to break with the fixed nominal exchange rate (relative to the French franc) and adopt a variable exchange rate paved the way for future reforms that would allow a greater role for prices in determining foreign exchange allocations.

Structural Adjustment

By 1984 a large measure of macroeconomic stability had been achieved in Madagascar. The current account deficit had fallen from 16.4 percent of GDP in 1980 to only −4.9 percent, and over the same period the government budget deficit was reduced from −14.9 to −3.6 percent of GDP. Overall inflation had subsided. Although the economy appeared to be near financial equilibrium in terms of its internal and external accounts, this new equilibrium was at a low level of national income. Real per capita incomes had fallen by 18 percent since 1979, the peak year for the invest-

Table 11.4. Monetary data, Madagascar

	1970	1971	1972	1973	1974	1975	1976	1977	1978	1979	1980	1981	1982	1983	1984	1985	1986	1987	1988	1989
M1 (billion FMG)	46.2	47.0	53.3	57.3	67.9	69.4	79.7	100.0	112.8	124.2	151.3	193.8	208.0	192.6	239.9	238.6	289.6	371.8	455.0	598.3
M1 (% change)	—	1.6	13.6	7.5	18.5	2.1	14.9	25.5	12.8	10.1	21.8	28.1	7.3	-7.4	24.6	-0.5	21.4	28.4	22.4	31.5
M2 (billion FMG)	55.2	58.6	65.4	68.0	82.2	83.7	97.1	116.4	136.3	140.6	169.6	210.2	228.9	208.1	258.1	292.4	367.3	435.0	532.6	712.0
M2 (% change)	—	6.1	11.8	4.0	20.8	1.9	16.0	19.9	17.1	3.2	20.7	23.9	8.9	-9.1	24.0	13.3	25.6	18.4	22.5	33.7
CPI (1984=100)	18.5	19.5	20.7	21.7	25.9	28.6	30.2	31.5	33.7	38.1	44.6	57.5	75.5	91.0	100.0	19.9	124.8	146.6	184.5	201.2
CPI inflation (%)	—	5.6	5.8	5.2	19.3	10.1	5.7	4.4	6.8	13.3	17.0	29.0	31.3	20.5	9.8	9.9	13.6	17.4	25.9	9.0
GDP deflator (1984=100)	18.3	19.0	19.5	21.9	26.8	28.1	30.9	33.5	35.8	40.0	45.8	58.1	74.7	90.7	100.0	111.7	127.6	156.7	191.8	211.4
Inflation—GDP deflator (%)	—	3.6	3.0	11.9	22.8	4.7	9.9	8.6	6.8	11.3	15.0	26.7	28.6	21.5	10.3	11.7	14.1	22.8	22.4	10.2
Discount rate—end of period (%)	—	5.5	5.5	5.5	5.5	5.5	5.5	5.5	5.5	5.5	5.5	8.0	12.5	13.0	13.0	11.5	11.5	11.5	11.5	11.5
Real interest rate (%)	—	1.9	2.5	-6.4	-17.3	0.8	-4.4	-3.1	-1.3	-5.8	-9.5	-18.7	-16.1	-8.5	2.7	-0.2	-2.6	-11.3	-10.9	1.3

Sources: IMF 1991c, and various years(b), Pryor 1988.
Notes: 1989 discount rate is through the third quarter. The CPI is a weighted average of the traditional and modern CPIs, with weights of 0.75 and 0.25, respectively. The traditional CPI reflects consumer prices for lower-income Malagasy households; the modern CPI reflects consumer prices for upper-income Malagasy and expatriate households.

ment boom, when in spite of inflationary domestic financing of government spending and unsustainable external borrowing per capita income was still 5 percent below the 1975 level. Madagascar's economic policy reforms thus came under sharp criticism from both national and foreign observers, not only for hardships endured by the population during the adjustment process but also for the apparent bleak prospects for growth if the policies continued. Clearly, the disappointing growth performance of the Malagasy economy since the 1970s suggested the need for reform.[9]

Beginning in 1985, a series of reforms in key sectors of the Malagasy economy were undertaken, supported by large structural adjustment loans from the World Bank and other donors. In general, these reforms aimed at removing perceived constraints on economic growth by creating competitive markets and increasing the role of price incentives in the process of allocation of resources. Dismantling the administrative controls on markets proved to be a difficult and slow process, however.

The industrial sector adjustment credit signed in August 1985 was designed to liberalize the rigid pricing system of formal sector industry in Madagascar. Various restrictions on prices, such as ex-factory price controls, controls on profit margins, and export taxes on industrial goods, were gradually eliminated from 1985 to 1989. A new investment code, also part of the intended policy package, was not enacted by the Malagasy government until December 1989.

Decontrol of agricultural producer prices was supported by the agricultural sector adjustment credit signed in May 1986. The official maximum prices for paddy set by the government had been replaced with floor prices in 1985, and private trade in rice was officially encouraged. At the local level, though, private rice traders still encountered government attempts to limit or tax trade (Berg 1989). Official prices for export crops remained in place until 1988. Up through 1990, vanilla prices continued to be fixed by the state.

The World Bank public sector adjustment credit, signed in June 1988, was tied to reform of public enterprises. Public enterprises had proliferated in the late 1970s during the investment boom, but through ineffective management and rigid pricing policies they had recorded substantial financial losses. Even though public enterprises accounted for only 2.3 percent

---

[9] There was considerable debate over whether price liberalization and reliance on market forces would be sufficient or even helpful. Duruflé (1989) argued that several important constraints blocked economic growth, including low savings and investment rates, weak internal articulation of the economy, poor growth of exports, and the pervasiveness of market "intermediaries" whose speculative and rent-seeking behavior "impede[d] all efforts to stimulate production." Thus a more appropriate policy response included a greater role for the state in marketing and direct investment. The actual reforms undertaken were of different nature.

of GDP and 7.1 percent of formal sector employment, they absorbed about 60 percent of total nongovernment short-term credit between 1980 and 1986 (Swanson and Wolde-Semait 1989). The reforms supported by the World Bank loan included a moratorium on the creation of new enterprises and reductions in the flow of government transfers and bank credit. Only limited progress had been achieved as of mid-1990 toward the goal of privatizing or liquidating half of the 170 public enterprises by mid-1991: only twenty-six had been dissolved or sold to private investors; actions on another twenty-four were still in progress.

Nonpayment of interest and principal by public enterprises seriously undermined the financial position of the three state commercial banks in Madagascar. A major restructuring of the state banks' portfolios was undertaken in 1988 and 1989 which involved the use of treasury transfers and counterpart funds from foreign aid to help counteract the write-off of nonperforming assets. More important, the enactment of a new banking law in 1988 ended the government's monopoly on the domestic banking system and allowed the participation of foreign capital. In August 1989 a private commercial bank, the Banque Malgache de l'Océan Indien (BMOI) was established (World Bank 1991c).

Perhaps the most important reforms of the late 1980s involved changes in trade and exchange rate policies in 1987 and 1988. Under the liberalized import regime begun in January 1987, a predetermined amount of foreign exchange was allocated each month according to each potential importer's share in total foreign exchange requested. Concurrent devaluation of the currency helped reduce the demand for foreign exchange. In the first six months of 1987, the nominal exchange rate was devalued from FMG 790 per dollar to FMG 1,362 per dollar, resulting in a 46 percent devaluation of the trade-weighted exchange rate (IMF 1991c, and various years). The transition to a market-determined allocation of foreign exchange was completed by July 1988, when the open general license system was instituted for all imports. Under this system, foreign exchange for imports was made available in unlimited quantities at the current exchange rate. Trade in export crops was also liberalized in 1988, allowing the effects of the devaluation of 1987 to have a direct positive impact on domestic producer prices of export crops.

In broad terms, the performance of Madagascar's economy was somewhat disappointing through 1987. Although a stable financial situation was achieved, including low budget deficits (averaging 3.7 percent of GDP from 1983 to 1987) and a small current account deficit (averaging 5.9 percent of GDP), real GDP growth averaged only 1.4 percent per year from 1983 to 1987 (Table 11.5).

This lack of positive per capita GDP growth was perhaps the most

Table 11.5. Macroeconomic summary, Madagascar

	1981–82	1983–87	1988–89
Real GDP (billion 1984 FMG)	1,667.00	1,719.00	1,880.00
Real GDP per capita (billion 1984 FMG)	183.60	170.30	164.60
Average GDP growth rate (%)	−5.80	1.40	4.40
Annual % change in GDP deflator			
Average	27.60	16.10	16.30
End of period	28.60	22.80	10.20
Trade deficit/GDP (%)			
Average	−10.40	−5.90	−5.90
End of period	−9.40	−4.60	−5.20
Budget deficit/GDP (%)[a]			
Average	−9.60	−3.70	−5.60
End of period	−7.10	−3.30	−7.90
Rice imports (thousand tons)			
Average	272.00	130.00	101.00
End of period	351.00	94.00	112.00
Exchange rate (FMG/$)			
Average	311.00	683.00	1,505.00
End of period	350.00	1,069.00	1,465.00
Industrial value added (billion 1984 FMG)			
Average	204.00	203.00	241.00
End of period	197.00	22.00	250.00

Sources: World Bank 1991c; IMF 1988b, 1991c.
[a]Budget deficit on a commitment basis.

glaring deficiency of the structural adjustment program. In this context, the GDP estimates for 1988 attracted more than the usual attention. The Ministry of Plan's estimates showed a drop in real GDP per capita of 1.2 percent in 1988.[10] Alternate World Bank figures, constructed using a preliminary version of the statistical office's (BDE) accounts for 1984, showed an increase in real GDP per capita of 0.8 percent for 1988. Moreover, the World Bank figures showed the level of real GDP to be approximately 20 percent higher every year between 1984 and 1988.[11] The wide divergence in estimates served to complicate policy analysis and discussions between the government and donors, in part because conditions for loans and policy recommendations are often expressed as shares of GDP and because

[10] Initial Ministry of Planning estimates for 1988 GDP showed a decline in real per capital GDP of 5.0 percent, due in large part to early pessimistic figures for rice production (EIU 1990, quoted in Dorosh, Bernier, and Sarris 1990, 59).
[11] The Banque des Données de l'Etat, technically responsible for producing the official estimates, was still in the process of constructing a detailed set of national accounts for 1984 which would serve as a base for subsequent years' estimates. Meanwhile, the Ministry of Planning produced annual estimates of real GDP per capita based to a large extent on the detailed national accounts for 1973. The differences in the growth rate estimates were mainly due to a larger share of services in GDP in the bank's base year.

Table 11.6. Foreign public debt, Madagascar (million US$)

	1977	1980	1982	1987	1988	1989
Total debt	296.0	1,257.0	1,920.0	3,728.0	3,637.0	3,607.0
Long-term total debt	224.1	965.5	1,679.0	3,113.0	3,289.0	3,345.0
Multilateral	108.0	220.0	326.8	857.2	931.0	1,023.0
Bilateral	86.8	390.3	905.0	1,981.6	2,014.0	2,021.0
Private	29.3	346.2	447.2	274.7	345.0	301.0
Total disbursements	43.7	391.8	245.6	278.8	227.0	160.0
Principal repayments	8.4	31.0	32.9	64.0	69.0	59.0
Interest payments	5.3	26.7	44.8	83.5	76.0	113.0
Total debt service	13.7	57.7	77.7	147.5	145.0	172.0
Net long-term transfers	30.0	318.0	168.0	144.0	82.0	−12.0
Debt/GDP (%)	9.5	23.7	47.6	120.2	131.5	131.8
Net transfers/GDP (%)	1.3	7.9	4.8	5.6	3.2	−0.5

Sources: World Bank 1991c, various years.
Note: GDP data are from World Bank 1991c, so debt/GDP ratios do not correspond to debt tables figures.

the GDP growth rate is often regarded as a key indicator of success or failure of reforms.

Despite technical differences in methodologies, both sets of GDP estimates showed positive growth in per capita GDP in 1989 and 1990.[12] These latest GDP figures are especially encouraging since they indicate positive growth despite a decline in the terms of trade caused by a sharp drop in world coffee prices. Of continued concern, though, is Madagascar's large long-term foreign debt, equal to $3.35 billion at the end of 1989, 132 percent of GDP (Table 11.6). The failed *investir à outrance* strategy of the late 1970s increased the country's foreign debt (including undisbursed) from $0.46 billion to $2.06 billion between 1977 to 1981. There was little change in the overall debt between 1981 and 1984, but the debt again increased substantially from 1985 to 1989 due to the large structural adjustment loans (Figure 11.1). Interest payments on the debt were equal to 36 percent of merchandise exports and 5 percent of GDP in 1989. Substantial debt forgiveness by France amounting to Fr 4 billion (approximately $700 million dollars at the exchange rate prevailing at the end of 1989), or 21 percent of the outstanding disbursed debt, was announced in 1989.[13] Madagascar still owes 15–20 percent of its large bilateral debts to

[12] Formal agreement by the Malagasy government to make the Banque des Données de l'Etat 1984 GDP estimates the base for current estimates of GDP was finally reached in 1991. The new official figures thus show that turnaround in real per capita income growth occurred one year earlier and indicate smaller ratios for standard performance indicators such as current account deficit/GDP.

[13] Technically, the debt is being forgiven on an annual basis, not written off all at once. Annual amortization and interest payments will be cancelled each year as they come due.

Figure 11.1. Total public debt and net transfers, Madagascar

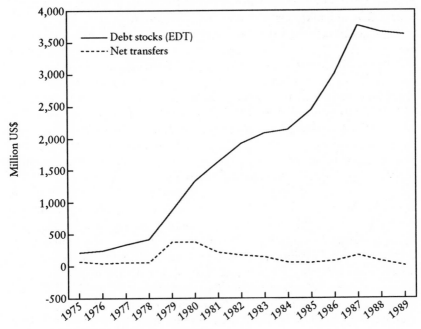

Source: Data from World Bank, undated.

non–Paris Club lenders (Soviet Union, China, and Arab countries) (IMF 1991c)[14]; about 50 percent of the remaining debt is owed to multilateral organizations such as the World Bank.

Economic and Social Outcomes on the Micro and Sectoral Levels

The slow growth rate of the Malagasy economy through 1987, in spite of successful macroeconomic stabilization, can largely be attributed to the uneven rate of structural reforms at the sectoral level. The multitude of constraints on increased production were removed only gradually, and a key reform of trade policy and the allocation of foreign exchange took place only in 1988.

Effects of changes in government policy were most noticeable on the rice

[14] The data on non–Paris Club debt are taken from IMF sources and do not correspond exactly to debt figures from the 1990 World Bank World Debt Tables.

sector, arguably the most important sector of the economy. Rice policy in Madagascar had largely been oriented to the needs of urban consumers: maintaining adequate supplies and low prices in urban areas.[15] To achieve this, the government relied on a monopoly control of purchases of domestically produced rice and subsidized sales of imported rice. Beginning in the mid-1970s, the Malagasy government gave monopoly control of domestic rice marketing to various parastatals and banned private trade in rice. Producer prices were suppressed in an effort to keep consumer prices of rice low: official producer prices fell by 33 percent in real terms from 1976 to 1982. Official market sales accounted for only about half of rice sales from 1975 to 1980; by 1982 only 5 percent of domestic production (20–25 percent of total domestic sales of paddy) was sold through official channels (AIRD 1984; Dorosh, Bernier, and Sarris 1990).

In part because of inadequate price incentives as well as shortages of inputs (fertilizer) and poor maintenance of existing irrigation systems, total production stagnated and annual production per capita dropped from 147.4 to 120.9 kilograms per person from 1976 to 1982 (Table 11.7). Consumption per capita was maintained by increasing reliance on imports, which increased from 5.9 percent of total availability in 1976 to 25.0 percent in 1982. Meanwhile, consumer rice prices (in the official market) fell by 46 percent in real terms between 1975 and 1981. The limited data available suggest that consumer prices in the parallel market were roughly equal to official market prices in 1977 but rose to between 20 and 70 percent above official market prices in 1982 in various urban centers. Thus prices in the parallel market appear to have declined by about 20 percent in real terms from 1976 to 1982 as the quantity of rice imports increased.

Rice policy reforms were begun in 1982 with a 28 percent increase in the official producer price of paddy (Figure 11.2). Although representing a 3 percent decline in real terms, the nominal price increase did slow the decline in real prices which had occurred in every year since 1977. More important, private trade in rice was again legalized in most regions of the country in 1983.[16] Initially, there was little improvement in real producer prices. Private market prices in 1983–85 were essentially unchanged in real terms from the official producer price of 1982. Later in the decade, as local administrative constraints on marketing were gradually removed, real producer prices rose, so that the average real producer price in the 1987–90 period was 30 percent higher than the official market price in 1982. Higher

---

[15] The bias in rice policy in favor of select urban groups continued throughout the 1980s. Rice policies of the Malagasy government are in many ways similar to policies adopted by several West African governments in the 1970s and early 1980s (see Hart 1982; Bates 1981).

[16] The government retained its monopoly on rice trade in the two major rice-surplus regions of the country, Lac Alaotra and Marovoay, until 1986.

Table 11.7. Rice production, availability, and prices, Madagascar

| | | | Quantity (thousand MT) | | | | | Product per capita (kg/yr) | Availability per capita[b,c] (kg/yr) | Imports/ Availability (%) | Index (1980=100) | |
	Area[a] (thousand ha)	Yield (MT/ha)	Paddy product[a]	Exports[b]	Imports[b,h]	Change in stocks[b]	Total availability[b]				Real consumer price[d,e]	Real producer price[f,g]
1970	935	1.99	1,865	68	20	0	1,002	155.7	148.6	2.0	146.1	80.2
1971	943	1.99	1,873	36	61	0	1,079	152.8	156.4	5.7	138.3	80.2
1972	1,008	1.67	1,687	26	49	0	972	134.4	137.7	5.0	131.2	76.0
1973	1,055	1.64	1,730	6	96	5	1,064	134.5	147.0	9.0	138.1	71.6
1974	1,134	1.63	1,844	7	129	0	1,155	139.9	155.7	11.2	184.6	97.8
1975	1,078	1.83	1,972	5	64	0	1,174	146.0	154.4	5.5	178.9	108.5
1976	1,064	1.92	2,043	4	72	5	1,213	147.4	155.5	5.9	170.4	120.5
1977	1,175	1.76	2,067	2	95	9	1,252	145.4	156.5	7.6	141.8	116.9
1978	1,133	1.70	1,922	1	153	7	1,236	131.6	150.4	12.4	131.2	109.7
1979	1,158	1.77	2,045	1	156	0	1,313	136.4	155.6	11.9	115.0	104.4
1980	1,178	1.79	2,109	1	177	2	1,361	136.9	156.9	13.0	100.0	100.0
1981	1,186	1.70	2,012	0	193	8	1,319	127.0	148.0	14.6	96.7	83.7
1982	1,188	1.66	1,970	0	351	66	1,402	120.9	152.9	25.0	121.7	81.1
1983	1,189	1.81	2,147	0	179	120	1,333	128.2	141.5	13.4	175.6	81.6
1984	1,163	1.82	2,112	0	111	30	1,390	122.7	143.4	8.0	231.6	82.5
1985	1,180	1.84	2,178	0	106	0	1,362	123.1	136.7	7.8	291.4	85.2
1986	1,187	2.06	2,230	0	162	28	1,417	122.4	138.2	11.4	344.4	159.1
1987	1,213	2.09	2,296	0	94	37	1,386	122.6	131.5	6.8	212.9	120.4
1988	1,189	1.88	2,235	0	90	-40	1,308	115.8	120.5	6.9	206.4	83.8
1989	1,221	1.91	2,332	0	112	-8	1,416	117.5	126.7	7.9	220.8	102.4

[a] 1970–85, Shuttleworth 1989; 1985–89 IMF 1988b, 1991c; MPARA 1989.
[b] Rice equivalent, equal to 0.67 kilogram of milled rice per kilogram of paddy times loss rate of 16 percent (loss rate from Hirsch 1986b).
[c] 1970–85, Hirsch 1986b; 1986/87, IMF 1988b, 1991c; World Bank 1990c.
[d] Consumer price, calendar year basis, deflated by traditional basket CPI; 1970–82 data are official consumer prices; 1983–89 data are free market consumer prices.
[e] AIRD 1984; Dorosh, Bernier, and Sarris 1990.
[f] Producer price deflated by traditional CPI; 1970–82 data are official producer prices for the crop year (e.g., the crop year t/t + 1 is shown as t + 1); 1983–89 data are free market producer prices, annual average.
[g] 1970–74, AIRD 1984; Dorosh, Bernier, and Sarris 1990.
[h] Net imports.

Figure 11.2. Rice prices, Madagascar

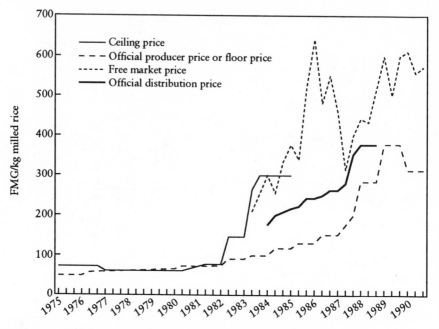

Sources: Ceiling price, Hirsch 1986b; free market and official distribution/floor price, IMF 1988b, 1991c; official producer price/floor price, ODI, forthcoming, IMF 1991c.
Note: A milling rate of 2/3 is assumed.

producer prices led to some gains in rice production. The annual average increase in rice production was 2.4 percent per year from 1982 to 1989, compared with no increase in production between 1975 and 1982. The production increases decelerated somewhat in the late 1980s, however, averaging only 1.7 percent per year from 1985 to 1989, indicating a need for further agricultural investments, more readily available inputs, and research on and extension of new technology.

The changes in consumer prices in the early 1980s were more dramatic. Official market consumer prices were raised by 91 percent to FMG 225 per kilogram between 1982 and 1985 (an increase of 31 percent in real terms), and nominal private market prices jumped by 186 percent to FMG 411 per kilogram in the same period. Higher sales prices in the official market sharply increased the marketing margin for the parastatals and helped to reduce their losses. Their marketing margin, which was actually negative in 1980 (FMG −8 per kilogram paddy), increased to FMG 16 (1980) per kilogram paddy in 1982 and FMG 40 (1980) per kilogram in 1985.[17] Estimated private marketing margins were very high in 1985 and 1986 due to a shortage of imported rice, which led to sharp increases in free market consumer prices.[18] Private market consumer prices in the period 1987–90 averaged FMG 118 (1980) per kilogram, 39 percent below the 1986 private market price in real terms but 72 percent above the official market price of 1982 in real terms. Marketing margins in the private sector averaged FMG 51 (1980) per kilogram from 1987 to 1990.

Official sales of imported rice continued through 1989, as part of a buffer stock (*stock tampon*) arrangement agreed on by the Malagasy government and the World Bank in 1985.[19] Although in principal the purpose of the buffer stock was to reduce seasonal variations in rice prices and guarantee adequate supplies of rice for urban areas, the imported rice was distributed to favored urban groups, much to the consternation of aid donors (World Bank 1991b). Nevertheless, reductions in the government's rice imports combined with declining per capita production (a 4.5 percent drop between 1985 and 1989) to reduce per capita consumption. From 1987 to 1989, per capita consumption of rice averaged 126.2 kilograms per annum, 7.7 percent less than in 1985 and 17.5 percent less than in 1982.

The higher domestic rice prices in the 1980s brought about by the market liberalization and reduced imports helped to lessen the implicit taxation on Madagascar's rice producers. From 1972 to 1982, the producer price was on average 39.6 percent below the import parity price, evaluated at the official exchange rate (Table 11.8). Taking into account

[17] The marketing margin reported here represents the costs of marketing and transformation (milling) between farmgate and consumer and is calculated as the difference between average consumer prices for rice expressed in paddy equivalents and the average producer price for paddy.

[18] In late 1985, open market consumer prices skyrocketed by 39 percent due to a lack of coordination between government ministries. Limits on total rice imports for the year had been set in an agreement between the Ministry of Finance and the IMF. Yet the Ministry of Transportation, Food and Tourism, responsible for arranging rice imports and distributing rice to consumers, continued to supply the previous year's ration of 400 grams per person per day through the official distribution channels. Given the import quota, however, the ration amount was excessive and the entire import quota was used up by August.

The price situation was further aggravated by domestic procurement policy, which mandated a constant sales price of rice throughout the marketing season. With no provision for storage costs, the parastatals had no incentive to hold stocks and disposed of their purchases soon after the harvest. Thus, by late 1986 the government also held no domestically procured stocks of rice (see Shuttleworth 1989 for further details).

[19] There was no physical buffer stock, in the sense of grain stored in Madagascar. Rather, contingent import contracts with world suppliers were set up and lines of credit were made available so that additional imports could be arranged at short notice in an emergency.

distortions in the real exchange rate caused by trade policies, the gap between domestic producer prices and import parity border prices was even larger.

Trade policies, especially import tariffs and quotas, raised the prices of protected importables (e.g., manufactured goods) relative to nontradables, thereby leading to increased demand for nontradable goods and encourag-

Table 11.8. Policy effects on producer prices, Madagascar

	World price ($/ton)	World price (FMG/kg)	Border price (FMG/kg)	Producer price (FMG/kg)	Real producer price (FMG/kg)	Direct effect (%)	Total effec (%)
Rice							
1972–77	290	67.8	50.2	26.7	43.1	−42.5	−67
1978–80	341	78.9	55.6	36.5	45.6	−33.8	−61
1981–82	402	121.9	89.6	53.5	36.0	−39.7	−65
1983–87	249	166.8	123.1	121.7	45.8	−2.5	−39
1988–90	299	448.3	331.2	198.3	38.5	−40.5	−63
1972–90	304	161.4	119.2	83.5	42.6	−30.0	−58
Coffee							
1972–77	1,434	342.1	296.0	158.3	262.9	−29.3	−64
1978–80	2,889	638.8	571.8	182.5	228.6	−67.9	−82
1981–82	1,840	567.6	449.9	255.0	173.9	−43.2	−71
1983–87	2,416	1,618.4	1,417.2	481.0	179.2	−66.7	−80
1988–90	1,237	1,844.6	1,487.2	781.3	153.9	−35.2	−67
1972–90	1,944	985.8	838.9	357.3	208.2	−47.3	−72
Cloves							
1972–77	3,706	876.1	785.9	299.2	499.2	−59.7	−78
1978–80	5,843	1,261.8	1,130.4	355.0	412.5	−68.2	−83
1981–82	7,950	2,452.9	2,222.6	412.5	280.4	−81.4	−90
1983–87	4,731	2,930.8	2,536.8	453.0	177.8	−80.7	−89
1988–90	2,244	3,321.0	2,621.5	614.0	120.8	−75.5	−85
1972–90	4,529	2,030.0	1,742.1	410.1	318.2	−71.3	−84
Vanilla							
1972–77	15,875	3,746.2	2,822.4	1,119.3	1,864	−69.8	−79
1978–80	29,878	7,557.7	5,287.4	1,909.0	2,362	−72.2	−81
1981–82	49,017	15,398.6	13,039.7	3,220.0	2,202	−79.5	−86
1983–87	64,040	44,492.6	40,457.5	4,876.0	1,913	−89.3	−92
1988–90	70,946	106,523.0	99,356.7	8,740.0	1,710.0	−93.0	−94
1972–90	43,769	32,525.0	29,579.4	3,701.8	1,988	−80.0	−86

Sources: World Bank 1980b, 1984c, 1986b, 1991c; IMF 1988b, 1991c; AIRD 1984; BDE 1988.

Notes: World price is derived as the quotient of the FMG value (f.o.b. Tomasina for export crops, c.i Tomasina for rice) of values and volumes at annual average exchange rates. Marketing cost derived from 198 marketing cost and the CPI. Border price is equal to world price less marketing and transport costs. The dire effect is the nominal rate of protection, defined as $[(Pi/Pna) - (Pi'/Pna)]/(Pi'/Pna) = (Pi - Pi')/Pi'$, where Pi the producer price of good i, Pi' is the border price of good i (at the official exchange rate), and Pna is the pri index on nonagricultural goods. The total effect measures the direct effect and the indirect effect of commerci and macroeconomic policies on the real exchange rate.

ing a shift of resources away from the production of nontradable goods toward the protected importable sectors.[20] The result was a decrease in the ratio of the price of exportables (and importable goods that were not protected by quotas or import tariffs) and the price of nontradables, that is, an appreciation of the real exchange rate. In Madagascar, the appreciation in the real exchange rate caused by trade policy averaged an estimated 57 percent from 1972 to 1982. Taking into account both the total effect of direct trade and pricing policies on rice and the real exchange rate distortion, domestic producer prices were an average 65.1 percent below border prices in this period.

Exchange rate devaluation in 1988 as well as higher world rice prices measured in dollars increased the border prices of rice such that implicit taxation measured at the official exchange rate actually increased in the 1988–90 period despite an increase in the real domestic producer price. Higher world rice prices also raised export parity border prices of rice such that domestic rice prices were approximately equal to export parity border prices evaluated at the equilibrium exchange rate. Madagascar was still importing small amounts of rice (3.7 percent of its total rice supply in 1988–89), and without these imports domestic market-clearing prices would have been somewhat higher. The above calculations suggest, however, that if remaining exchange rate distortions were removed, Madagascar might again be able to compete on world rice export markets, particularly in years of high world rice prices.

Macroeconomic reforms, including exchange rate devaluations, had little impact on the export crop sector until the adoption of reforms in the marketing of export crops in 1988. Real export crop prices fell by 4 percent between 1981 and 1986 and rose by 13.5 percent in 1987. Despite a 48 percent depreciation of the real exchange rate between 1981 and 1987, export crop prices rose in real terms by only 9 percent (Dorosh, Bernier, and Sarris, 1990).

Until 1987 the major export crops, coffee, vanilla, and cloves, were taxed heavily through the government marketing system. Producer prices were set at low levels, allowing the government and parastatals to capture the large margin between world prices and domestic producer prices. Although domestic producer prices were insulated from the large swings in world prices, domestic producers were heavily taxed. Between 1972 and 1987, coffee producers received on average only 50 percent of the border price evaluated at the official exchange rate and only 26 percent of the

[20] A formal presentation of the theory is found in Dornbusch 1974. See Krueger, Schiff, and Valdés 1988 for the methodology of calculating direct and total effects of trade policies. Details of the calculations for Madagascar are found in Dorosh, Bernier, and Sarris 1990.

border price evaluated at the equilibrium exchange rate; clove and vanilla producers received only 29 percent of the border price evaluated at the official exchange rate on average in the same period (Table 11.8).

Not only were producer prices kept substantially below border prices, producer prices declined sharply in real terms. Producer prices for coffee fell by 45 percent in real terms between 1978 and 1983, as the government allowed only small increases in nominal producer prices in the midst of accelerating domestic inflation. Although real coffee prices did finally increase by 39 percent in 1986, they still remained slightly below the levels of the 1970s in 1987. Private trade in coffee was legalized in 1988, but almost immediately afterward the world coffee price dropped by 27 percent. Not surprisingly, with very little improvement in real coffee prices for producers, production of coffee has stagnated.

Clove and vanilla exports were likewise hindered by low producer prices and declining or unstable world demand. Domestic producers of cloves saw a continuous decline in real producer prices from 1978 to 1990 by a total of 76 percent. World clove prices measured in dollars also fell by 60 percent between the late 1970s and the late 1980s as Indonesia, which accounted for 66 percent of world imports in the 1976–80 period, increased its domestic production and greatly reduced its import demand for cloves.

To push up world prices of vanilla, Madagascar and other members of the small cartel of vanilla exporters reduced quantities sold and built up their domestic stocks in the 1980s. Vanilla prices rose on world markets by 176 percent between 1978 and 1990, and Madagascar's average annual exports in the period were 47 percent lower in quantity terms than in 1977. Domestic producer prices of vanilla were kept low to discourage overproduction, and real prices of vanilla fell by 25 percent between 1979 and 1989. Although the cartel's strategy produced higher incomes in the short run, it also resulted in a loss of 10 percent of the cartel's market share in 1990 to Indonesia, not a member of the cartel.

Exports of shellfish emerged as a major export category during the 1980s, to some extent offsetting the decline in revenues from clove exports. Shellfish exports grew by 57 percent in value terms between 1980 and 1989 to $29.1 million, representing 9 percent of total exports in 1989.

Higher real rice prices accompanying the reduction in rice imports led to increased demand and higher real prices for other food crops. Real prices of cassava and potatoes rose by about 100 percent between 1980 and 1986 but fell 79 percent between 1986 and 1989. In response to higher real prices, production increased by growing 4.7 and 7.1 percent for cassava and potatoes, respectively, during the 1981–86 period. As real prices fell after 1986, production growth fell to 1.3 and 3 percent per year in each case.

Overall, agricultural value added grew in real terms by 10 percent between 1981 and 1984 and by 8 percent between 1987 and 1989, due to price policy, institutional reform, and favorable weather conditions. Given an annual average population growth rate of 2.8 percent during the 1980s, and an annual sectoral growth rate of 3 percent between 1981 and 1989, per capita growth in agriculture was only slightly positive (World Bank 1991c).

Madagascar's industry was adversely affected by the policy reforms of the 1980s. The 60 percent decline in imports of capital and intermediate goods between 1979 and 1983 contributed to a sharp decline in industrial value added, which fell by 34 percent between 1980 and 1982 (Figure 11.3). Between 1983 and 1989, however, there was a 29 percent increase in industrial value added despite a 47 percent decline in intermediate and capital goods imports. Reduced industrial sector output had little effect on

Figure 11.3. Indices of output and input use, Madagascar

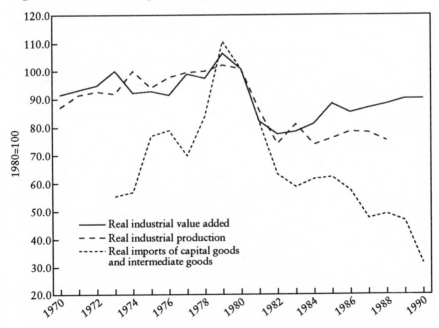

Sources: Pryor 1988; IMF 1988b, 1991c; World Bank 1980b, 1986b, 1991c; USAID 1988.

Figure 11.4. Real minimum wage and private and public employment, Madagascar

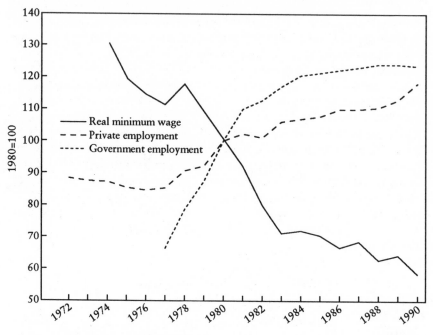

Sources: Pryor 1988; IMF 1988b, 1991c.

employment, though. Employment in formal industry grew by 7.5 percent between 1980 and 1984 while industrial output was falling (Figure 11.4).

Real incomes of formal sector workers apparently declined, however. Real wages fell throughout the 1980s, by 30 percent between 1980 and 1988 as increases in nominal wages failed to keep up with inflation. Since formal sector employment grew more slowly than did the labor force during the period of economic decline, the number of people seeking employment in the informal sector rose. Given falling aggregate demand during the post-1980 period, the increase in informal sector earnings was small, especially in the first half of the decade. By 1986, however, incomes in the informal sector were increasing more rapidly, by an estimated 4 percent in 1986 and 5 percent in 1987.

Approximately 10,000 jobs were created in the formal sector between 1987 and June of 1990. Capacity utilization rates increased from 45–60 percent to 60–75 percent from 1987 to 1990 for a sample of thirty manufacturing firms surveyed. Although the domestic market was relatively weak, the increase in production was enabled by increased exports (World

Bank 1990d). The garment industry has been particularly successful in the wake of the structural adjustment reforms. In addition to highly publicized examples of successful informal sector enterprises,[21] after 1986 ten new export-oriented garment manufacturers began production in the Antananarivo region.

The sharp decline in economic activity in the early 1980s led to a growing concern over the welfare of poor households in Madagascar, in particular the most visible low-income households—the urban poor. Purchases of subsidized rice by the urban poor represented an income transfer of FMG 8,400 per person per year in 1982/83, or 22 percent of total expenditures by the poor in Antananarivo. After the removal of the subsidy, per capita consumption fell by 53 percent for the lowest expenditure quartile in Antananarivo by 1986/87 (Table 11.9).[22] The rice subsidy had been poorly targeted, however. Although most of the poor lived in rural areas, rural households consumed only 4 kilograms of subsidized rice per person; the urban poor consumed on average 105 kilograms per capita. Higher-income urban groups consumed even more subsidized rice (132 kilograms per capita). As rice and other prices rose, real wages fell. The rice purchasing power of the minimum wage fell from a 1975–80 average of 5.4 kilograms per day to 2 kilograms per day in 1984 (Table 11.10). Real wages had been falling before 1980, though: their decline had begun in 1974. But rice prices fell in real terms between 1974 and 1981, helping to maintain the purchasing power of the minimum wage.

Food items make up a relatively larger share of the household expenditures of the lowest income groups (74 percent for the first quartile compared to 59 percent for the fourth quartile in 1982/83). As a result, an increase in the ratio of food to nonfood prices would have a more adverse effect on the poor. The ratio of food to nonfood prices in Antananarivo changed little over time, however.

Some data on malnutrition collected at public feeding clinics suggest higher rates of child malnutrition in the early 1980s in the midst of the initial stabilization period, but data do not provide a basis for statistically valid comparison over time. Programme National de Surveillance Alimen-

---

[21] The World Bank (1989k) cites the case of a Malagasy woman who penetrated the European market for high-quality children's clothing by taking advantage of the embroidery made in Madagascar and the large devaluation of the Malagasy franc. Almost all the primary inputs are produced locally, including the packing materials, which are made from the natural fibers used in basketry. Export sales exceed $1 million, and the woman employed three hundred embroiderers.

[22] Increases in the official rice price had an adverse effect on consumers who had enjoyed the benefits of the official distribution network, but few consumers consumed only subsidized rice. For the large majority of consumers, their marginal purchases came from the parallel market, both before and after the legalization of the private rice trade.

Table 11.9. Rice subsidy transfer, Madagascar

	First quartile	Second quartile	Third quartile	Fourth quartile	Average
1982/83					
Official rice consumption (kg)	106	135	144	151	132
Total transfer (FMG)	8,400	10,800	11,520	12,000	10,560
Total expenditure (FMG)	39,000	71,000	108,000	229,000	103,000
Transfer as % of expenditure	21.5	15.2	10.7	5.2	10.3
Official distribution price (FMG)	140				
Free market price (FMG)	220				
Per kg transfer (FMG)	80				
1986/87					
Official rice consumption (kg)	49	68	67	82	64
Total transfer (FMG)	14,918	20,703	20,399	24,966	19,485
Total expenditure (FMG)	32,000	61,000	90,000	165,000	84,000
Transfer as % of expenditure	46.6	33.9	22.7	15.1	23.2
Official distribution price (FMG)	251				
Free market price (FMG)	556				
Per kg transfer (FMG)	304				

Source: Calculated from World Bank 1989j.

taire et Nutritionelle (1989) points out that the increased numbers of children treated in some clinics does not necessarily indicate an increase in actual malnutrition, since the sample may not be representative of the entire population. Moreover, the considerable month-to-month variation in the numbers of malnourished at the clinics does not seem to correspond to fluctuations in economy activity, weather, or other known factors.[23]

In addition to the rice subsidy, government expenditures had other important effects on the welfare of lower-income groups in Madagascar—particularly personnel, health, and education expenditures. Government personnel expenditure declined by 21 percent between 1982 and 1987 and by 37 percent between 1978, the peak year, and 1987. Real expenditures fell 11 percent between the 1983–87 period and 1988/89 (Figure 11.5). Since only the lower-skilled public sector employees would be among the low-income earners in Madagascar, only some of these expenditure cuts fell on the poor.

Real current expenditures on education and health peaked in 1978–80 at FMG 65.5 billion and FMG 19.7 billion (1984 prices), respectively, representing 26 (education) and 8 (health) percent of the total current

[23] Among the other problems with the data was the lack of a fixed criteria for diagnosing malnutrition.

Table 11.10. Real wages and food prices, Madagascar

	Minimum wage nonagricultural[a] (FMG/mo.)	Consumer price for rice[b] (FMG/kg)	Real wage index[c]	Real rice index[c]	Purchasing power of rice[d]	Ratio of food to nonfood prices[e]
1974	8,839	62.0	136.7	184.6	4.75	106.1
1975	8,839	65.0	126.4	178.9	4.53	102.6
1976	8,839	65.0	120.3	170.4	4.53	99.6
1977	8,839	55.8	116.7	141.8	5.28	95.6
1978	9,625	55.0	119.3	131.2	5.83	98.0
1979	10,018	55.0	108.8	115.0	6.07	98.9
1980	10,877	56.5	100.0	100.0	6.42	100.0
1981	13,038	71.3	91.8	96.7	6.10	102.9
1982	14,742	118.3	78.8	121.7	4.15	101.8
1983	15,591	203.7	69.8	175.6	2.55	95.2
1984	17,369	295.2	70.8	231.6	1.96	97.5
1985	19,000	410.5	70.1	291.4	1.54	102.1
1986	20,699	555.6	66.7	344.4	1.24	110.5
1987	24,399	394.9	68.3	212.9	2.06	94.6
1988	29,032	485.6	64.1	365.3	1.99	79.3
1989	32,250	564.3	65.3	220.0	1.91	78.9

[a]1974–83, Toro Ochoa 1988; 1983–89, IMF 1988b, 1991c.
[b]Spliced consumer price is based on crop year May–April; data for 1970–74 from Bérthelemy 1988; data for 1975–83 from "Riz de Consommation," ceiling price, as cited in Hirsch 1986b; 1983–87 from IMF 1988b, 1991c.
[c]Using traditional household basket CPI as the price deflator (1980=100).
[d]Purchasing power of rice calculated by the following formula: (minimum wage/consumer price for rice)/30 days.
[e]Ratio of the indices of the prices of food and the prices of nonfood items from the traditional basket CPI (1980=100).

budget (Table 11.11). Between 1978–80 and 1981/82, real expenditures fell by 23 percent for education and by 38.6 percent for health. Real expenditures on education were still 86 percent higher than the 1975–77 level, in real terms, and maintained a slightly higher overall share in the budget. Real health expenditures, however, were only 3.4 percent higher in 1981–82 than during the 1975–77 period. Real expenditures on education and health fell by a further 3 and 8.5 percent, respectively, between 1983–87 and 1988–89, whereas overall current expenditures grew by 0.4 percent in real terms. The access by the poor to education and health services would ultimately determine the extent to which expenditure reductions hurt the poor. Little information exists on the actual beneficiaries of social expenditures in Madagascar, but it is likely that the poor suffered some reduction in health services as a result of the cutbacks.

Figure 11.5. Public employment, Madagascar

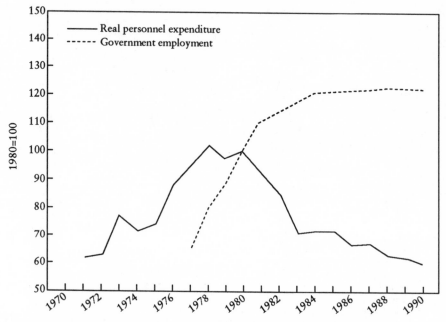

Sources: Pryor 1988; IMF 1988b, 1991c; World Bank 1980b, 1986b, 1991c.

Conclusions

Macroeconomic stabilization in Madagascar has been a clear success in terms of its primary goals: restoring internal and external equilibria. Between 1981 and 1984, the government budget deficit was reduced from 12.1 to 3.6 percent of GDP, and the current account deficit fell from 11 to 6 percent of GDP. These results were achieved largely through reducing aggregate demand, however.

Increasing aggregate supply and promoting economic growth have proved much more difficult. The dismantling of extensive government controls on prices and marketing of rice, export crops, and industrial goods proceeded comparatively slowly with policy reforms staggered through the mid-1980s. In part, this sluggishness may have been due to reluctance on the part of the Malagasy authorities to relinquish control, or to a basic mistrust of market forces (a suspicion of private traders manipulating markets and a fear that liberalization would result in increased concentration *ble II.II.* Real government expenditures, Madagascar (billion 1984 FMG)

	1975–77	1978–80	1981–82	1983–87	1988–89
)TAL CURRENT EXPENDITURE	111.5	247.9	187.8	171.6	172.3
:neral public services	40.5	100.4	79.4	61.7	59.8
lucation	27.0	65.5	50.3	44.6	40.8
ealth	11.7	19.7	12.1	12.9	11.3
)cial/comm. services	2.3	0.0	0.0	1.9	1.8
:onomic services	18.8	34.0	18.2	11.7	11.0
of which agriculture	9.9	18.0	7.8	6.0	5.6
of which public works	5.9	8.5	3.0	2.6	1.8
nallocable	11.2	28.3	27.7	38.9	47.6
of which interest on govt. debt	0.0	0.0	0.0	25.9	43.7
.are of total expenditure					
Education (%)	24.3	26.4	26.8	26.0	23.7
Health (%)	10.5	7.9	6.4	7.5	6.6

Sources: Pryor 1988; IMF 1988b, 1991c; World Bank 1980b, 1986b, 1991c.

of wealth). Faster reforms may not have been politically feasible either. There was considerable debate and conflicting policy advice over whether the outward-looking development strategy based on market forces advocated by the World Bank and others was in fact the best policy.

The overall results of structural adjustment programs through 1987 were rather bleak: slow overall growth, in spite of moderate increases in rice production; clear adverse effects on some of the poorest urban households; and mounting foreign debt. Increases in per capita GDP in recent years despite adverse changes in the external terms of trade are hopeful signs that suggest the structural adjustment effort is bearing fruit. Two of the necessary conditions for rapid, sustained growth now appear to be in place: macroeconomic stability and restored economic incentives for increased production. Yet impediments to faster growth and alleviation of poverty in Madagascar remain.

Although price incentives for agricultural producers have improved and market liberalization has opened up greater trading opportunities, high real marketing costs are a major constraint on increased real output. The country's infrastructure remains weak, especially in rural areas, in part because it has deteriorated over time, in part because little rural infrastructure ever existed. Investments in roads and small bridges are needed to better integrate rural households with the urban economy. A lack of physical security in many rural areas also discourages private investment.

Concerns about income distribution persist. The harsh criticisms of stabilization and structural adjustment in the mid-1980s were largely based on changes in rice policies and their effects on lower-income urban groups. The vast majority of the poor (88.2 percent) who live in rural areas were little affected by removal of the subsidy and may finally be enjoying the fruits of improved price incentives reflected in increased agricultural production. Madagascar's experience suggests that targeted aid to the urban poor during the initial phases of a macroeconomic adjustment program may be required to mitigate the adverse effects of aggregate demand–reducing policies for the most vulnerable groups. The dangers that should be avoided remain expensive government subsidies that threaten macroeconomic stability and misguided attempts to keep the cost of the subsidies down through implicit taxation of domestic farmers (which destroys producers' price incentives).

With market liberalization, a growing economy, and privatization of state assets, there is also increasing concern about the distribution of benefits of growth and a possible increase in the concentration of wealth. Special care needs to be taken especially in the sale of parastatal capital assets to ensure an authentically competitive bidding process and avoid a situation in which capital assets are purchased at below-market values by small groups of the richest households.

Other daunting problems lie ahead. Despite the write-off of 15–20 percent of Madagascar's external debt in 1989, total foreign debt exceeds GDP. Unless considerably more debt forgiveness is offered, payments of interest and principal will continue to absorb a large share of export earnings. Of special concern is the preservation of Madagascar's fragile natural environment in the context of increasing population pressure. On a more positive note, further investment in the transportation network and hotels could spur a large increase in "ecological tourism" revenues.

In a sense, a decade of macroeconomic and sectoral adjustment policies has succeeded in putting Madagascar's economy is a position similar to that of the early 1970s. Slow but positive growth in per capita income has once more been achieved, and the policy debate is again focusing on long-term development strategies and how best to accelerate economic growth. The control of assets and the distribution of the economic gains from growth remain central issues. Although the external terms of trade are not as favorable as in the mid-1970s, there again appear to be substantial amounts of foreign capital available for investment. Unlike the late 1970s, however, when Madagascar relied mostly on short-term credits from commercial banks for its big investment push, there are now realistic prospects of significant direct investments and increases in aid flows. The challenge facing Madagascar now is to accelerate income growth and improve the distribution of earnings and wealth while maintaining the hard-won macroeconomic stability. One can only hope that the lessons of the past will be heeded: the need for realistic goals, the importance of financial discipline, and an understanding of the limits of the government's role in the economy.

12

Economic Crisis and Policy Reform in Africa: Lessons Learned and Implications for Policy

David E. Sahn

The ten case studies in this book together present a comprehensive picture of African economies in crisis. Broadly dichotomizing Africa's post-independence economic experience into the periods before and after structural adjustment, the authors discuss the evolution of the economic decline, the measures taken in response to the crisis, and the outcomes of these measures. Perhaps no lesson is more manifest than that of the diversity of experience in sub-Saharan Africa and the perils of generalization when it comes to such a large region. The causes and severity of the economic crisis, the nature of policy reforms conceived and implemented in response, and the consequences of those changes vary widely from one country to the next. Nonetheless, it is important to extract common themes and lessons, and to garner the broader perspective that is crucial to advance our knowledge of how economies work, both under the stress of negative external conditions and destructive policies and during periods of reform and recovery.

The story of economic deterioration in Africa, as told in the preceding chapters, is predominantly a consequence of the failure of domestic policy and of the institutions the state helped to develop and sustain. Exogenous factors have also been harmful, especially given the structure of the economies as inherited after independence. Policy, however, determines the degree to which shocks can be absorbed and the degree to which economic transformation occurs. On both counts, examples of state failures abound.

The Colonial Legacy

Many of the chapters help us appreciate the economic difficulties that ensued from the struggle for independence. For example, the abrupt and contentious departures of the French from Guinea and the Portuguese from Mozambique, when these colonial powers endeavored to destroy as much as possible of the countries' infrastructure and productive capacity, stand as a testament to the extremes of vindictiveness and of the hardship brought by the fight for freedom. On the other hand, one observes an accommodation of former colonial powers and emergent nation states in many other instances, such as Niger and Cameroon in the franc zone, or Britain's former colonies, Malawi and Tanzania. But regardless of the circumstances of separation, all the countries had in common a fragility of institutions that had been set up by colonial powers to accommodate their outward-looking export orientation. For example, the British introduced into Malawi a system of individualized tenure parallel to the traditional customary tenure in order to promote tobacco production and facilitate marketing. Export agriculture and the related emergence of public enterprises in the West African nations of Ghana and Cameroon were designed to foster the production of coffee and cocoa. Mineral exploitation in Guinea and Zaire likewise had important implications for the structure of the economy and indigenous institutions, as did the slave trade and subsequently export agriculture in Mozambique. Similarly in Madagascar, the expropriation of large tracks of land, to be given to French commercial companies and settlers, had important long-term effects on the structure of the rural economy.

In general, then, the nature of land tenure rules, the roles of the community, tribe, and household in providing services and generating output, the terms of exchange, and so forth had begun to be redefined by the colonial authorities. The colonialists effectively managed the new institutional structures and incentive framework they imposed, extracting a considerable surplus while generating respectable levels of economic growth, albeit largely limited to specific sectors or enclaves that often had limited linkages with the rest of the economy. It was after independence, however, that public sector intervention in the economy became pervasive, government grew larger, and the state assumed roles that proved most disruptive and counterproductive.

Post-Independence Policy Failures

The contention by leaders of the newly independent nations based on a combination of ideology and popular, new paradigms of economic development was that new institutions, primarily state-controlled, were needed to guide and sustain an economic structure that had been transformed during the colonial period. There was a need to cope with the vulnerability implied by the concentration of exports in a few minerals or primary agricultural products, whether it be coffee and vanilla in Madagascar, cocoa in Ghana, copper in Zaire, tobacco in Malawi, bauxite in Guinea, or groundnuts in Gambia. The new economies were vulnerable not only to the shock of falling world prices for primary exports but to rising prices for fuel, consumer goods, and imported intermediates as well. There was some validity, therefore, in the assertion that the health of the post-independence economies was dependent on diversifying the countries' exports and managing the development of import-substituting manufacturing. In practice, however, the newly developed, state-run institutions charged with achieving these laudable objectives served mainly to sustain and legitimize the state rather than to achieve the espoused developmental objective.

At the time of independence, policymakers in Africa inherited trade-dependent economies, performing reasonably well, although fragile in terms of the institutional structures on which the export orientation was built. These colonial structures, which were inherited and transformed by the newly independent states, failed to promote economic growth and technological advancement. Rather, they, along with the new institutions set up after independence, facilitated the abuse of power and contributed to the devastating impact of policy on the economies discussed in the preceding chapters. This now seems almost predictable, especially given the wisdom of hindsight that comes from having witnessed the dissolution of centrally planned economies in which state endeavored to occupy the commanding heights of decisionmaking. But the actions of the newly independent nations must be placed in the context of the times: new ideas about the potential of the development state were gaining currency. In particular, several states moved quickly toward their own brand of African socialism. The usefulness of the command economy was an article of faith for most leaders of the new African states. It was viewed as a means of quickly transforming their economies, catapulting them it into the twentieth century.

This faith in the state was taken to the extreme in the cases of Guinea, Mozambique, and even Tanzania and Madagascar, where what may have been valid ideas had been distorted by men lacking in ideals, prone to corruption, and with limited management capability. State controls in these countries included taking over banks and trading companies, supplying inputs, and marketing output. In some cases, government control went so far as to redefine the structure and organization of production. State-operated farms in Mozambique and Guinea were to be the vanguard of agriculture. Communal villages in Tanzania, as decreed by the Arusha declaration, were to revolutionize not only agriculture but the whole institutional fabric of society. This was deemed necessary in order to galvanize all the country's resources in the quest to alleviate poverty, develop human capital, and ensure technological advancement. These initiatives involved outlawing markets, trade, entrepreneurship, and so forth, imposing in their place an economic order orchestrated by the party and the state. That such policies were an unmitigated failure, at least as measured by economic aggregates, is clear from the information contained in this book. But it is of equal importance to note that such egregious forms of state interference as found in Guinea, Madagascar, Mozambique, and Tanzania were not required to impede economic growth; less overt and comprehensive forms of interference by governments in their quest to be directly involved in resource allocation and production also led to economic failure.

Some countries discussed in this book assumed a more tolerant position vis-à-vis the activities of the private sector, thereby restraining to a greater degree the role of the state in directly productive activities. Still, the results were not favorable: whenever government did try to assume a role in the allocation of goods and services, the desired outcomes were not achieved. For example, the pervasive interference of government in pricing and allocating foreign exchange in the non-CFA countries discussed in this book resulted in grossly overvalued exchange rates (with the possible exception of Malawi). These policies, which stifled exports, went hand in hand with the import-substitution strategies directed by the state, which were clearly not sustainable. Capacity utilization plummeted in countries such as Ghana, Madagascar, Tanzania, and Zaire. Gross inefficiencies and a simple lack of foreign exchange to purchase needed intermediate inputs, coupled with the suppression of markets to which output could be sold, proved devastating in most of the countries examined. Even in those instances where government left the smallholder sector to engage in food crop production and marketing without interference, the combined effects of heavy taxation of export agriculture and poor management of the industrial sector sent the economy into a tailspin. Stagnation of the rural economy occurred commensurate with the decline of the formal sector. The extremes of *ujaama* (the villagization scheme in Tanzania) or FAPAs (Guinea's state farms) were not necessary to bring the African economies to their knees.

Of those countries discussed in this book, the best economic performers after independence were Malawi and Cameroon, where the state's role was also most limited and world market prices were relatively more important for both producers and consumers. Even here, however, underlying features of the government intervention made the economy extremely susceptible to a decline in the terms of trade. Malawi's macroeconomy eventually came to suffer the deleterious consequences of technological stagnation, largely because of government policies which, among other things, neglected agricultural research and actively discriminated against smallholder agriculture. Allowing the leasehold estate sector special privileges to sell their product at remunerative world market prices initially propelled the economy into reasonable growth rates, but the discrimination against the smallholder sector eventually contributed to the faltering economy. Such bias against smallholders was also manifested in terms of chronic malnutrition among Malawi's rural population, which is unparalleled in sub-Saharan Africa.

In Cameroon, previously lauded for its prudent macroeconomic policies and nondiscriminatory treatment of agriculture, the economic hardships that followed the decline in its terms of trade exposed some serious underlying problems of state policy. These included the large proportion of nonperforming assets of the state-controlled banking system, which resulted in a liquidity crisis, and a financially unsound and inefficient public enterprise sector that had incurred large payments arrears to domestic suppliers, who in turn were unable to service their obligations to banks, exacerbating their liquidity problem. But the failure of state-run enterprises was perhaps nowhere more apparent than in the complex system of rules and regulations concerning marketing of agricultural inputs and outputs, a system that fostered inefficiencies and rent seeking. The latter was exemplified by the government-operated, subsidized fertilizer distribution system, which, despite high fiscal costs to the treasury, failed to help farmers significantly. Most of the value of the subsidy accrued to inefficient public enterprises with high marketing costs and often an inability to provide farmers with fertilizer of appropriate nutrient composition.

The socialist ideals and inappropriate economic theory that led to the nationalization of industries and the formation of inefficient public enterprises, distortionary subsidies, and disincentives to production were also often manifested in inappropriately large investment pushes. In Madagascar, for example, unproductive capital investment was followed by the need to stabilize an economy with acute account imbalances, which contributed to the subsequent decline in real income per person. The cases of Gambia and Niger also amply illustrate the adverse effects of poorly conceived and implemented investment strategies, bringing about little more than acute imbalances in the internal and external accounts. The story of the enormous Inga hydroelectric dam in Zaire, generating capacity that was far beyond effective demand, is yet another illustration of the state systematically overestimating its ability to bring about rapid transformation to a modern economy.

In apportioning blame for the excessive foreign borrowing that boosted investment and had short-lived Keynesian effects, or in considering the development of state-run institutions such as marketing parastatals that were to cripple agriculture, the international community also deserves mention. The development push of the late 1970s discussed in many of the chapters in this book was consistent with the advice of donors and international financial institutions. Indeed, the skepticism of the donors over indigenous institutions and the expectation of market failures spurred the growth of public enterprises, just as development theory espoused by international financial institutions commended increased capital spending to generate rapid economic growth. But the advice received by African policymakers went beyond good theory being misapplied. In reality, cheap credit on international markets and the excess of petrodollars in the world financial system also motivated the counsel of the donors which encouraged African governments to borrow and invest. In practice, instead of generating growth, the excessive borrowing and investment was to prove detrimental and unsustainable, thereby setting the stage for economic decline and the need for adjustment.

The failure of the state, often resulting from policy implemented with at least the tacit approval and occasionally the explicit encouragement of international financial institutions and donors, is therefore a common theme throughout the case studies in this book. But perhaps of equal interest are the often heroic efforts of the population to circumvent the heavy hand of the state, especially when it was directly involved in producing and marketing goods and services. In fact, there is some support for the assertion that parallel market activities were more extensive and critical in sectors and countries where distortions were more acute and the government was trying harder to occupy commanding heights of their economies. One area where parallel markets were pervasive was in the foreign exchange market. The gross overvaluation of the domestic currency in the countries studied resulted in thriving parallel markets in which a large share of transaction took place. Parallel markets were also extensive in urban food markets. With the exception of Madagascar, where the official subsidized food distribution seems to have been able to meet demand in urban areas during the late 1970s (although not so in rural areas, where 93 percent of the poor reside), extensive rationing of food at official prices occurred. This is most clearly illustrated by the cases of Guinea, Mozambique, Niger, and Tanzania. Consequently, in these countries the parallel market was of critical importance in ensuring availability of product and determining the real price faced by most consumers.

The case studies indicate that parallel markets were not limited to foreign exchange and urban food markets. Where fertilizer was under the direct control of the state, such as in Cameroon and Malawi, considerable diversion occurred. In the case of the former, the fertilizer officially destined for export crop producers was ultimately used by food crop producers. In Malawi, 25–50 percent of the subsidized fertilizer targeted by the government to smallholders ended up in the hands of estate farms.

Another pervasive form of parallel market activity was illegal cross-border trade to circumvent any of the number of state controls over exports. Ghana's farmers responded to the high level of taxation by diverting cocoa to neighboring countries. In Gambia, groundnuts were regularly bough and sold illegally across the border with Senegal, the direction of trade depending on the prevailing price differential. In Niger, illegal trade in livestock and cowpeas comprised more than 20 percent of official exports and, despite the intentions of the state, government parastatals had little influence on the producer price of the staple grains, millet, and sorghum. Similarly, in Tanzania's controlled economy it was estimated that around 30 percent of the GDP was in the second economy.

Parallel markets, in which prices were free to rise above official levels, were paramount in maintaining incentives for production. They were likewise central in generating incomes and making available consumer goods so that people could survive. Relative to a competitive and efficient free market situation, however, high transaction costs were characteristic of the parallel markets. In fact, the transaction costs associated with parallel markets were greater the more government, failing to see them as performing basic and necessary functions, attempted to crack down on these markets. These costs were paid by the state treasury as well as by private sector producers and consumers. In the case of the losses incurred by the state, the attempts to increase the tax rate by administering prices and various other forms of direct intervention in markets contributed to a shrinking of the tax base as producers and consumers endeavored to avoid participation in the official or formal economy. The irony is that, the more the state tried to impose controls, and the more official prices were out of line, the greater the distortions were and the more the government's actual control of the economy dissipated as parallel markets flourished and the informal economy expanded.

In the private sector, the major losses from state controls were in the form of transaction costs and the dead-weight losses associated with avoiding participation in the official economy, as well as in the expansion of rent seeking behavior. Thus, state controls slowed economic growth, as much of the economic activity went underground and expended resources to avoid the heavy hand of state controls. Such state intervention also furthered income inequality as the elite, primarily in urban areas, effectively captured the rents (i.e., pecuniary returns not derived from producing goods and services but instead associated with profit seeking from unproductive activities; Bhagwati 1987) associated with market repression and the emergence of parallel markets. By all accounts, in fact, the actions of the elite went beyond rent seeking, beyond responding to the difficulties encountered by government in its earnest attempt to administer prices in hopes of generating economic growth. Instead, a great deal of evidence, nowhere more overt than in Zaire, suggests that those with access to power were in fact rent creators, using their influence to create distortions from which they prospered. This behavior was manifested in privileged access to cheap food in the cities, subsidized credit, and foreign exchange through official windows where the domestic currency was grossly overvalued. This same elite in turn provided the support for the state's existence, legitimizing government obtrusiveness, corruption, and ineptitude which in combination further amplified income inequalities and impeded growth.

The welfare losses to non-elites resulting from this perversion of the developmental state were palpable. This was especially the case for the rural smallholders, who faced low official prices for their product. Some stopped producing for the market altogether; those who sold output through officially sanctioned enterprises or found alternative channels paid a price in terms of low financial awards or the high costs of circumventing official policy. In any event, productivity stagnated, especially in agriculture, the backbone of most African economies. Likewise, as shown in the cases of Guinea, Mozambique, and Tanzania, urban consumers often lacked access to rationed official market goods and instead relied on parallel markets, although the transaction costs meant that the prices they paid were in excess of world market prices. It is arguable that most net consumers were best off in countries such as Ghana, where government stayed out of the business of food grain marketing, or Niger, where their intervention was too limited to have any appreciable effect on the prices paid by most households. Likewise, urban households also suffered, especially as real wages fell and inflation eroded nominal incomes in the years before reform. Few good jobs were available in government or publicly controlled or operated enterprises, and few good employment opportunities existed in the small and repressed private sector in urban areas. Instead, the informal sector, or second economy, was generally able to provide subsistence incomes, but it could not serve as the basis for the investment and technological progress essential to promote economic growth.

Policy Reform: Variable Commitment and Mixed Results

The attempt by the state to control prices and allocate resources proved to be an unmitigated failure, which was most obviously manifested in unsustainable account imbalances. Most of the countries examined in this volume had little recourse but to adjust to the adverse economic climate as tax revenues dwindled, budget resources to pay salaries, maintain infrastructure, and deliver services were greatly diminished, and the ability to command foreign exchange for vital imports disappeared. Indeed, it was generally when all domestic policy options for dealing with the repercussions of negative shocks and irresponsible policy had been exhausted that government turned to international financial institutions for help. The process of donor-financed adjustment thus began only when the economic decline reached crisis proportions, with official economies generally on the verge of total collapse. For many of the countries in this volume, in fact, the onset of donor-financed adjustment found them in a worse position, in terms of both national income per capita and infrastructure development, than at the time of independence.

Measured against the failed policies that predated the reforms, considerable progress has since been made in the countries included in this volume, both in policy change and performance. In many instances, however, policy change has lagged behind rhetoric; implementation of reforms has often proved more perilous than planning them. For example, fiscal deficits have fallen in several countries, including Gambia and Tanzania, but others such as Zaire and Niger have met with little success. In either case, the deficits have been moderated by the infusion of external financing, which also helped sustain the size of government relative to GDP and was responsible for the majority of capital expenditures. And when data on discretionary government expenditures (i.e., net of interest payments) are examined, a slightly different story emerges. The increasing burden of debt on the treasury in most countries in the post-adjustment era portends future problems.

Another area of progress reported on in this volume is trade policy. The non-CFA countries for the most part dealt effectively with their overvalued exchange rates. The spread between official and parallel rates declined dramatically in some countries, most notably Gambia, Ghana, Guinea, and Madagascar. Of equal importance, the move to market-oriented exchange rate regimes, something facilitated by adjustment lending's balance-of-payments support, has allowed enough flexibility so that parallel rates have not deviated significantly from official rates since initial devaluation occurred. Concurrently, the adjustment lending has moderated the degree of currency depreciation, at least relative to what it would have been in the absence of the considerable external financing. And though official rates remain overvalued in Mozambique and Tanzania, the divergence from parallel rates has been reduced substantially as part of adjustment efforts. Furthermore, complementary to exchange rate reforms, the countries included in this volume have made great strides in commercial policies. The elimination of quantitative restrictions and adoption of simplified tariffs have met with considerable progress in Gambia, Ghana, Madagascar, Guinea, Malawi, and Zaire, for example.

The net result of trade policy reforms, both on indicators such as inflation and the trade balance and on the incomes of producers and consumers, has been variable. One encouraging finding is that liberalizing exchange rate systems need not be accompanied by large increases in inflation. And even in those cases, such as Gambia, where inflation jumped in the year immediately after the exchange rate liberalization, this consequence was temporary and ultimately was effectively controlled by appropriate credit, monetary, and fiscal policies. Although there is little question that problematic inflation persists in many of the countries included in this volume, especially Zaire, there is no compelling evidence that an overvalued exchange rate as opposed to prudent fiscal and monetary policy, is the appropriate mechanism to control inflation.

In regard to export performance in the wake of trade reforms, the evidence is mixed. Countries such as Tanzania, where trade and related price policy reforms have been slow in coming, have witnessed exports falling short of target levels, particularly in agriculture. Donor support has nonetheless helped ensure that the level of imports has remained high, at least relative to exports. In contrast, the volume of merchandise exports soared in Ghana, which along with the foreign aid inflows contributed to a noteworthy increase in import volume as well. In Guinea, too, official market exports increased dramatically, although not as fast as the increase in imports, which generally were financed through the adjustment program's foreign exchange auction. It is also interesting that a large portion of Guinea's auction-financed imports were accounted for by consumer goods. This substantial increase in the availability of consumer goods subsequent to reforms is not unusual, since prior to adjustment in countries such as Ghana and Mozambique the availability of consumer goods had been dramatically reduced as a result of trade policy distortions and related foreign exchange shortages. There is little doubt, however, that sustaining import levels, both for consumer goods and intermediate imports, is predicated on the continued support of the IMF, World Bank, and bilateral donors, which finance a significant share of the current account deficits of most of the countries examined. Furthermore, increasing rescheduling of ODA debt could make a major contribution to closing the financing gap of the countries discussed in this book. In addition, efforts to reduce and not just reschedule commercial debt are required to ameliorate Africa's economic crisis. Such initiatives could be accomplished through, for example, cash buybacks, which in turn must rely on donor funding. This tactic would save foreign exchange for those countries that pay, and it would foster improved relationships with commercial creditors, thereby helping to attract capital inflows and foreign investment and ease access to trade credits.

But perhaps of greatest importance is that, in those cases where trade reforms have proceeded, there are indications that the incomes of producers of exportables in general, and of export-oriented farmers in particular, have benefited. This opportunity has been realized in many of the countries discussed in this volume, including Ghana, where the reform of trade policies afforded the opportunity to raise the producer price for cocoa despite declining world prices. Likewise, reforms in Guinea and Mozambique, among others, increased the price paid to farmers for their export crops. In contrast, however, is the slow pace of progress in other countries. Malawi's smallholder farmers remain heavily taxed, and devaluation, if not accompanied by other reforms in marketing, will likely show limited success in terms of its impact on incentives and incomes. The limits to exchange rate reforms to restore producer incentives in the absence of complementary trade reforms are perhaps no better illustrated than in the case of Madagascar, where the real value of export crop price increases was less than one-fifth the real exchange rate depreciation through 1987. It was not until reforms of marketing institutions were belatedly undertaken that farmers captured a larger share of the world price for their products.

In addition to trade and exchange rate measures, a variety of other types of policy changes has been adopted, indicating considerable progress in the process of reform. Whether it be the reduction in the size of the civil service and subsequent increase in the real wage from absurdly low levels (which forced workers to seek outside employment and lowered worker productivity) in Gambia, Ghana, and Guinea, or the progress in public enterprise reforms and privatization in Gambia, Ghana, Guinea, Madagascar, Mozambique, and Niger, there is little question that serious, albeit not always complete efforts at economic restructuring are well under way.

The domain where the most important policy changes have occurred, however, is in agricultural price and marketing reforms. These are of special importance primarily because of agriculture's central role in generating value added and employment, coupled with the fact that earlier extensive government controls had all but eliminated competitive markets and disassociated domestic from border prices. The improvement in the incentive structure and efficiency of markets has been partly achieved through macropolicy reforms, such as the reduction of indirect taxation due to exchange rate overvaluation. Institutional reforms, as exemplified by liberalizing markets and limiting parastatal involvement in price determination, have been of equal importance.

In examining the experiences to date, it is useful to distinguish between reforms that apply to food crops and those to export crops. In general, reforms of the food crop sector have proceeded more quickly. This is especially true in Gambia, Guinea, Madagascar, Malawi, Mozambique, Niger, Tanzania, and Zaire. In Ghana and Cameroon, where there was relatively little interference in food crop pricing and marketing before the reforms, major policy reversals were not yet required. Perhaps the most important aspect of the marketing reforms in the food crop sector is that, where they have proceeded, there is evidence that the cost of raising prices paid to farmers has generally not been borne by consumers. Lower marketing margins owing to increased competition and greater efficiency have served the interests of those at both ends of the marketing chain. Similarly, in many of those cases (e.g., Guinea, Niger, Malawi, Tanzania, but not in urban areas of Madagascar) where liberalization involved abandoning efforts to maintain low retail prices, consumers have generally not been losers, because of the ineffectiveness of government intervention in that regard before reform.

Concerning export crops, progress has been much more sporadic in the countries included in this volume. Guinea, Madagascar, and Niger have moved to allow private traders to engage in marketing export crops. Other countries, including Ghana and Mozambique, maintained controls over exports through the 1980s but have raised producer real prices. Still others, such as Malawi (for smallholders) and Tanzania, moved very slowly in terms of improving the marketing environment for export crops. The continued high level of taxation resulted, not unexpectedly, in a poor response by producers. In Cameroon, where relatively high prices prevailed before the reforms, the fall in world prices for the major export crops necessitated a similar decline in prices paid to producers in order to avoid the government assuming the untenable burden of high subsidy payments.

So, though there has been considerable variability in the degree to which price incentives were restored and producers responded, the underlying issue of the appropriate nature and extent of state involvement in agricultural markets was rarely explicitly addressed in formulating a strategy to reinvigorate agriculture. Indeed, all indications suggest that the driving paradigm for agricultural sector reform was the neoclassical approach to border pricing. This is not only a consequence of a faith in markets to provide signals that will result in the optical allocation of resources but additionally an artifact of the disillusionment of prior efforts at state intervention, generally in the name of equity and stability.

Left unresolved is to what extent wholehearted adoption of the free market is in the best long-term interests of African agriculture. It is clear that, relative to the failures of market interventions in the past, the failures of the unfettered market will in most cases represent a vast improvement. This should not, however, be confused with the suggestion that no intervention in pricing is the best medium- and long-term strategy. Given the instability in world prices, the imperfect information that hinders decision-making by producers and marketing agents, the absence of insurance markets, the weaknesses of credit markets and financial intermediation, and the externalities of a robust and stable agriculture and an adequate and stable food supply to feed consumers, the argument is compelling that the role of the state should go beyond the requisite activities of simply building infrastructure, performing agricultural research, and engaging in agricultural extension. But the risks of more direct intervention as is followed in both developed countries and other regions of the developing world— for example, stabilizing prices or setting price margins in which private sector agents can operate—need to be considered in light of past policy failures. Specifically, the pervasive corruption coupled with analytical limits that presently constrain policymaking in African agriculture represents a challenge that needs to be confronted.

The overall picture of the process of African economic reform, in terms of both policy change and the ensuing response, has been one of progress on many fronts—although the commitment to reform has differed markedly from one country to the next, just as the requirements for policy change vary, reflecting differences in the conditions before adjustment. These differences have ultimately manifested themselves in terms of the sequencing, pace, and orderliness of reforms, about which the experiences of the countries included in this volume suggest a few generalizations. First, in many of the countries, the priority in the early stages of adjustment was stabilizing the economy. Consequently, the goal of restoring growth in many cases assumed a secondary role. This was especially the case in Gambia, Madagascar, Niger, and Zaire, which before adjustment undertook massive investment pushes characterized by a loss of fiscal and monetary discipline. The case of Cameroon also illustrates an adjustment program oriented to reducing absorption. Its membership in the franc zone left little recourse other than tight fiscal policy to stabilize the economy when account deficits burgeoned in the wake of the drop in oil revenues. Although there is some merit to the assertion that stabilization is an important prerequisite to establishing an environment in which growth-oriented policies can succeed, there are also compelling reasons to move forward quickly to promote the structural and institutional changes that offer the possibility of enhancing output and productivity. Cutting government expenditures is easier than identifying constraints to growth and bringing about the changes required to ensure their relaxation. This explains in part why a period of stabilization measures may precede attempts at structural change. It is also likely that the loss of privileges that accrue to the elite is less immediate in a program that is expenditure-reducing rather than expenditure-switching, so the resistance to the latter may have been stronger. In any case, it is unfortunate, but not inevitable, that years transpired before discernable growth-oriented policies were implemented in many countries and positive results began to emerge.

Moving more quickly to increase output and exports through trade liberalization, restructuring public enterprises, shifting the internal terms of trade in favor of agriculture, and related institutional reforms such as of the formal financial sector, therefore, need to be given the priority that fiscal, monetary, and exchange rate measures have received. It was in fact exactly these types of output-enhancing measures that were aggressively pursued early in adjustment in Ghana and Guinea, where the response has been more promising than in those countries that pursued such reforms more slowly. This, of course, raises the issue of pace and timing of reforms. As noted, macroeconomic stability is rightfully viewed as an early objective of reform. It is also true that certain aspects of institutional reforms that improve the incentive framework may work against that objective. For example, trade liberalization may adversely affect the balance of payments in the short term. Nonetheless, moving more quickly to reduce taxation of farmers or liberalize marketing (e.g., in Madagascar and Malawi) would likely have served to reinvigorate the economy earlier in the adjustment process. The improved economic performance in Tanzania in the wake of reforms would also have materialized quicker if the pace of structural reform has been accelerated. There was little economic or distributional justification, for example, for waiting until 1991 to truly liberalize grain marketing. Likewise, the magnitude of the response to reform in Tanzania could have been enlarged if the market for export crops, still heavily controlled, had been liberalized. Thus the full extent of the economic recovery still awaits long overdue reforms in many cases.

There remains ample room for research on what is in fact the optimal pace and sequencing of reform. It is nonetheless beyond contention that the issue of orderliness and commitment are paramount. This can be no better illustrated than with the case of Zaire. The fact that the adjustment process was chaotic, showing signs of success where conditions were adhered to in the first two years of adjustment and setbacks when politics derailed the programs, shows how a government attempting to maintain the allegiance of the elite and civil servants will adopt ruinous policies such as monetary and fiscal policy distortions and licensing and control of economic activities.

Zaire's story of failure is an extreme—one that contrasts with the economic upturn occurring in countries such as Ghana which were serious about reform. Overall, however, progress in restoring growth has generally been slow in coming. The magnitude and pace of supply responses have not been strong enough to give comfort to the participants in, and observers of, the process. This weak response is explained by factors other than the sequencing, pace, and orderliness of economic reforms. For example, conditions before reform are critical to determining outcomes. In the case of Cameroon, exogenous shocks applied in the prevailing policy context precipitated the need for adjustment and made it virtually certain that the adjustment would be demand-reducing by nature. In Guinea, in contrast, policy deserves the bulk of the blame for the crisis. There, unlike Cameroon, the dismal economic standing before reform made it almost certain that any reduction in state controls and any change in policy, regardless of the external environment, would have positive growth implications.

Another aspect of the importance of prior conditions is represented by Niger, whose disappointing economic performance after adjustment is simply due to the fact that most of the adjustment measures boosted sectors of the economy that contribute relatively little to the country's GDP. Large portions of the economy, such as the livestock sector, were only marginally affected by the measures. The uranium sector, in particular, the most important source of exports, was untouched by the reforms, its performance instead being subject to the vagaries of world markets and the level of subsidies provided by France. It thus comes as no surprise that there was little response to government measures to reform economic policies.

But perhaps the most important aspect of prior conditions is the nature and strength of existing indigenous, informal, or even state-run institutions that are needed to respond to price-oriented adjustment measures and related efforts at state disengagement. A robust response by existing and newly emerging institutions is a prerequisite to restoring incentives and efficiency in markets. On this count, the evidence suggests a great deal of variability and amply elucidates the limitations of the reforms undertaken. There is little question that the outcomes of reforms have been disappointing in some cases because of an absence of indigenous or other civil institutions to fill the vacuum left by the state. A combination of old and emerging formal and informal sector institutions strengthening fledgling or repressed markets, as the state disengages and loosens controls, is essential to successful adjustment. The failures in that regard are shown, for example, by the fertilizer story told for Gambia or the rice marketing reforms for Guinea. In the case of the former, a combination of factors—including the lack of access to credit, imperfect information, and uncertainty as to the actions of the state—resulted in the private sector not making a market for fertilizer. In the case of Guinea, despite the improvement in producer incentives, infrastructure, credit, and information constraints impeded significant increases in domestic rice availability in the capital and consequently resulted in a precarious situation in which the inflow of foreign exchange in the form of adjustment lending increased the reliance on food imports. Similarly, in Zaire, supply response to food crop price and marketing liberalization has been greatly hindered by the extremely dire condition of the nation's road system and the weakness of government agencies for agricultural research and extension.

Stories of these types are found throughout the chapters in this volume. Even in the star performer, Ghana, the lack of private investment in the wake of a successful adjustment program has limited the progress to date. Consideration of such shortcomings brings to the fore the issue of what can be done in these differing contexts to restore more rapidly and thereafter sustain growth.

Restoring Growth and the Role of the State

Most African economies have yet to achieve the critical mix of policy reforms, institutional development, and investment in physical and human capital necessary for sustainable and accelerated growth. Meeting this challenge is clearly critical to raising incomes, moderating consumer prices, and enhancing the quality and quantity of social and economic services and infrastructure. What is also clear is that the state will inevitably play a crucial role in these areas, albeit one that must depart markedly from its role in most economies since independence. The need to define a new role for government—not simply a disengagement from directly intervening in markets but an assumption of critical responsibilities in delivering public services and providing public goods—is paramount. The case studies in this book suggest that exports and production will marginally increase, and more consumer goods will be available, as state disengagement is accompanied by market liberalization and privatization proceeds. But it is not so clear that simply letting prices clear markets and adopting the principles of border pricing are sufficient to promote rapid sustainable growth or recovery from a serious external shock. Rather, there remains a crucial role for government in promoting development and leading a nation out of crisis.

Several fundamental functions can be identified for the state, and all of them require care in implementation to preclude them from becoming means for the elite to serve their own interests. One need only look to see who benefited from subsidies—whether for food in Guinea and Tanzania, fertilizer in Malawi and Cameroon, overvalued foreign exchange in Mozambique and Zaire, civil service jobs in Niger and Gambia, or education and health services in virtually all countries in the region—to be concerned over the prospects of new or continued opportunities for rent seeking. This suggests that strong and fair-minded political leadership is required to prevent new or changed means of state intervention from empowering special interests. And though the issue of governance is not explicitly addressed in this book, it is clear that it underlies much of what has been wrong in Africa, and what needs to be done in the future.

Perhaps the foremost government role is in fostering investments in human capital. Markets cannot be relied on to ensure that adequate resources are directed toward education, health, and other basic needs. The evidence in this book suggests that adjustment has brought no general pattern of declining spending for health and education, either in real terms or as a percentage of total expenditures. In some countries, such as Malawi, these expenditures remained largely unchanged: in others, such as Gambia, they fell; in still others, such as Ghana, spending on education and health increased markedly. The major issue is not, however, simply whether adjustment affected the level of spending; but whether adjustment affected the way available resources were being allocated within the social sector. Too much money was spent on secondary and university education, subsidizing the urban elite; too much of the health budget was for curative, hospital-based services for the urban population. Thus, there was considerable room for adjustment programs to encourage reallocation of spending within the social sector to increase technical efficiency and equity.

Experience from most of the countries examined in this book suggest no major strides made in reallocating social sector spending. Some positive developments were observed, however. For example, Niger's and Ghana's education adjustment credits were partially successful in redirecting government priorities. They also illustrate the feasibility and importance of social sector restructuring occurring commensurate with macroeconomic adjustments. Nonetheless, there remains much to be done to redefine the role of the state in areas such as expanding the quality and quantity of available services, improving the technical training of service delivery personnel, and building social service infrastructure. More careful management of existing financial resources and cost recovery to maintain and expand the resource base are key elements to achieving such objectives, with the understanding that certain expenditures with high externalities such as vector control or immunizations receive justly deserved subsidies.

In addition to investing in people, however, the state must also help regulate and guide, as distinguished from controlling, markets. The fact is that, left to their own devices, markets do occasionally fail and often distribute rents. In the absence of state controls, there may be different winners and losers of market failures. Promoting an environment in which competitive price-clearing markets function requires a range of government attention and function. And the complementarities of public and private investment must also be recognized. In particular, government must assume a central role in "crowding in" investments in physical infrastructure, such as ports, roads, and communications. Other "crowding-in" investments in research, particularly in agriculture, are essential. Without the diffusion of new technology, there is little basis for growth in productivity and the returns to complementary human capital investments are lessened.

Collecting and disseminating information are also necessary state functions, since market imperfections are often a consequence of imperfect and asymmetric access to information. Developing a set of rules and regulations that foster competition, fairness, and openness is essential, too. These must include an enlightened legal framework that is respected and enforced, governing everything from property rights to investment codes to banking regulations.

These roles for the state, although far more restricted than the all-encompassing posture of direct involvement in producing and allocating goods and services, still may have perilous consequences in the absence of an enlightened system of governance. The same types of selective credit policy used in Korea and Japan to guide and promote economic development, for example, would indeed be corrupted by the rent-seeking behavior found in Zaire and other countries which led to investment being channeled in the direction of nonperforming assets. Likewise, arguments in favor of some degree of price stabilization are certainly pervasive. For farmers, reducing market instability may encourage investments in output-enhancing technology and equipment as well as in marketing and processing services; for food consumers, it may reduce some of the transaction costs associated with changing expenditure patterns in response to changes in relative prices; for other sectors in which stability in wage goods encourages investments in labor-using technology, some degree of price stability may also prove valuable (Timmer 1989). But, once again, the perils of inappropriate intervention, as has been the case through most of sub-Saharan Africa, are pervasive. In particular, when such objectives translate into supplanting the private sector through too much intervention, the costs to the budget, efficiency, and generally macrostability become untenable.

All this is not to suggest that some degree of intervention in credit, factor, and product markets cannot be successful in sub-Saharan Africa, or that Africans are not ready for government to assume such a role. It is to suggest, however, that assuming such roles is predicated on (1) proper analytical foundations and the related human capital and management resource base; (2) adequate financial resources to invest, for example, in needed infrastructure and to support requisite budgetary expenditures implied by certain stabilization activities; and (3) extensive financial controls and fiscal discipline. A corollary of these points is that the political process needs to be sufficiently open to prevent state controls so excessive as to suppress private sector agents and destabilize markets and the macroeconomy.

Challenges Ahead

In the final analysis, as Africa moves into its second decade of adjustment, it is too simple to suggest that success in this process will result solely by the strengthening of market forces through disengagement of the state. Instead, three complementary challenges must be confronted by donors, policymakers, political leaders, and their communities. First, structural adjustment has been slow to build on and foster growth in indigenous institutions that have evolved in response to the economic crisis or existed before it. There is a need to give greater thought and emphasis on how to legitimize and strengthen these traditional or newly developed civil institutions in order to help induce technological change and economic recovery. This is paramount given that the preadjustment era of forced institutional change, under the sponsorship of the state, failed so miserably.

A second challenge begins with the recognition that formal sector institutions, including government, also must be strengthened. In some cases this will simply involve a reorientation of state roles without an expansion of government's share of GDP; in others, where state functions contracted considerably before reform, government size will have to increase to engage in public capital formation that "crowds in" private investment and provides the needed regulatory and legal framework to promote the dual objectives of growth and poverty alleviation. In either case, the developmental state needs the support of donors and international financial institutions. The challenge is enormous, requiring major financial and moral support to ensure that government is restrained yet sufficiently strong to sustain the process of economic development. Simply put, just as the state has been at the forefront of abusing power and threatening individual rights and the effective functioning of markets, so too is it essential for protecting individual rights and setting the parameters for efficient and competitive markets.

A third challenge is to alleviate poverty and improve income distribution. This topic, and in particular the effect of adjustment on the poor, has been discussed only tangentially in this book; the complexity of the subject demands special treatment, which we intend to supply in a complementary volume. Nonetheless, it is clear that implications of economic reform on poverty and equity will be substantial. In fact, the impact of adjustment will be felt both through the effects of policy on factor and product markets in which the poor are engaged as well as more directly through the provision of services and transfers from the state.

In essence, these two pathways are analogous to the direct and indirect routes to poverty reduction, which have been persistently debated by development economists. Basically, the debate has centered on the merits of striving to alleviate poverty by creating new income through improved resource allocation and market efficiency versus the advantages of redistributing existing income through subsidies, in-kind transfers, and so forth. The optimal mix of the growth-oriented indirect path and the targeted direct path must be determined on a country-by-country basis. It is wrong, however, to suggest that the use of scarce financial resources to support investment and economic growth, rather than directly provide subsidies to the poor, will neither raise living standards nor provide opportunities for all economic agents in the society. Instead, it is more correct to assert that the needs of the poor will be addressed primarily by a growth-oriented adjustment program that emphasizes inclusion of the poor as well as the other groups in society in productive activities and economic revitalization. Ideally, participation of the poor in the growth process can be achieved by policies that promote the redistribution of assets and investment in human capital. Policies may need to be designed toward raising the returns to factors owned and controlled by the poor while attempting to redirect constrained budgetary resources toward services used by low-income households. Thus the marked distinction between a laissez-faire approach to development and active public policy designed to promote growth while improving the living standards of low-income groups must be recognized. Understanding this distinction and particularly the effects of various paths to adjustment on poverty and income distribution are key areas for future research.

References

Abbott, Richard. 1991. *Privatization of Fertilizer Marketing in Cameroon: A Case Study.* The Agricultural Marketing Improvement Strategies Project. University of Idaho/Postharvest Institute for Perishables and Abt Associates. Washington, D.C.: USAID.

———. 1990. *Privatization of Fertilizer Marketing in Cameroon: A Second-Year Assessment of the Fertilizer Sub-Sector Reform Program.* The Agricultural Marketing Improvement Strategies Project. University of Idaho/Postharvest Institute of Perishables and Abt Associates. Washington, D.C.: USAID.

Acharya, S. N., and B. F. Johnston. 1978. *Two Studies of Development in Sub-Saharan Africa.* World Bank Staff Papers. Washington, D.C.: World Bank.

Alderman, Harold. 1991a. *Downturn and Economic Recovery in Ghana: Impacts on the Poor.* Cornell Food and Nutrition Policy Program, Monograph no. 10. Ithaca, N.Y.: CFNPP.

———. 1991b. "Incomes and Food Security in Ghana." Washington, D.C.: CFNPP. Photocopy.

———. 1990. *Nutritional Status in Ghana and Its Determinants.* Social Dimensions of Adjustment in Sub-Saharan Africa Working Paper no. 3. Washington, D.C.: World Bank.

Alderman, Harold, David E. Sahn, and Jehan Arulpragasam. 1991. "Food Subsidies and Exchange Rate Distortions in Mozambique." *Food Policy* (October): 395–404.

Alderman, Harold, and Gerald Shively. 1991. *Prices and Markets in Ghana.* Cornell University Food and Nutrition Policy Program, Working Paper no. 10. Ithaca, N.Y.: CFNPP.

Arulpragasam, Jehan, Carlo del Ninno, and David E. Sahn. 1992. *Household Consumption in Conakry: Expenditures, Calories, and Prices.* ENCOMEC

"Processed" means produced in an informal way such as mimeographed.

Findings Bulletin no. 5. Washington, D.C.: Cornell Food and Nutrition Policy Program.

Arulpragasam, Jehan, and David E. Sahn. Forthcoming. *Economic Reform in Guinea*. New York: New York University Press.

Associates for International Resources and Development (AIRD). 1990. "Madagascar Irrigated Sub-Sector Review: An Economic and Financial Analysis." Somerville, Mass.: AIRD. Draft.

———. 1989. *Agricultural Sector Assessment, Republic of Guinea*. Somerville, Mass.: AIRD.

———. 1985. *Impact de la stratégie d'autosuffisance en riz*. Rapport pour le Ministère de la Production Agricole et de la Réforme Agraire. Somerville, Mass.: AIRD.

———. 1984. *Etude du secteur rizicole: Rapport Final*. Somerville, Mass.: AIRD.

Atukorala, V., A. Batchelder, G. Gardner, and T. Ware. 1990. *The Malawi Fertilizer Subsidy Reduction Program: The Impact of the African Economic Policy Reform Program*. Washington, D.C.: USAID.

Azam, Jean-Paul. n.d. *Les difficultés d'ajustement au Niger: Le rôle de la naira*. Clermont-Ferrand: CERDI.

Azam, Jean-Paul, and Tim Besley. 1989. "General Equilibrium with Parallel Markets for Goods and Foreign Exchange: Theory and Application to Ghana." *World Development* 17 (12): 1921–30.

Azam, Jean-Paul, and J. J. Faucher. 1988. *The Supply of Manufactured Goods and Agricultural Development: The Case of Mozambique*. OECD Development Centre Papers. Paris: OECD.

Bagachwa, M. S. D., N. E. Luvange, and G. D. Mjema. 1990. "Tanzania: A Study of Non-Traditional Exports." Report prepared for the Ministry of Finance, Dar es Salaam.

Banco de Moçambique. 1989. *Balance of Payments Statistics*. Mozambique: Banco de Moçambique, Office of External Debt.

Bank of Zaire. Various years. *Annual Report*. Kinshasa: Bank of Zaire.

Banque Central de la République de Guinée (BCRG). Various years. *Bulletin trimestriel d'études et de statistiques*. Nos. 4, 5, 7, 8, 9, and 15. Conakry: BCRG.

———. 1989. *Rapport annuel d'activités au 31 décembre 1989*. Conakry: BCRG.

Banque des Données de l'Etat, Direction Générale (BDE). 1989. *Situation économique au 1 er Janvier 1989*. Antananarivo: Présidence de la République.

———. 1988. *Inventaire socio-economique, 1976–1986*, Pts. 1–2. Antananarivo: Présidence de la République.

———. 1987a. *Enquête sur les budgets des ménages: Revenu/milieu rural et centres urbains secondaires*. Antananarivo: Présidence de la République.

———. 1987b. *Enquête sur les budgets des ménages: Dépenses/centres urbains secondaires*. Antananarivo: Présidence de la République.

———. Undated. *Recensement industriel, année 1983 et 1984*. Antananarivo: Présidence de la République.

Barghouti, Shawki, and Gary Le Moigne. 1990. *Irrigation in Sub-Saharan Africa: The Development of Public and Private Systems*. World Bank Technical Paper no. 123. Washington, D.C.: World Bank.

Barker, Jonathan. 1985. "Gaps in the Debates about Agriculture in Senegal, Tanzania, and Mozambique." *World Development* 13(1): 59.

Bateman, Merrill, Alexander Meeraus, David Newberry, William Okyere, and Gerald O'Mara. 1990. *Ghana's Cocoa Pricing Policy.* Agricultural and Rural Development Working Paper no. 429. Washington, D.C.: World Bank.

Bates, Robert K. 1981. *Markets and States in Tropical Africa: The Political Basis of Agricultural Policies.* Berkeley: University of California Press.

Ben-Senia, Mohamed. 1991. *Agricultural Production, Poverty, and Consumption in Rural Bandundu.* Washington, D.C.: Cornell Food and Nutrition Policy Program.

Berg, Elliot. 1989. "The Liberalization of Rice Marketing in Madagascar." *World Development* 17(5).

Berg, Elliot, and Associates. 1983. *Joint Program Assessment of Grain Marketing in Niger.* Report prepared for USAID/Niger and the Government of Niger. Alexandria, Va.: Elliot Berg and Associates.

Bernier, René, and Paul A. Dorosh. 1993. *Constraints on Rice Production in Madagascar: The Farmer's Perspective.* Working Paper no. 34. Ithaca, N.Y.: Cornell Food and Nutrition Policy Program.

Bérthelemy, J. C. 1988. Offre de biens manufacturés et production agricole: Le cas de Madagascar: Textes du Centre de Développement. Paris: Centre de Développement de l'OCDE.

Bevan, David, Arne Bigsten, Paul Collier, and Jan Willen Gunning. 1987. *East African Lessons on Economic Liberalization.* Thames Essay no. 48. Trade Policy Research Center, London: Gower.

Bevan, David, Paul Collier, Jan Willen Gunning, Peter Horsnell, and A. Bigsten. 1989. *Peasants and Governments: An Economic Analysis.* Oxford: Clarendon Press.

Bhagwati, Jagdish. 1987. "Directly Unproductive Profit-Seeking Activities." In *The New Palgrave,* ed. John Eatwell, Murray Milgate, and Peter Newman. London: Macmillan.

Bhatia, Rattan J. 1985. *The West African Monetary Union.* Occasional Paper no. 35. Washington, D.C.: IMF.

Biermann, W., and J. Campbell. 1989. "The Chronology of Crisis in Tanzania, 1974–1986." In *The IMF, the World Bank, and the African Debt: The Economic Impact,* ed. B. Onimode. London: Zed.

Blandford, David, and Sarah Lynch. 1990. *Structural Adjustment and the Poor in Cameroon.* Washington, D.C.: CFNPP.

Blane, Dianne, Michael Fuchs-Carsch, David Hess, and Jane Seifert. 1990. *Cameroon: The Fertilizer Sub-Sector Reform Program.* Yaoundé: USAID.

Boateng, Oti, Kodwo Ewusi, Ravi Kanbur, and Andrew McKay. 1989. "A Poverty Profile for Ghana, 1987–1988." Washington, D.C.: World Bank. Photocopy.

Bond, Maureen. 1983. "Agricultural Response to Prices in Sub-Saharan African Countries." *IMF Staff Papers* (December): 703.

Borsdorf, Roe. 1979. *Marketing Profile: Cereals and Cash Crops.* Vol. 2, P. F. Niger Agricultural Sector Assessment. Niamey: USAID.

Caputo, E. 1991. *Rapport d'évolution d'un programme national d'appui à la filière riz en Guinée.* Conakry: CCCE.

Chazan, Naomi. 1983. *An Anatomy of Ghanaian Politics: Managing Political Recession, 1969–1982.* Boulder, Colo.: Westview Press.

Chemonics International. 1986. "Investment Opportunities in the Guinean Rice Industry." Washington, D.C.: Chemonics International. Photocopy.

Chhiber, Ajay, and Nimat Shafik. 1990. "Exchange Reform, Parallel Markets, and Inflation in Africa: The Case of Ghana." Washington, D.C.: World Bank. Photocopy.

Christiansen, R. E., and V. R. Southworth. 1988. "Agricultural Pricing and Marketing Policy in Malawi: Implications for a Development Strategy." Paper presented at the Symposium on Agricultural Policies for Growth and Development, Magnochi, Malawi.

CILSS, Assistance Technique Italienne. 1989. *L'impact de l'urbanisation sur les modèles de consommation alimentaire de base au Niger*. Niamey: Direction Statistique et Informatique, Ministère du Plan.

Collier, Paul, and Jan Willen Gunning. 1991. "Money Creation and Financial Liberalization in a Socialist Banking System: Tanzania 1983–1988." *World Development* 19(5): 533–38.

Comissão Nacional do Plano. 1988. *Informacão estatistica 1987*. Maputo: Comissão Nacional do Plano, Direccão Nacional de Estatistica.

Cook, Andy. 1988. *Niger's Livestock Export Policy*. Government of Niger/Tufts University/USAID Integrated Livestock Production Project. Medford, Mass.: Tufts University.

Coulson, Andrew. 1989. "Nigerian Markets for Livestock and Meat: Prospects for Niger." Niamey: USAID. Mimeo.

———. 1982. *Tanzania: A Political Economy*. Oxford: Oxford University Press.

Covell, Maureen. 1989. *Madagascar: Politics, Economics, and Society*. London: Frances Pinter.

Cowitt, P.P., ed. 1985. World Currency Yearbook: International Currency Analysis. Brooklyn, N.Y.: Picks.

David, Paul. 1985. "Clio and the Economics of QWERTY." *American Economic Review* 75 (May): 332–37.

del Ninno, Carlo, David E. Sahn. 1990. "Survey Methodology and Preliminary Results of Household Welfare in Conakry: A Progress Report." Cornell University Food and Nutrition Policy Program. Ithaca, N.Y.: CFNPP. Draft.

Devarajan, Shantayanan, and Jaime de Melo. 1990. *Membership in the CFA Franc Zone: Odyssean Journey or Trojan Horse?* PPR Working Papers, no. 482. Washington, D.C.: World Bank.

———. 1987. "Adjustment with a Fixed Exchange Rate: Cameroon, Côte d'Ivoire, and Senegal." *World Bank Economic Review* (May).

Dittus, Peter. 1987 "Structural Adjustment in the Franc Zone: The Case of Mali." CPD Discussion Paper no. 1987–11. Draft.

Dornbusch, Rudiger. 1974. "Tariffs and Nontraded Goods." *Journal of International Economics* 4 (May): 177–85.

Dorosh, Paul A., René E. Bernier, Armand Roger Randrianarivony, and Christian Rasolomanana. 1991. *A Social Accounting Matrix for Madagascar: Methodology and Results*. Cornell Food and Nutrition Policy Program, Working Paper no. 6. Ithaca, N.Y.: CFNPP.

Dorosh, Paul A., René E. Bernier, and A. H. Sarris. 1990. *Macroeconomic Adjustment and the Poor: The Case of Madagascar*. Cornell Food and Nutrition Policy Program, Monograph no. 9. Ithaca, N.Y.: CFNPP.

Dorosh, Paul, and Boniface Essama Nssah. 1991. *A Social Accounting Matrix for Niger: Methodology and Results*. Cornell University Food and Nutrition Policy Program, Working Paper no. 18. Washington, D.C.: CFNPP.

Dorosh, Paul, and Alberto Valdés. 1990. *Effects of Exchange Rates and Trade Policies on Agriculture in Pakistan*. International Food Policy Research Institute Report no. 84. Washington, D.C.: IFPRI.

Duruflé, Gilles. 1989. "Structural Disequilibrium and Adjustment Programmes in Madagascar." In *Structural Adjustment in Africa*, ed. Bonnie K. Campbell and John Loxley. New York: St. Martin's Press.

Economist Intelligence Unit (EIU). 1991a. *Cameroon, CAR, Chad*. Country Report no. 3 1991. London: EIU.

———. 1991b. *Cameroon Country Profile, 1990–1991*. London: EIU.

———. 1990. *Madagascar, Mauritius, Seychelles, Comoros*. Country Report no. 1 1990. London: EIU.

Edwards, Sebastian. 1989. *Real Exchange Rates, Devaluation, and Adjustment*. Cambridge, Mass.: MIT Press.

Ellis, Frank. 1983. "Agricultural Marketing and Peasant State Transfers in Tanzania." *Journal of Peasant Studies* 10(4).

Ephson, Ben. 1990. "Wage Low Not High Enough." *West Africa*: 484–85.

Filippi-Wilhelm, Laurence. 1988. *Circuits de commercialisation et de distribution en Guinée*. Vol. 2. Conakry: UNCTAD/UNDP.

———. 1987. *Circuits de commercialisation et de distribution en Guinée*. Vol. 1. Conakry: UNCTAD/UNDP.

Finnegan, W. 1989. "The Emergency." *New Yorker* (May): 43–76.

Food and Agriculture Organization (FAO). 1984. *Food Balance Sheets, 1979–1981 Average*. Rome: FAO.

———. Various years (a). *FAO Production Yearbook*. Rome: FAO.

———. Various years (b). *Food Outlook Statistical Supplement*. Rome: FAO.

Forbeau, Francis, and Yannick Meneux. 1989. *Riz local ou riz importé en Guinée?* Montpellier: Institute de Recherches Agronomique Tropicales.

Franco, Robert. 1981. "The Optimal Producer Price of Cocoa in Ghana." *Journal of Development Economics* 8: 77–92.

Garcia, Jorge Garcia, and Gabriel Montes Llamas. 1988. *Coffee Boom, Government Expenditure, and Agricultural Prices: The Colombian Experience*. Research Report no. 68. Washington, D.C.: International Food Policy Research Institute.

Ghana, Government of Ministry of Agriculture. n.d. National Directorate of Agricultural Economics data bank.

Ghana, Government of, and World Bank. 1989. *Ghana Living Standards Survey, First Year Report*. Accra: Ghana Statistical Service, and Washington, D.C.: World Bank, Social Dimensions of Adjustment Unit.

Gould, David. 1980. *Bureaucratic Corruption and Underdevelopment in Third World: The Case of Zaire*. Elmsford, N.Y.: Pergamon Press.

Gould, David, and Joe A. Amaro-Reyes. 1983. *The Effects of Corruption on Administrative Performance*. World Bank Staff Papers no. 580. Washington, D.C.: World Bank.

Green, H. R. 1987. *Stabilization and Adjustment Policies and Programmes: Ghana*. Helsinki: World Institute for Development Economics Research.

Green, H. R., D. G. Rwegasira, and Brian van Arkadie. 1980. *Economic Shocks and National Policy Making: Tanzania in the 1970s*. The Hague: Institute of Social Studies.

Guillamont, Patrick, and Sylviane Guillamont. 1984. *Zone franc et développement Africain*. Paris: Economica.

Guillaumont, Sylviane Jeanneney. 1985. "Foreign Exchange Policy and Economic Performance: A Study of Senegal, Madagascar, and Guinea." In *Crisis and Recovery in Sub-Saharan Africa*, ed. Tore Rose. Paris: OECD.

Guinea, Government of (GOG). 1991a. *Rapport économique et social 1990*. Conakry: Ministère du Plan et de la Coopération Internationale.

——. 1991b. *La sécurité alimentaire composante de la politique sociale du gouvernement*. Conakry: Ministère du Plan et de la Coopération Internationale.

——. 1990a. *Loi de finances pour 1990*. Conakry: Ministère de l'Economie et des Finances.

——. 1990b. *Rapport économique et social 1989*. Conakry: Ministère de Plan et de la Coopération Internationale.

——. 1990c. *Résultats du recensement national de l'agriculture*. Conakry: Ministère du Plan et de la Coopération Internationale and Direction Nationale de la Statistique et de l'Informatique.

——. 1989a. *Rapport annuel 1988: Situation et perspectives économiques, 1988–91*. Conakry: Ministère du Plan et de la Coopération Internationale.

——. 1989b. *Recensement general de la population et de l'habitat, 1983, analyse de résultats définitifs*. Conakry: Ministère du Plan et de la Coopération Internationale and Direction Nationale de la Statistique et de l'Informatique.

——. 1988. *Analyse économique et financière des dépenses du secteur éducation en République de Guinée*. Conakry: Ministère de l'Education Nationale.

——. 1986. "Enquête légère sur la consommation de ménages de la ville de Conakry, 30/09/84–3/11/84." Conakry: Ministère du Plan et de la Coopération Internationale. Photocopy.

——. Undated (a). "Etude sur la filière-riz en Guinée Maritime." Conakry: Ministère du Plan et de la Coopération Internationale and Ministère du Développement Rural. Photocopy.

——. Undated (b). *Enquête filière-riz Haute Guinée, 1987–87*. Conakry: Ministère de l'Agriculture et Ressources Animales.

Hanrahan, Charles E., and Steven Block. 1988. *Food Aid and Policy Reform in Guinea*. Prepared for USAID/Conakry. Cambridge, Mass.: Abt Associates.

Harrigan, Jane. 1988. "Malawi: The Impact of Pricing Policy on Smallholder Agriculture, 1971–1988." Paper presented at the Symposium on Agricultural Policies for Growth and Development, Mangochi, Malawi.

Harris, Laurence. 1980. "Agricultural Cooperatives and Development Policy in Mozambique." *Journal of Peasant Studies* 7(3): 338.

Hart, Keith. 1982. *The Political Economy of West African Agriculture*. Cambridge: Cambridge University Press.

Hirsch, R. 1986a. *Rapport d'une mission préliminaire sur le secteur rizicole Guinéen*. Paris: Caisse Centrale de Coopération Economique.

——. 1986b. *Rapport final d'une mission de réflexion sur le secteur rizicole Malgache*. Paris: Département d'Appui aux Opérations, Caisse Centrale de Coopération Economique.

Hopkins, Jane, and Thomas Reardon. 1990. *Cereal and Pulse Consumption of*

Rural Households in Sahelian and Sudanian Zones of Western Niger in 1988–89: Preliminary Results. Washington, D.C.: International Food Policy Research Institute.

———. 1989. *Annual Crop and Livestock Transactions Patterns for Rural Households in Western Niger.* Washington, D.C.: International Food Policy Research Institute.

Horowitz, Michael, et al. 1983. *Niger: A Social and Institutional Profile.* Binghamton, N.Y.: Institute for Development Anthropology.

Houyoux, Joseph. 1986. *Consommation de produits vivriers a Kinshasa et dans les grandes villes du Zaire, 1986.* Kinshasa: Bureau d'Etudes d'Amenagement et d'Urbanisme.

Hugon, Philippe. 1987. "La crise économique à Madagascar." *Afrique Contemporaine* 144 (October–December).

Institut National de la Statistique. 1984. *Zaire: Recensement scientifique de la population—Juillet 1984.* Kinshasa: Government of Zaïre.

International Fertilizer Development Center (IFDC). 1989. *Malawi Smallholder Fertilizer Marketing Survey.* Prepared for Ministry of Agriculture, Government of Malawi. Muscle Shoals, Ala.: IFDC.

———. 1986. *Cameroon: Fertilizer Sector Study.* Washington, D.C.: USAID.

International Labour Office. 1982. *Basic Needs in Danger: A Basic Needs Oriented Development Strategy for Tanzania.* Jobs and Skills Programme for Africa. Addis Abbaba: ILO.

International Monetary Fund (IMF). 1991a. *Cameroon: Staff Report.* Washington, D.C.: IMF.

———. 1991b. *International Financial Statistics Yearbook.* Washington, D.C.: IMF.

———. 1991c. "Madagascar: Statistical Appendix." Unpublished.

———. 1991d. *Malawi: Recent Economic Developments.* Washington, D.C.: IMF.

———. 1990a. *Cameroon: Statistical Annex.* Washington, D.C.: IMF.

———. 1990b. *Cameroon: Staff Report.* Washington, D.C.: IMF.

———. 1990c. *Government Finance Statistics Yearbook.* Washington, D.C.: IMF.

———. 1990d. *Guinea: Statistical Annex.* Washington, D.C.: IMF.

———. 1990e. *International Financial Statistics Yearbook.* Washington, D.C.: IMF.

———. 1990f. *Niger: Recent Economic Developments.* Washington, D.C.: IMF.

———. 1989a. *Cameroon: Second Review under Standby Arrangement.* Washington, D.C.: IMF.

———. 1989b. *The Gambia: Statistical Annex for Article IV Consultation.* Washington, D.C.: IMF.

———. 1989c. *Zaire: Recent Economic Developments.* Washington D.C.: IMF.

———. 1988a. *The Gambia: Recent Economic Developments.* Washington, D.C.: IMF.

———. 1988b. "Madagascar: Statistical Annex." Unpublished.

———. 1988c. *Malawi: Recent Economic Developments.* Washington, D.C.: IMF.

———. 1988d. *Niger: Statistical Annex.* Washington, D.C.: IMF.

———. 1988e. *Zaire: Recent Economic Developments.* Washington, D.C.: IMF.

———. 1987a. *The Gambia: Recent Economic Developments.* Washington, D.C.: IMF.

———. 1987b. *Government Finance Statistics Yearbook.* Washington, D.C.: IMF.

———. 1987c. *Guinea: Recent Economic Developments.* Washington, D.C.: IMF.

———. 1987d. *Niger: Recent Economic Developments*. Washington, D.C.: IMF.
———. 1987e. *Zaire: Recent Economic Developments*. Washington, D.C.: IMF.
———. 1985. *Niger: Recent Economic Developments*. Washington, D.C.: IMF.
———. 1980. *Government Finance Statistics Yearbook*. Washington, D.C.: IMF.
———. 1971. *Surveys of African Economies*. Vol. 4. Washington, D.C.: IMF.
———. Various years (a). *Balance of Payments Statistics*. Washington, D.C.: IMF.
———. Various years (b). *International Financial Statistics*. Washington, D.C.: IMF.
Isaacman, Allen, and Barbara Isaacman. 1983. *Mozambique: From Colonialism to Revolution, 1900–1982*. Boulder, Colo.: Westview Press.
Jabara, Cathy L. 1991. *Structural Adjustment and Stabilization in Niger: Macroeconomic Consequences and Social Adjustment*. Cornell Food and Nutrition Policy Program, Monograph no. 11. Ithaca, N.Y.: CFNPP.
———. 1990a. *Economic Reform and Poverty in the Gambia: A Survey of Pre- and Post-ERP Experience*. Cornell Food and Nutrition Policy Program, Monograph no. 8. Washington, D.C.: CFNPP.
———. 1990b. "Zaire: Transport and Marketing Costs for Bandundu Subsector Model." Washington, D.C.: CFNPP. Photocopy.
Jamal, Vali. 1986. "Economics of Devaluation: The Case of Tanzania." *Labor and Society* 2(3).
Jeffries, Richard. 1991. "Leadership Commitment and Political Opposition to Structural Adjustment in Ghana." In *Ghana: The Political Economy of Recovery*, ed. Donald Rothchild. Boulder, Colo.: Lynne Reinner.
Kandoole, B. F. 1990. "Structural Adjustment in Malawi: Short-Run Gains and Long-Run Losses." Paper presented at a workshop on the effects of the structural adjustment program in Malawi, Club Makokola, Mangochi, Malawi.
Kennedy, Eileen, and Harold Alderman. 1987. *Comparative Analysis of Nutritional Effectiveness of Food Subsidies and Other Food Related Interventions*. Washington, D.C.: International Food Policy Research Institute and UNICEF.
Khan, Haider A., and Erik Thorbecke. 1988. *Macroeconomic Effects and Diffusion of Alternative Technologies within a Social Accounting Matrix Framework*. Aldershot, England: Gower.
Killick, Tony. 1978. *Development Economics in Action: A Study of Economic Policies in Ghana*. London: Heinemann.
Körner, Peter. 1991. "Zaire." In *Indebtedness and Kleptocracy in the Poverty of Nations: A Guide to the Debt Crisis from Argentina to Zaire*, ed. Elmar Altavater et. al. English translation, Biddles Limited: Guildford and King's Lynn.
Kraus, Jon. 1991. "The Political Economy of Stabilization and Structural Adjustment in Ghana." In *Ghana: The Political Economy of Recovery*, ed. Donald Rothchild. Boulder, Colo.: Lynne Reinner.
Kristjanson, Patrick, Mark D. Newman, Cheryl Christensen, and Martin Abel. 1990. *Export Crop Competitiveness: Strategies for Sub-Saharan Africa*. Agricultural Policy Analysis Project, Phase II, Technical Report no. 109. Bethesda, Md.: Abt Associates.
Krueger, Anne, Maurice Schiff, and Alberto Valdés. 1988. "Agricultural Incentives in Developing Countries: Measuring the Effect of Sectoral and Economywide Policies." *World Bank Economic Review* 2 (3): 255–72.
Krugman, Paul, and Lance Taylor. 1978. "Contractionary Effects of Devaluation." *Journal of International Economics* 8(3).

Kydd, J. 1989. "Maize Research in Malawi: Lessons from Failure." *Journal of International Development* 1(1): 112–44.

Kyle, Steven. 1991. "Economic Reform and Armed Conflict in Mozambique." *World Development* 19(6): 637–49.

Lancaster, Carol. 1991. *African Economic Reform: The External Dimension.* Washington, D.C.: Institute of International Economics.

Langan, Glenn E. 1987. "An Assessment of Agricultural Input Marketing in The Gambia within the Context of the Economic Recovery Program." Report prepared for USAID. Banjul. Mimeographed.

Leith, Clark. 1974. *Foreign Trade Regimes and Economic Development: Ghana.* New York: Columbia University Press.

Lele, Uma, Robert E. Christiansen, and Kundhave Kadirsesan. 1989. *Fertilizer Policy in Africa: Lessons from Development Programs and Adjustment Lending, 1979–87.* Washington, D.C.: World Bank.

Leslie, Winsome J. 1987. *The World Bank and Structural Transformation in Developing Countries: The Case of Zaire.* Boulder, Colo.: Lynne Rienner.

Lewis, Stephen R., Jr. 1984. *Taxation for Development Principles and Applications.* New York: Oxford University Press.

Lipton, Michael. 1989. "State Compression: Friend or Foe of Agricultural Liberalization?" *Indian Society of Agricultural Economics,* Golden Jubilee volume.

———. 1988. *Report on the National Situation: Zaire.* Reprint no. 140. Washington, D.C.: International Food Policy Research Institute.

Lofchie, M. F. 1989. *The Policy Factory: Agricultural Performance in Kenya and Tanzania.* Boulder, Colo.: Lynne Rienner.

———. 1988. "Tanzania's Agricultural Decline." In *Coping with Africa's Food Crisis,* ed. Naomi Chazan and T. M. Shaw. Boulder, Colo.: Lynne Rienner.

Louis Berger International. 1989a. *Etude sur les mesures d'incitation à l'industrie.* Report prepared for the Government of Niger, Ministère du Commerce de l'Industrie et de l'Artisanat. East Orange, N.J.: Louis Berger International.

———. 1989b. *Final Evaluation Agriculture Sector Development Grant.* Report submitted to USAID/Niger. East Orange, N.J.: Louis Berger International.

Lowdermilk, Melanie. 1989. *Food Needs Assessment, 1989–90.* Washington, D.C.: Energy/Development International.

Loxley, John. 1989. "The Devaluation Debate in Tanzania." In *Structural Adjustment in Africa,* ed. Bonnie K. Campbell and John Loxley. New York: St. Martin's Press.

Lucas, Robert E. B. 1987. "Emmigration to South Africa's Mines." *American Economic Review* 77(3): 313.

MacGaffey, Janet. 1991. *The Real Economy of Zaire: The Contribution of Smuggling and Other Unofficial Activities to National Wealth.* Philadelphia: University of Pennsylvania Press.

———. 1987. *Entrepreneurs and Parasites: The Struggle for Indigenous Capitalism in Zaire.* Cambridge: Cambridge University Press.

Macintosh, Maureen. 1988. "Mozambique." In *Agricultural Pricing Policy in Africa,* ed. Harvey Charles. London: Macmillan.

———. 1986. "Agricultural Marketing and Socialist Accumulation: A Case Study of Maize Marketing in Mozambique." *Journal of Peasant Studies* 14(2): 243.

McPherson, Malcolm, and Steven C. Radelet. 1989. "Economic Reform in the

Gambia: Policies, Politics, Foreign Aid, and Luck." Draft prepared for Harvard Institute for International Development. Photocopy.

Malawi Government. 1990. *Economic Report 1990.* Lilongwe: Office of the President and Cabinet, Department of Economic Planning and Development.

———. 1989a. *Economic Report 1989.* Lilongwe: Office of the President and Cabinet, Department of Economic Planning and Development.

———. 1989b. *Guide to Agricultural Production.* Lilongwe: Ministry of Agriculture.

———. 1988. *Malawi Statistical Yearbook 1986.* Zomba: National Statistical Office.

———. 1987. *Malawi Statistical Yearbook 1985.* Zomba: National Statistical Office.

———. Various years. *Annual Statement of External Trade.* Zomba: National Statistical Office.

Maliyamkono, T. L., and M. S. D. Bagachwa. 1990. *The Second Economy in Tanzania.* London: James Currey.

Marketing Development Bureau. 1985. *Import Intensity of Major Crops in Mainland Tanzania 1985/86 (July).* Dar es Salaam: Ministry of Agriculture and Livestock Development, Marketing Development Bureau.

May, Ernesto. 1985. *Exchange Controls and Parallel Market Economies in Sub-Saharan Africa: Focus on Ghana.* Staff Working Paper no. 711. Washington D.C.: World Bank.

Mbelle, A. V. Y., 1991. "Tariffs and Quantitative Restrictions in Tanzania." Paper prepared for the ERB/Cornell Food and Nutrition Policy Program. May. Processed.

———. 1990. "Anatomy of Current Institutional and Production Constraints." ERB/Cornell Food and Nutrition Policy Program. December. Processed.

Ministère de la Production Agricole et de la Réforme Agraire (MPARA). 1989. *Chroniques statistiques agricoles, 1981–1984.* Antananarivo: Direction de la Programmation, Service de la Méthodologie et du Traitement des Informations Statistiques.

———. 1988a. *Caractéristiques générales du milieu rural, campagne agricole, 1984/1985.* Projet Recensement National de l'Agriculture et Système Permanent des Statistiques Agricoles, Pt. 2. Antananarivo: MPARA.

———. 1988b. *Cheptel et équipement des exploitations agricoles, campagne agricole, 1984/1985.* Projet Recensement National de l'Agriculture et Système Permanent des Statistiques Agricoles, Pt. 5. Antananarivo: MPARA.

———. 1988c. *Cultures et superficies des exploitations agricoles, campagne agricole, 1984/1985.* Projet Recensement National de l'Agriculture et Système Permanent des Statistiques Agricoles, Pt. 4. Antananarivo: MPARA.

———. 1988d. *Généralités et méthodologie, campagne agricole, 1984/1985.* Projet Recensement National de l'Agriculture et Système Permanent des Statistiques Agricoles, Pt. 1. Antananarivo: MPARA.

———. 1988e. *Main-d'oeuvre des exploitations agricoles, campagne agricole, 1984/1985.* Projet Recensement National de l'Agriculture et Système Permanent des Statistiques Agricoles, Pt. 3. Antananarivo: MPARA.

———. 1988f. *Les rendements des cultures et estimation de la production campagne agricole, 1984/1985.* Project Recensement National de l'Agriculture et Système Permanent des Statistiques Agricoles, Pt. 6. Antananarivo: MPARA.

———. 1987a. *Etude méthodologique de détermination des coûts de production: Le cas du paddy.* Project Recensement National de l'Agriculture et Système Permanent des Statistiques Agricoles. Antananarivo: MPARA.

———. 1987b. *Statistiques agricoles annuaire, 1984–86.* Direction de la Programmation. Service de la Méthodologie et du Traitement des Informations Statistiques. Antananarivo: MPARA.

———. 1986. *Chroniques statistiques agricoles.* Direction de la Programmation. Service de la Méthodologie et du Traitement des Informations Statistiques. Antananarivo: MPARA.

———. 1978. *Chroniques statistiques agricoles, 1977.* Direction de la Programmation. Service de la Méthodologie et du Traitement des Informations Statistiques. Antananarivo: MPARA.

Ministère du Plan, Direction de l'Analyse Economique et de la Planification. 1991. *Etudes et conjoncture.* Niamey: Ministère du Plan (March).

———. Undated. Unpublished national accounts data tables.

Ministère du Plan, Bureau Central du Recensement. 1990a. *2ème recensement général de la population, 1988.* Niamey: Ministère du Plan.

———. 1990b. *Etudes et conjoncture.* Niamey: Ministère du Plan (March).

———. 1989a. "Banque De Données Macroéconomiques." Niamey: Ministère du Plan. Mimeo.

———. 1989b. *Elaboration du tableau entrée-sorties 1987.* Comptes Nationaux du Niger, no. 7. August. Niamey: Ministère du Plan.

———. 1989c. *Etudes et conjoncture.* Niamey: Ministère du Plan (April).

Minot, Nicholas. 1991. *Impact of the Fertilizer Sub-Sector Reform Program on Farmers: The Results of Three Farm-Level Surveys.* The Agricultural Marketing Improvement Strategies Project. University of Idaho/Postharvest Institute of Perishables and Abt Associates. Washington, D.C.: USAID.

Mkandawire, R., S. Jaffee, and S. Bertoli. 1990. *Beyond "Dualism": The Changing Face of the Leasehold Estate Sub-Sector of Malawi.* Lilongwe: Bunda College and the Institute for Development Anthropology.

Morna, Colleen. 1989. "An Exercise in Educational Reform." *Africa Report* (November): 34–37.

Morris, R. 1985. *Tanzania's Productivity Crisis: A Social and Institutional Profile.* Tanzania: USAID.

Mukonoweshuro, Eliphas G. 1990. "State 'Resilience' and Chronic Political Instability in Madagascar." *Canadian Journal of African Studies* 24(3): 376–98.

Munslow, Barry. 1984. "State Intervention in Agriculture: The Mozambican Experience." *Journal of Modern African Studies* 22(2): 199.

Ndulu, B. J. 1988. *Stabilization and Adjustment Policies and Programmes.* Country Study no. 17, Tanzania. Helsinki: World Institute for Development Economics Research.

———. 1986. "Governance and Economic Management." In *Strategies for African Development,* ed. Robert J. Berg and Jennifer Seymour Whitaker. Berkely: University of California Press.

Ndulu, B. J., W. M. Lyakurwa, J. J. Semboja, and A. E. Chaligha. 1987. "Import Tariff Study." Report submitted to the Ministry of Finance, Economic Affairs and Planning, Dar es Sallam, March. Processed.

Nellum, A. L., and Associates. 1980. "Guinea Agricultural Sector Report." Mimeo.

Newton, Alex. 1988. *West Africa: A Travel Survival Guide.* Berkeley: Lonely Planet Press.

Odegaard, K. 1985. "Cash Crop versus Food Crop Production in Tanzania: An Assessment of the Major Post-Colonial Trends." *Lund Economic Studies* 33.

Overseas Development Institute (ODI) and International Fund for Agricultural Development. Forthcoming. *Madagascar: The Impact of Economic Recovery Programs on Smallholder Farmers and the Rural Poor.* Working Paper no. 4. London: ODI.

Peterson, E. Wesley. 1989. *Niger: Monitoring the Effect of Policy Reform.* Technical Report no. 105. Bethesda, Md.: Abt Associates.

Pick's Currency Yearbook. Various years. New York: Picks.

Plane, Patrick. 1989. "Financial Crises and the Process of Adjustment in the Franc Zone: The Experience of the West African Monetary Union." Presented at the USAID/World Bank/IMF Conference on Policy Reform and Financial Systems in Sub-Saharan Africa, Washington, D.C., June 20–22.

Programme National de Surveillance Alimentaire et Nutritionelle. 1989. *Document de travail: Synthèse des données existantes sur l'état nutritionnel.* Document de Travail. Antananarivo: Ministère de la Production Agricole de la Réforme Agraire.

Project PNUD/BIT. 1980. *Enquête sur le secteur informel dans le Grand Antananarivo.* Antananarivo: Ministère auprès de la Présidence de la République Chargé des Finances et du Plan.

Pryor, Frederic L. 1988. *Income Distribution and Economic Development in Madagascar: Some Historical Statistics.* World Bank Discussion Papers no. 37. Washington, D.C.: World Bank.

———. 1963. *The CFA Franc System.* IMF Staff Papers no. 10. Washington, D.C.: IMF (November).

Puetz, Detlev, and Joachim von Braun. 1988. "Parallel Markets and the Rural Poor in a West African Setting." Paper prepared for Harvard Institute of International Development Workshop on Parallel Markets, Nov. 11-12. Mimeographed.

Radelet, Steven Charles. 1990. "Economic Recovery in the Gambia: The Anatomy of an Economic Reform Program." Ph.D. diss. presented to the Committee on Public Policy. Cambridge, Mass.: Harvard University.

Raikes, Philip. 1986. "Eating the Carrot and Wielding the Stick: The Agricultural Sector in Tanzania." In *Tanzania: Crisis and Struggle for Survival,* ed. J. Boesen, K. J. Havnevik, J. Koponen, and R. Odegaard. Uppsala: Scandinavian Institute for African Studies.

———. 1985. "Food Policy and Production in Mozambique." In *Food Systems in Central and Southern Africa,* ed. Johan Pottier. London: University of London, London School of Oriental and African Studies.

Ramahatra, Olivier. 1989. *Madagascar: Une économie en phase d'ajustement.* Paris: L'Harmattan.

Rao, J. Mohan. 1989. "Agricultural Supply Response: A Survey." *Agricultural Economics* 3(1): 1.

Rassas, Bechir, and Thierry Loutte. 1989. *Niger: Rice and Cotton Policy.* Prepared for USAID/Niger. Technical Report no. 106. Bethesda, Md.: Abt Associates.

Republic of Zaire. 1990. Dossier du Programme de Révision des Comptes Nationaux. Kinshasa: Ministère du Plan, Institut Nationale de la Statistique.

——. 1989a. *Les prix au producteur du manioc, du mais et des arachides dans la région de Bandundu et les marges de commercialisation pour le marché de Kinshasa.* Direction des Marchés, Prix et Credits de Campagne. Project "Commercialization des Produits Agricoles." Publication no. 1. Kinshasa: Département de l'Agriculture, et Leuven: AGCD–Katholieke Universiteit.

——. 1989b. *Analyse des prix des produits vivriers à Kinshasa pendant la période 1961–1989.* Direction des Marchés, Prix et Credits de Campagne. Projet "Commercialisation des Produits Agricoles." Publication no. 12. Kinshasa: Département de l'Agriculture, et Leuven: AGCD–Katholieke Universiteit.

——. 1987. *Situation actuelle de l'Agriculture Zairoise.* Kinshasa: Département de l'Agriculture.

——. Various years. *Conjoncture économique.* Kinshasa: Département de l'Agriculture.

Reserve Bank of Malawi. 1988. *Financial and Economic Review* (Lilongwe) 20 (4).

——. 1987. *Financial and Economic Review* (Lilongwe) 19 (2).

Rimmer, Douglas. 1992. *Staying Poor: Ghana's Political Economy, 1950–1990.* New York: Pergamon Press.

Robert R. Nathan Associates. 1989. *Final Evaluation of USAID/Kinshasa's African Economic Policy Reform Program Structural Adjustment Support Grant (SASG 660–0121).* Prepared for USAID. Washington, D.C.: Robert R. Nathan Associates.

——. 1987. *The Impact of the Fertilizer Subsidy Removal Program on Smallholder Agriculture in Malawi: Annexes.* Presented to Ministry of Agriculture, Malawi Government and USAID.

Roe, A. R., and M. Johnston. 1988. "Malawi: The Social Dimensions of Structural Adjustment—Background Paper." Photocopy.

Roesch, Otto. 1984. "Peasants and Collective Agriculture in Mozambique." In *The Politics of Agriculture in Tropical Africa,* ed. Randolph Barker. Beverly Hills, Calif.: Sage.

Sahn, David, and Harold Alderman. 1987. *The Role of Foreign Exchange and Commodity Auctions in Trade, Agriculture, and Consumption in Somalia.* Washington D.C.: International Food Policy Research Institute.

Sahn, David, Jehan Arulpragasam, and Lemma Merid. 1990. *Policy Reform and Poverty in Malawi: A Review of a Decade of Experience.* Cornell Food and Nutrition Policy Program, Monograph no. 7. Ithaca, NY: CFNPP.

Sarris, A. H., and Rogier Van den Brink. 1993. *Economic Policy and Household Welfare during Crisis and Adjustment in Tanzania.* New York: New York University Press.

SCETAGRI. 1986. "Etude de restructuration des services, agricoles, et de schémas directeurs regionaux de développment rural programmes nationaux."

Scobie, Grant, 1989. *Macroeconomic Adjustment and the Poor: Toward a Research Strategy.* Cornell Food and Nutrition Policy Program, Monograph 89–1. Ithaca, N.Y.: CFNPP.

SEDES. 1988. *Etude du secteur agricole du Niger: Bilan-diagnostic—Phase 1.* Paris: SEDES.

——. 1987. *Etude du secteur agricole du niger: Les politiques—Phase 2.* Paris: SEDES.

Shapiro, K., and Elliot Berg. 1988. "The Competitiveness of Sahelian Agriculture." Elliot Berg Associates. Photocopy.

Sharpley, Jennifer. 1985. "External versus Internal Factors in Tanzania's Macro-economic Crisis, 1973–1982." *Eastern African Economic Review* 1 (December).

Shuttleworth, Graham. 1989. "Policies in Transition: Lessons from Madagascar." *World Development* 17(3).

Sines, Richard, Christopher Pardy, Mary Reintsma, and E. Scott Thomas. 1987. *Impact of Zaire's Economic Liberalization Program on the Agricultural Sector: A Preliminary Assessment.* Washington, D.C.: Robert Nathan Associates and USAID.

Singh, Ajit. 1986. "Tanzania and the IMF: The Analytics of Alternative Adjustment Programs." *Development and Change* 17.

Skarstein, R., and S. Wangwe. 1986 *Industrial Development in Tanzania: Some Critical Issues.* Uppsala: Scandinavian Institute of African Studies in cooperation with Tanzania Publishing House, Dar es Salaam.

Streeten, Paul. 1987. *What Price Food? Agriculture Price Policy in Developing Countries.* New York: St. Martin's Press.

Stryker, D. 1988. *A Comparative Study of the Political Economy of Agricultural Pricing Policies.* Washington, D.C.: World Bank.

Svedberg, P. 1991. "The Export Performance of Sub-Saharan Africa." *Economic Development and Cultural Change* 39(3): 549–66.

Swanson, Daniel, and Teferra Wolde-Semait. 1989. *Africa's Public Enterprise Sector and Evidence of Reforms.* World Bank Technical Paper no. 95. Washington, D.C.: World Bank.

Tabatabai, Hamid. 1990. "Poverty, Food Consumption, and Nutrition in Zaire: Review of Selected Studies." Kinshasa. Photocopy.

———. 1988. "Agricultural Decline and Access to Food in Ghana." *International Labour Review* 127 (6): 703–34.

Tarp, Finn. 1984. "Agrarian Transformation in Mozambique." *Land Reform, Land Settlement, and Cooperatives* nos. 1 and 2: 1.

Thenevin, Pierre. 1988. "Politique de relance de la filière rizicole et approvisionnement en riz local de la Guinée: Identification et faisibilité de quelques actions." Photocopy.

Thomas, Scott, and Mary Reintsma. 1989. "Zaire's Economic Liberalization and Its Impact in the Agricultural Sector." *Development Policy Review* 7: 29–50.

Timmer, C. Peter. 1989. "Food Price Policy: The Rationale for Government Intervention." *Food Policy* (February).

Tinguiri, Kiari Liman. 1990. *Adjustement structurel et satisfaction des besoins essentiels: Une étude de cas à partir de l'éxperience du Niger (1982–89).* Prepared for the Centre International pour le Développement de l'Enfant de l'UNICEF, version préliminaire (January). Niamey: Université de Niamey.

Toro Ochoa, Javier. 1988. *Propositions pour un programme d'aide alimentaire ciblée au profit des couches les plus démunies à Madagascar.* UNICEF/République Démocratique Malgache.

Trail, Susan. 1991/92. *The Africa Review.* 15th ed. Edison, N.J.: Hunter.

Truong, Tham. 1989. *Privatization of Cameroon's Subsidized Fertilizer Subsector: The Implementation Process.* Yaoundé: USAID/Cameroon.

Tshishimbi, wa Bilenga, and Peter Glick. Forthcoming. *Stabilization and Structural*

Adjustment in Zaire since 1983. Cornell Food and Nutrition Policy Program, Monograph no. 16. Washington D.C.: CFNPP.

UNDP and World Bank. 1989. *African Economic and Financial Data.* Washington, D.C.: World Bank.

UNICEF. 1990. "Plan quinquennal de cooperation: Programme santé et nutrition." Programme de Cooperation Gouvernement–UNICEF. Conakry: UNICEF.

United Nations Development Programme (UNDP). 1991. *Human Development Report 1991.* Oxford: Oxford University Press.

United Nations, Subcommittee on Nutrition. 1989. *SCN News* (4): 3.

University of Michigan, Technical Assistance Team. 1989. "Illustrated Methodology for the Impact Assessment of Policy Reforms: Agricultural Sector Development Grant to Niger." Prepared for USAID/Niamey. Niamey: USAID. Mimeo.

——. 1988a. "Policy Reform Area: Agricultural Inputs." Niamey: Ministère de l'Agriculture. Mimeo.

——. 1988b. "Policy Reform Area: Cross-Border Trade." Niamey: Ministère de l'Agriculture. Mimeo.

——. 1988c. "Pricing Policy and Cereals Marketing." Niamey: Ministère de l'Agriculture. Mimeo.

U.S. Agency for International Development (USAID). 1990a. *Program Assistance Authorization Document (PAAD), Transport Reform Program.* Kinshasa: USAID/Kinshasa.

——. 1990b. "USAID/Cameroon's Response to AID/W Impact Assessment of the Fertilizer Sub-Sector Reform Program (FSSRP)." Unpublished memo.

——. 1989. *Guinea Grant Food Assistance Programs Second Mid-Term Evaluation.* Conakry: USAID.

——. 1988. *Madagascar: Agricultural Export Liberalization Program (MAELP).* Washington, D.C.: USAID.

——. 1987. "An Evaluation of United States Food Aid in Guinea." Conakry: USAID. Photocopy.

——. 1986a. *Niger Health Sector Support Program: Program Assistance Approval Document,* Vol. 2: *Technical Annexes.* Washington, D.C.: USAID.

——. 1986b. *Niger: Current Macroeconomic Situation and Constraints.* Country Development Strategy Statement. Washington, D.C.: USAID.

——. 1986c. *Program Assistance Approval Document (PAAD), Zaire Private Sector Support Program.* Kinshasa: USAID/Kinshasa.

Université de Clermont I Faculté des Sciences Economiques (CERDI). 1990. *Politique macro-économique et pauvreté au Niger.* Clermont-Ferrand, France: Université de Clermont. Preliminary.

——. 1989. *La politique d'ajustement au Niger, 1982–87.* Prepared for the Government of Niger. Clermont-Ferrand, France: Université de Clermont.

van Arkadie, Brian. 1983. "The IMF Prescription for Structural Adjustment in Tanzania: A Comment." In *Monetarism, Economic Crisis and the Third World,* ed. Kalel Jansen. London: Frank Cass.

van de Walle, Nicholas. 1990. "The Politics of Non-Reform in Cameroon." In *Hemmed In: Responses to Africa's Economic Decline,* ed. John Ravenhill and Thomas Callaghy. New York: Columbia University Press.

von Braun, Joachim, Ken Johm, Sambou Kinteh, and Detlev Puetz. 1990. *Structural*

Adjustment, Agriculture, and Nutrition: Policy Options in the Gambia. Working Papers on Commercialization of Agriculture and Nutrition no. 4. Washington, D.C.: International Food Policy Research Institute.

Wangwe, Samuel. 1983. "Industrialization and Resource Allocation in a Developing Country: The Case of Recent Experiences in Tanzania." *World Development* 11(6): 483–92.

Weaver, James H., and A. Anderson. 1981. "Stabilization and Development of the Tanzanian Economy in the 1980s." In *Economic Stabilization in Developing Countries,* ed. William R. Kline and Sidney Weintraub. Washington, D.C.: Brookings Institution.

Weaver, Robert D. 1987. *Comparative Advantage in Food Production in Guinea: A Study of Smallholders.* Washington, D.C.: World Bank.

Willame, Jean-Claude. 1972. *Patrimonialism and Political Change in the Congo.* Stanford, Calif.: Stanford University Press.

World Bank. 1991a. *Ghana: Progress on Structural Adjustment.* Washington, D.C.: World Bank.

———. 1991b. "Madagascar: Agricultural Sector Adjustment Credit: Evaluation of the 'Buffer Stock'." Unpublished.

———. 1991c. *Madagascar: Beyond Stabilization to Sustainable Growth.* Washington, D.C.: World Bank.

———. 1991d. *World Development Report 1991.* New York and Oxford: Oxford University Press.

———. 1991e. *World Tables.* Washington, D.C.: World Bank.

———. 1991f. *Zaire: A Review of Government Expenditure.* Report no. 8995-ZR. Washington, D.C.: World Bank.

———. 1990a. *Cameroon: Public Investment Review.* Washington, D.C.: World Bank.

———. 1990b. *Cameroon: Social Dimensions of Adjustment: Country Assessment Paper.* Washington, D.C.: World Bank.

———. 1990c. *Food Security and Nutrition in Madagascar.* Washington, D.C.: World Bank.

———. 1990d. *Madagascar: Adjustment in the Industrial Sector.* Report no. 7784-MAG. Washington, D.C.: World Bank.

———. 1990e. *Republic of Guinea: Country Economic Memorandum.* Vols. 1–2. Washington, D.C.: World Bank.

———. 1990f. *World Debt Tables 1989/90.* Washington, D.C.: World Bank.

———. 1990g. *World Development Report.* Washington, D.C.: World Bank.

———. 1989a. *African Economic and Financial Data.* New York: United Nations Development Programme and the World Bank.

———. 1989b. *Cameroon Agricultural Sector Report.* Vols. 1–2. Washington, D.C.: World Bank.

———. 1989c. *Cameroon Agricultural Sector Report.* Vol. 2: Statistical Volume. Washington, D.C.: World Bank.

———. 1989d. *Ghana Public Expenditure Review, 1989–91.* Washington, D.C.: World Bank.

———. 1989e. *Ghana: Structural Adjustment for Growth.* Washington, D.C.: World Bank.

———. 1989f. *Guinea: Public Investment Review, 1986–1991.* Washington, D.C.: World Bank.

———. 1989g. "Malawi Country Economic Memorandum: Growth through Poverty Reduction." Washington, D.C.: World Bank. Draft.

———. 1989h. *Memorandum and Recommendation of the President of the IBRD to the Executive Directors on a Proposed Loan of US$ 150 Million Equivalent to the Republic of Cameroon for a Structural Adjustment Program.* Washington, D.C.: World Bank.

———. 1989i. *Mozambique: Food Security Study.* Washington, D.C.: World Bank.

———. 1989j. *Poverty Alleviation in Madagascar: Country Assessment and Policy Issues.* Washington, D.C.: World Bank.

———. 1989k. *Sub-Saharan Africa: From Crisis to Sustainable Growth. A Long-Term Perspective Study.* Washington, D.C.: World Bank.

———. 1989l. *Tanzania Public Expenditure Review.* Washington, D.C.: World Bank.

———. 1989m. *World Development Report 1989.* New York and Oxford: Oxford University Press.

———. 1989n. *World Tables Update.* Washington, D.C.: World Bank.

———. 1988a. *Agricultural Sector Memorandum Zaire: Toward Sustained Agricultural Development.* Report no. 7356-ZR. Washington, D.C.: World Bank.

———. 1988b. *Country Briefs.* Washington, D.C.: World Bank.

———. 1988c. *Guinea: Structural Adjustment Credit.* Washington, D.C.: World Bank.

———. 1988d. *Human Resources Development in Malawi: Issues for Consideration: Labor and Employment.* Washington, D.C.: World Bank.

———. 1988e. *Report and Recommendation of the President of the International Development Association to the Executive Directors on a Proposed Industrial and Trade-Policy Adjustment Credit in an Amount Equivalent to US$ 85 Million to the Government of Malawi.* Washington, D.C.: World Bank.

———. 1988f. *Report and Recommendation of the President of the International Development Association to the Executive Directors on a Proposed Credit in an Amount Equivalent to US$ 70 Million to the Republic of Malawi for an Industrial and Trade Policy Adjustment Program.* May 25. Washington, D.C.: World Bank.

———. 1988g. *Sector Review: Zaire Population, Health, and Nutrition.* Report no. 7013-ZR. Washington, D.C.: World Bank.

———. 1988h. *Urban Sector Report.* Washington, D.C.: World Bank.

———. 1988i. *World Development Report 1988.* New York: Oxford University Press.

———. 1988j. *World Tables.* Washington, D.C.: World Bank (tapes).

———. 1987a. *Cameroon: Country Economic Memorandum.* Washington, D.C.: World Bank.

———. 1987b. *Côte d'Ivoire Country Economic Memorandum.* Washington, D.C.: World Bank.

———. 1987c. *Ghana: Policies and Issues of Structural Adjustment.* Washington, D.C.: World Bank.

———. 1986a. *Cameroon: Financial Sector Report.* Washington, D.C.: World Bank.

———. 1986b. *The Democratic Republic of Madagascar: Country Economic Memorandum.* Report no. 5996-MAG. Washington, D.C.: World Bank.

———. 1986c. *Guinea: Population, Health, and Nutrition Sector Review.* Washington, D.C.: World Bank.

———. 1986d. *Guinea: Structural Adjustment Credit, Credit Summary.* Washington, D.C.: World Bank.

———. 1986e. "Improving Agricultural Marketing and Food Security Policies and Organization: A Reform Proposal." Washington, D.C.: World Bank. Draft.

———. 1986f. *Zaire: An Interim Assessment of Prospects and Preconditions for the Five Year Agricultural Development Plan.* Washington, D.C.: World Bank.

———. 1985a. *The Gambia: Development Issues and Prospects.* Washington, D.C.: World Bank.

———. 1985b. *Malawi Economic Recovery: Resource and Policy Needs—An Economic Memorandum.* Washington, D.C.: World Bank.

———. 1985c. *Zaire Country Economic Memorandum: Economic Change and External Assistance.* Washington, D.C.: World Bank.

———. 1984a. *Guinea: Agricultural Sector Review.* Washington, D.C.: World Bank.

———. 1984b. *Guinea: The Conditions for Economic Growth, a Country Economic Memorandum.* Washington, D.C.: World Bank.

———. 1984c. *Madagascar: Export Crops Sub-Sector Review.* Washington D.C.: World Bank.

———. 1984d. *World Development Report 1984.* New York: Oxford University Press.

———. 1982a. *Malawi Growth and Structural Change: A Basic Economic Report—Statistical Appendix.* Washington, D.C.: World Bank.

———. 1982b. *Zaire Country Economic Memorandum.* Washington, D.C.: World Bank.

———. 1981a. *Accelerated Development in Sub-Saharan Africa: An Agenda for Action.* Washington, D.C.: World Bank.

———. 1981b. *Report and Recommendation of the President of the International Bank for Reconstruction and Development to the Executive Directors on a Structural Adjustment Loan to the Republic of Malawi.* Washington, D.C.: World Bank.

———. 1981c. *Revolutionary People's Republic of Guinea: Country Economic Memorandum.* Washington, D.C.: World Bank.

———. 1981d. *Revolutionary People's Republic of Guinea: Survey of the Public Enterprise Sector.* Washington, D.C.: World Bank.

———. 1980a. *Cameroon Economic Memorandum.* Washington, D.C: World Bank.

———. 1980b. *Madagascar: Recent Economic Developments and Future Prospects.* World Bank Country Study. Washington, D.C.: World Bank.

———. 1979. *The Zairean Economy: Current Situation and Constraints.* Report no. 2158-ZR. Washington, D.C.: World Bank.

———. 1978. *External Public Debt of Zaire.* Report no. 2159a-ZR. Washington, D.C.: World Bank.

———. Various years. *World Debt Tables.* Washington, D.C.: World Bank.

Young, Crawford. 1986. "Africa's Colonial Legacy." In *Strategies for African Development,* ed. Robert J. Berg and Jennifer Seymour Whitaker. Berkeley: University of California Press.

Young, Crawford, and Thomas Turner. 1985. *The Rise and Decline of the Zairien State.* Madison: University of Wisconsin Press.

Younger, Stephen. 1991a. *Monetary Management in Ghana.* Cornell University Food and Nutrition Policy Program, Working Paper no. 8. Washington, D.C.: CFNPP.

——. 1991b. "Testing the Link between Devaluation and Inflation: Time-Series Evidence from Ghana." Washington, D.C.: CFNPP. Photocopy.

Zaire Trading and Engineering (ZTE/SOCFINCO). 1987. *Etude sur les cultures perennes au Zaire.* Kinshasa: ZTE/SOCFINCO.

Zaire Trading and Engineering and COGEPAR (ZTE/COGEPAR). 1987. *Etude de la compétitivité de l'agriculture Zaïroise face aux produits agricoles importes.* Kinshasa: ZTE/COGEPAR.

Index

Food Systems and Agrarian Change

Edited by Frederick H. Buttel, Billie R. DeWalt,
and Per Pinstrup-Andersen